Palgrave Studies in World Environmental History

Series Editors
Vinita Damodaran
Department of History
University of Sussex
Brighton, UK

Rohan D'Souza
Graduate School of Asian and African Area Studies
Kyoto University
Kyoto, Japan

Sujit Sivasundaram
University of Cambridge
Cambridge, UK

James John Beattie
History
Victoria University of Wellington
Wellington, New Zealand

The widespread perception of a global environmental crisis has stimulated the burgeoning interest in environmental studies and has encouraged a range of scholars, including historians, to place the environment at the heart of their analytical and conceptual explorations. An understanding of the history of human interactions with all parts of the cultivated and non-cultivated surface of the earth and with living organisms and other physical phenomena is increasingly seen as an essential aspect both of historical scholarship and in adjacent fields, such as the history of science, anthropology, geography and sociology. Environmental history can be of considerable assistance in efforts to comprehend the traumatic environmental difficulties facing us today, while making us reconsider the bounds of possibility open to humans over time and space in their interaction with different environments. This series explores these interactions in studies that together touch on all parts of the globe and all manner of environments including the built environment. Books in the series come from a wide range of fields of scholarship, from the sciences, social sciences and humanities. The series particularly encourages interdisciplinary projects that emphasize historical engagement with science and other fields of study.

More information about this series at
http://www.palgrave.com/gp/series/14570

Simo Laakkonen • J. R. McNeill
Richard P. Tucker • Timo Vuorisalo
Editors

The Resilient City in World War II

Urban Environmental Histories

Editors
Simo Laakkonen
University of Turku
Turku, Finland

J. R. McNeill
Georgetown University
Washington, DC, USA

Richard P. Tucker
University of Michigan
Ann Arbor, MI, USA

Timo Vuorisalo
Department of Biology
University of Turku
Turku, Finland

Palgrave Studies in World Environmental History
ISBN 978-3-030-17438-5   ISBN 978-3-030-17439-2   (eBook)
https://doi.org/10.1007/978-3-030-17439-2

© The Editor(s) (if applicable) and The Author(s) 2019
This work is subject to copyright. All rights are solely and exclusively licensed by the Publisher, whether the whole or part of the material is concerned, specifically the rights of translation, reprinting, reuse of illustrations, recitation, broadcasting, reproduction on microfilms or in any other physical way, and transmission or information storage and retrieval, electronic adaptation, computer software, or by similar or dissimilar methodology now known or hereafter developed.
The use of general descriptive names, registered names, trademarks, service marks, etc. in this publication does not imply, even in the absence of a specific statement, that such names are exempt from the relevant protective laws and regulations and therefore free for general use.
The publisher, the authors and the editors are safe to assume that the advice and information in this book are believed to be true and accurate at the date of publication. Neither the publisher nor the authors or the editors give a warranty, express or implied, with respect to the material contained herein or for any errors or omissions that may have been made. The publisher remains neutral with regard to jurisdictional claims in published maps and institutional affiliations.

Cover illustration: Photo 12/Universal Images Group/Getty Images

This Palgrave Macmillan imprint is published by the registered company Springer Nature Switzerland AG
The registered company address is: Gewerbestrasse 11, 6330 Cham, Switzerland

Foreword

This book is not merely about the fate of cities during World War II—although that is an important topic unto itself—but a comment on the potential resilience of cities within a wartime setting. Such a focus has to contend with resilience within a specific time frame and over a wide geographic expanse. And like most studies of cities and urbanization, it has to confront endless differences in demography, spatial location, environmental conditions, and the particular ways in which war came to the cities (and then left).

Within a spectrum of experiences—from isolation from the war to destruction and annihilation—are an infinite number of possible outcomes and specific impacts. In this sense, *The Resilient City in World War II* is more speculative than exhaustive. Indeed, the variety of topics touched upon (or suggested) should open many conversations about war, the environment, and cities. For example, can a city prepare for war by securing necessities such as food, water, and other resources, or must it improvise as conditions deteriorate? How can city leadership protect human and animal life while under attack? How can a population adjust to a wartime footing in general? Is it possible to deal with the physical destruction of infrastructure within a limited time span? What are the emotional and psychological responses of the citizenry that shape the war experience?

These questions and others are to be understood within the context of World War II—an industrial-dependent, highly mechanized, aerial-influenced, massively human-scaled event—covering five or more years, and ranging over several oceans and continents. Because of the nature and intensity of the war, were preparations possible and to what degree could cities avoid destruction, given their central place in war strategy?

Adding to the complexity of the war and its many effects is the great disparity in impacts even within the war zones themselves. Wide swaths of Europe and Asia were overwhelmed by soldiers and machines, but even there most of Scandinavia and even parts of the United Kingdom were untouched by combat. The same goes for parts of Asia and the Middle East, much of Africa, and all of North and South America. But the question remains: Were cities which sustained some or no physical destruction spared the repercussions of war? We know what happened in Hiroshima, Nagasaki, and Tokyo, but what about Yokohama, Osaka, Kyoto, or Kumamoto?

The question of resilience itself, so central to the themes of this book, has a universal as well as a more specific connotation. In one sense, the focus on resilience is a forward-looking, almost optimistic, tip of the hat to human survival in the wake of such catastrophe as war, and more typically speaks to the future. Can it be applied to the concurrent state of affairs as cities went through the wartime experience in real time? It also presumes that humans have the capacity to repair the damages wrought by war. Indeed, any of the memorials in the myriad of bombed cities throughout the war zone is a graphic reminder of the conflict. But if we were to stroll down the streets of Berlin, or Nuremberg, or London today, we might see some signs of wartime destruction (like the Kaiser Wilhelm Memorial Church in Berlin with its ruined West Tower), but otherwise we'd find thriving communities. The same might be said for Manila, Warsaw, or St. Petersburg (Leningrad).

To be further explored is the dystopian, and thus less optimistic, side of the assault on cities during World War II beyond the atomic bombings in Hiroshima and Nagasaki, and the firebombings in Hamburg. An appropriate question is: How did cities come out of the war? The apocalyptic despair of post-nuclear warfare widely represented in visions like those in films like *A Boy and His Dog* (1975) and books like Cormac McCarthy's *The Road* (2006) may be inappropriate in looking back at World War II beyond the obvious atomic events in Japan. But Kurt Vonnegut's *Slaughterhouse-Five* (1969) presents its powerful anti-war themes in the hellish terrain of Dresden, which was destroyed by "conventional" weapons. Hopelessness sometimes trumped resilience in cases like these.

That *The Resilient City in World War II* is so evocative to me, raising thunderously important questions about the impacts of war, makes it an important read. Its greatest contribution will be to prompt lively debates over very important questions.

University of Houston Martin V. Melosi
Houston, TX, USA

Preface

This book is based on long-term research on the environmental history of war at the University of Turku in Finland and the University of Michigan in the United States. The roots of cooperation between us extend to August 2012 when the first international workshop on the environmental history of World War II was organized on Magpie Island in Helsinki, Finland. In this workshop, urban issues concerning, for example, the cities of Leningrad, Gdansk, London, and Helsinki were addressed and discussed. However, the resulting book, *The Long Shadows: A Global Environmental History of the Second World War* (Corvallis: Oregon State University Press, 2017), did not include any chapter specifically on urban issues. Although it is an independent collection of articles, this new book continues and extends the themes of the earlier book by (nearly) the same editorial team, and by focusing on urban areas provides a novel and fruitful perspective to the rapidly expanding research on the environmental history of World War II.

Why urban areas? Simple demographic facts make it clear that it is not possible to understand the environmental history of the greatest violent conflict in the history of Earth without paying serious attention to the fate and role of towns and cities. Already at the turn of the nineteenth and twentieth centuries the rapidly proceeding concentration of human population in cities was considered the most remarkable contemporary social phenomenon.[1] While in 1800 under 3 percent and 1900 about 18 percent of humans lived in cities, the proportion of the world population, which was about 2.3 billion in 1940, had increased to 25 percent. The industrialized regions were already urbanized: in North America 59 percent, in

Europe and Oceania 53 percent, in Japan 38 percent and in the Soviet Union 32 percent of the population was urbanized. By 1940, the number of multimillion cities (with 2.5 million inhabitants or more) outside Europe had grown to nine, while the number in Europe remained at four.[2] Urbanization still proceeds, especially in the developing countries, and according to the United Nations' estimate 68.4 percent of the human population will live in cities by 2050.[3] Earth will become an urban planet.

From the perspective of military actions, the degree of urbanization was (and is) vitally important because it largely determined not only the geographical distribution of economic activities and thus resources of warfare of the nations involved in conflict, but also to some extent the targets of military offensives. Although attempts to include civilian populations in armed conflicts have a long history, total warfare gained unprecedented dimensions during World War II and was most severely manifested and experienced in the densely populated urban areas. The prewar urban growth had in many countries created a new spatial economic structure with clearly defined heartland (urban cores) and hinterland (or periphery) whose interaction could obviously be severely damaged by military actions.[4] Urban warfare that devastated cities such as Stalingrad, Dresden, and Hiroshima represented some of the most sinister chapters in the history of World War II.

However, in spite of extensive or even catastrophic damage, life continued in its various forms in wartime cities. Our collection of essays has an explicit focus on the patterns of urban resilience, defined here as the wartime capacity of towns and cities to function and maintain realistic living opportunities for the urban population in spite of extreme hardship. Resilience means very practical things. In the Syrian Civil War, for example, in Aleppo, which has been called "Stalingrad of the Middle East" due to destructive urban warfare against ISIS, urban inhabitants have increasingly started to grow their own food.[5] People have looked for alternatives for damaged urban infrastructure by means of decentralizing and improvising. In Deir ez-Zor, which became known as "Syria's Leningrad" because siege of both cities lasted about three years, and some other Syrian towns, alternative water supply networks, consisting mainly of newly dug public or private wells, or re-discovered old wells, have expanded. Due to the damage to central power plants and electrical grids, people have resorted to private or commercial diesel generators as well as car batteries. Alternative power sources, such as small-scale devices harnessing the energy of wind and solar power, have also been constructed.[6] Similarly, practical innovations were widespread during World War II. Perhaps it is worth taking a look at some previous practices in order to think creatively about how to arrange living in towns and cities during the coming wars and crises.

This book is the first attempt to synthesize the environmental impacts of World War II on urban areas. However, the challenge to triangulate urban history, environmental history, and military history is considerable, because despite being a turning point in global environmental history, World War II signified highly different issues for different kinds of towns and cities on various continents. In many cities life continued as before; some even benefited from warfare while others were almost entirely wiped out. Towns and cities were also in very different positions before, during, and after the war. Therefore we decided to focus in this book mainly on wartime experiences. As yet it is hard to draw any general conclusions on the urban environmental history of World War II. Nevertheless, in the Epilogue we address some main themes of the book in a wider framework. Urban resilience is today as important an issue as it was yesterday in Athens, Nanking, Freetown, Montevideo, or Tokyo.

The editorial quartet is grateful for the original grant from the Foundation for Baltic and East European Studies (Östersjöstiftelse) that laid the base also for this second publishing project. More specifically we would like to thank University of Turku and Pori University Center, Finland, for funds needed in revising language and the Degree Program of Cultural Production and Landscape Studies for providing research leave to edit the book manuscript. Above all we would like to thank the contributors for their excellent articles and the anonymous reviewers for their constructive comments. Finally, we would like to thank the staff of Palgrave Macmillan for their invaluable help and the editors of Palgrave Studies in World Environmental History for accepting our book in their distinguished series.

Helsinki, Finland	Simo Laakkonen
Washington, DC, USA	J. R. McNeill
Ann Arbor, MI, USA	Richard P. Tucker
Turku, Finland	Timo Vuorisalo
March 1, 2019	

Notes

1. Adna Ferrin Weber, *The Growth of Cities in the Nineteenth Century: A Study in Statistics* (New York: Macmillan, 1899).
2. United Nations, *Growth of the world's urban and rural population, 1920–2000*. Population Studies No. 44, Department of Economic and Social Affairs (New York: United Nations, 1969), Table 39. Estimated percentages of urban population as nationally defined, in the total population of the world and major areas, 1920, 1940, 1960, 1980, and 2000.
3. United Nations, Population Division, Department of Economic and Social Affairs, *World Urbanization Prospects 2018*, https://esa.un.org/unpd/wup/Download/, visited 4 June 2018.
4. Brian J. L. Berry, "Urbanization," in John M. Marzluff et al., eds., *Urban Ecology. An International Perspective on the Interaction Between Humans and Nature* (New York: Springer, 2008), 36.
5. "The war in Syria: Syrians growing their own food to survive in eastern Aleppo," *TRT World*, 1 November 2016, https://www.youtube.com/watch?v=K9CTNwr2F_c
6. Raja Rehan Arshad and Joy-Fares Aoun, *Syria damage assessment of selected cities Aleppo, Hama, Idlib* (Washington, DC: World Bank Group, 2017), viii, 50; The World Bank, *The Toll of War: The Economic and Social Consequences of the Conflict in Syria* (Washington, DC: World Bank Group, 2017), 30, 33, 48–49.

Contents

Section I Introduction   1

1 Environmental History, the Second World War, and Urban Resilience   3
   Simo Laakkonen

Section II Urban Environment   21

2 **Critical Networks**   23
   Santiago Gorostiza

3 **Fortress City**   47
   Katherine Macica

4 War and Urban-Industrial Air Pollution in the UK and the US   69
   Peter Brimblecombe

5 **Imagined Resilience**   81
   Sarah Frohardt-Lane

Section III Urban Nature — 103

6 Guerrilla Gardening? — 105
Rauno Lahtinen

7 Gaining Strength from Nature — 127
Thomas J. (Tom) Arnold

8 Resilience Behind Bars — 151
Mieke Roscher and Anna-Katharina Wöbse

9 Where Have All the Pigeons Gone — 177
Timo Vuorisalo and Mikhail V. Kozlov

Section IV Urban Society — 201

10 Partial Resilience in Nationalist China's Wartime Capital — 203
Nicole Elizabeth Barnes

11 Japanese-Occupied Hanoi — 219
Geoffrey C. Gunn

12 The Esteros and Manila's Postwar Remaking — 237
Michael D. Pante

13 Apocalyptic Urban Future — 259
Kimmo Ahonen, Simo Laakkonen, and William M. Tsutsui

Section V Conclusions — 279

14 Epilogue: What Makes a City Resilient? — 281
Simo Laakkonen, J. R. McNeill, Richard P. Tucker, and Timo Vuorisalo

Index — 303

Notes on Contributors

Kimmo Ahonen PhD, is working as a coordinator of international affairs at Tampere University, Finland. His principal research interests are film history and the cultural history of the Cold War. Ahonen's doctoral dissertation addressed alien invasion in American science fiction films of the 1950s. He is co-editor of *Frontiers of Screen History: Imagining European Borders in Cinema, 1945–2010* (2013).

Thomas J. (Tom) Arnold PhD, is an instructor of history at San Diego Mesa College, California. He studies the impacts of warfare on the relationship between humans and nature, and focuses on how war changes the character of cities and their connections to natural resources. He has taught courses in World, European and Environmental History, the History of World War II, and the Holocaust.

Nicole Elizabeth Barnes got her PhD in 2012 from University of California at Irvine. She is assistant professor of Chinese History at Duke University. Her most recent book is *Intimate Communities: Wartime Healthcare and the Birth of Modern China, 1937–1945* (2018). Barnes researches the histories of medicine, public health, and gender in twentieth-century China.

Peter Brimblecombe was born in Australia and educated in New Zealand. He is an atmospheric chemist whose research also includes the history of air pollution, the regulation of air pollutants, and their impact on material heritage. He is currently a chair professor in the School of Energy and Environment at City University of Hong Kong.

Sarah Frohardt-Lane is assistant professor of History and Environmental Studies at Ripon College. Her main fields of interest are race and ethnicity, environmental history, and twentieth-century United States history. Her articles have been published in the *Journal of American Studies* and the *Journal of Transport History*.

Santiago Gorostiza is a post-doctoral researcher at the Institute of Environmental Science and Technology at the Universitat Autònoma de Barcelona (ICTA-UAB). He is an environmental historian working on modern and early-modern Spanish history, with training as an historian, environmental scientist, and political ecologist. His doctoral research, concluded in 2017, examined how the Spanish Civil War transformed the country's socioecological relations and landscapes.

Geoffrey C. Gunn is emeritus professor at Nagasaki University, Japan. He first visited Indochina during the "American war," subsequently completing a PhD at Monash University, Australia, on Vietnamese communism/Lao nationalism. He is author of such works as *Rice Wars in Colonial Vietnam: The Great Famine and the Viet Minh Road to Power* (2014) and the edited collection *Wartime Macau: Under the Japanese Shadow* (2016).

Mikhail V. Kozlov is a senior research fellow in the Department of Biology at the University of Turku, Finland. He graduated from St. Petersburg University, Russia, in 1984 and obtained a PhD in entomology in 1986. Kozlov has written several books, including *Impacts of Point Polluters on Terrestrial Biota: Comparative Analysis of 18 Contaminated Areas* with Elena L. Zvereva and Vitali Zverev (2009). His current focus is on trophic interactions in terrestrial ecosystems.

Simo Laakkonen is senior lecturer of Landscape Studies at the University of Turku, Finland. He has explored the environmental history of cities and of World War II and the Cold War. He has co-edited *Ecology of War: Environmental History of Modern Warfare* (in Finnish, 2017) with Timo Vuorisalo and *The Long Shadows: A Global Environmental History of the Second World War* with Richard Tucker and Timo Vuorisalo (2017).

Rauno Lahtinen is a cultural historian working as a researcher at the University of Turku, Finland. In his doctoral thesis "Environmental debate in the city: Urban environment and environmental attitudes in Turku, Finland, 1890–1950" (in Finnish) he studied, among other things, the history and effects of wartime urban agriculture. Since then he has authored several books and articles on various urban issues.

Katherine Macica is a doctoral candidate in Loyola University Chicago's joint doctoral program in American history and public history. Her research focuses on environmental history, Western US history, military history, and urban history.

J. R. McNeill is a university professor at Georgetown University. He teaches world history, environmental history, and international history. He has published more than 50 scholarly articles in professional and scientific journals. McNeill's most recent book (with Peter Engelke) is *The Great Acceleration: An Environmental History of the Anthropocene since 1945* (2016).

Michael D. Pante is an assistant professor in the Department of History, Ateneo de Manila University, Philippines. He obtained his PhD in area studies from the Graduate School of Asian and African Area Studies, Kyoto University, Japan. He is also the associate editor of *Philippine Studies: Historical and Ethnographic Viewpoints*. His research interests include transportation history, environmental history, and Southeast Asian cities.

Mieke Roscher is assistant professor of Social and Cultural History and the History of Human-Animal Relations at the University of Kassel, Germany. She acquired her PhD in 2008 with a thesis on the history of the British animal rights movement, published in 2009 as *A Kingdom for Animals* (in German). Since then she has moved on to zoo history and animal historiography. Her current research project focuses on writing a political history of animals in the Third Reich.

William M. Tsutsui is president and professor of History at Hendrix College, Arkansas. A specialist in the business, environmental, and cultural history of modern Japan, he is the author or editor of eight books, including *Manufacturing Ideology: Scientific Management in Twentieth-Century Japan* (1998) and *Godzilla on My Mind: Fifty Years of the King of Monsters* (2004).

Richard P. Tucker is adjunct professor of Environmental History at the University of Michigan. His research centers on the global environmental history of warfare and military operations. He is author and editor of three volumes on World Wars I and II.

Timo Vuorisalo is senior lecturer of Environmental Science in the Department of Biology at the University of Turku, Finland. He completed his PhD in Turku in 1989, and worked as a postdoctoral fellow at Indiana

University, Bloomington, between 1989 and 1990. He studies evolutionary theory, urban ecology, and environmental history, and has written and edited several books in these fields.

Anna-Katharina Wöbse is an environmental historian, curator, and researcher at the Justus Liebig University of Giessen, Germany. She has published extensively on media and the environment, environmental biographies and the history of international environmental movements, and the environmental role of League of Nations and the United Nations (*Weltnaturschutz*, 2012). Her current research focuses on the European environmental history of wetlands.

List of Figures

Fig. 2.1	"Madrid will not lack water." Guards, soldiers, militiamen, and women in front of El Villar dam. Published in *Ahora*, July 30, 1936, p. 22. (Source: Hemeroteca Digital de la Biblioteca Nacional de España. Photo by Contreras and Vilaseca)	27
Fig. 2.2	The Lozoya water supply system and military operations, Madrid region, November 1936. (Source: Own elaboration based on J. M. Macías and C. Segura, eds., *Historia del abastecimiento y usos de agua en la Villa de Madrid* (Madrid: Confederación Hidrográfica del Tajo y Canal de Isabel II, 2000) and J. M. Martínez Bande, *La lucha en torno a Madrid en el invierno de 1936–1937* (Madrid: Librería Editorial San Martín, 1968))	28
Fig. 2.3	Swimming pool at the workshop n°33 of the Industry of Socialized Wood (Galileu street, Sants neighbourhood, Barcelona), a collectivization run by the CNT. Summer 1936. (Source: Merletti collection, Arxiu de l'Institut d'Estudis Fotogràfics de Catalunya, reference ACP-1-2799p)	31
Fig. 2.4	Bombing damages in Carrera de San Jerónimo, Madrid city center. (Source: Ministerio de Cultura y Deporte. Archivo General de la Administración, Fondo "Archivo Fotográfico de la Delegación de Propaganda y Prensa de Madrid durante la Guerra Civil", F-04042-53501-001r, ANTIFAFOT)	33
Fig. 2.5	A present of the city of Barcelona to Madrid: Six water-tank trucks to keep supply going in the bomb-torn neighbourhoods. December 20, 1936 (Source: Pérez de Rozas, Arxiu Fotogràfic de Barcelona)	35

xviii LIST OF FIGURES

Fig. 2.6   Poster by José Bardasano: "Water in bad conditions causes more casualties than shrapnel", 1937. (Source: Pavelló de la República CRAI Library) 39

Fig. 3.1   This resource map, produced by the State of Washington for school children in 1945, illustrates the diversification of the state's economy as a result of the war. (Source: Office of the Secretary of State of Washington, Washington State Archives) 50

Fig. 3.2   Grand Coulee dam began powering Northwest war industries in 1941, facilitating the growth of shipbuilding, aircraft manufacturing, and light metals processing in the region. (Source: Library of Congress, Prints & Photographs Division, FSA/OWI Collection, LC-USW33–035035-C) 53

Fig. 3.3   Workers assemble B-17Gs inside Boeing Plant 2. Subassemblies and component parts, such as wings, tail assemblies, and engines, arrived in Seattle via railroad from around the Puget Sound industrial region and across the country. (Source: U.S. Air Force photo, National Museum of the United States Air Force, VIRIN 060517-F-1234S-027) 58

Fig. 3.4   The Army Corps of Engineers' Passive Defense Division fabricated an artificial town with houses and streets on the roof of Boeing Plant 2 to protect the factory from Japanese aerial attack. The camouflage was made of chicken wire, spun glass, painted canvas, burlap, feathers, and local lumber. Newly assembled bombers are visible on the right side of the plant. (Source: Records of the Office of the Assistant Chief of the Air Staff, A-4, R & D Branch, Case Histories, 1941–1946, Box 31, Entry 22, Record Group 18, National Archives at College Park) 58

Fig. 3.5   The war militarized Seattle's seascape as well as landscape. Warships became a common sight in Puget Sound as they deployed overseas or returned from battle for repairs. Here the destroyer USS *Little* (DD-803), built by the Seattle-Tacoma Shipbuilding Corporation, departs Seattle for Pearl Harbor in November 1944. *Little* supported the invasions of Iwo Jima and Okinawa before being sunk by kamikazes in May 1945. (Source: Naval History & Heritage Command, photo NH 107134) 61

Fig. 4.1   Average annual deposit of (**a**) tar, (**b**) dissolved matter, (**c**) insoluble ash and (**d**) the concentrations of sulphur dioxide and smoke at sites from England in the first half of the twentieth century. (Note: Victoria Park, Park Square, Finsbury Park and smoke at Stoke-on-Trent are shown as light bars. The units for sulphur dioxide are parts per million and smoke mg m^{-3}. Source: Data from A. R. Meetham, *Atmospheric Pollution: Its Origins and Prevention* (London: Pergamon Press, 1952)) 74

List of Figures

Fig. 5.1	This poster by Henry Koerner (1943) printed for the Office of War Information illustrates the message that household waste fat helped the war effort. (Source: National Archives poster (NWDNS-44-PA-380), via Wikimedia Commons)	85
Fig. 5.2	Myriad materials were collected in salvage drives all over the world. Women had a crucial role in recycling due to their increased importance in both households and working places. (Source: San Francisco History Center, San Francisco Public Library)	88
Fig. 5.3	"DOWN WITH HITLER!" San Francisco school children dramatize the impact of their scrap paper drive on the war's outcome. (Source: San Francisco History Center, San Francisco Public Library)	94
Fig. 6.1	Public institutions supported urban agriculture in all belligerent countries. Victory gardens located across from San Francisco City Hall during World War II. (Source: San Francisco History Center, San Francisco Public Library)	107
Fig. 6.2	Elderly people, women, and children took care of wartime urban agriculture. Mrs. Rodney George irrigates her victory garden in Golden Gate Park, with her daughter Eleanor looking on. Scores of urban youngsters learned a lot about gardening during war years. (Source: San Francisco History Center, San Francisco Public Library)	110
Fig. 6.3	A potato war? A map based on newspapers and archival sources shows the extension of potato fields in Turku town during World War II. (Source: Rauno Lahtinen)	113
Fig. 8.1	Pandas were naturally on the Allied side. (Source: Poster of the London Zoo 1940. Reprinted from: R. and D. Morris, *Men and Pandas*, London 1966)	156
Fig. 8.2	Interspecies resilience in action. A zoo chimpanzee is helping to outfit an air raid shelter in London. (Source: London Zoological Society, Animals and Zoo Magazine, June 1941. ZSL Archive)	158
Fig. 8.3	At the end of the war, the Berlin Zoo also became a battlefield. A monumental flak tower built in the vicinity of the zoo, was surrounded by burned trees, stables and destroyed tanks in May 1945. (Source: Wikipedia Commons, Berlińska Victoria, E. Kmiecik, "Ruch" 1972)	166
Fig. 9.1	Viljo Erkamo (1912–1990) was a Viipuri-born Finnish botanist who studied plant species and birdlife in the war-damaged Viipuri in the summer and fall 1942. (Source: Pekka Kyytinen, Collection of The Finnish Heritage Agency)	180

Fig. 9.2	Viipuri Castle is the symbol of Viipuri, a city founded in the thirteenth century as a military outpost against Novgorod. This photo of burning Viipuri was taken during the Winter War, on March 7, 1940. (Source: Photographer probably Otso Pietinen, Finnish Wartime Photographic Archives)	182
Fig. 9.3	Vegetables grown in 1942 in the center of Leningrad, next to the Saint Isaac's Cathedral. (Source: M. Trakhman, National Library of Russia)	184
Fig. 9.4	The poster "Early spring wild herbs suitable for salad," published in Leningrad in 1943. (Source: National Library of Russia)	186
Fig. 9.5	The population growth of the black redstart in the inner London (minimum number of breeding pairs). (Source: Sara Vuorisalo, after The London Natural History Society, *The Birds of the London area*, 252)	193
Fig. 10.1	A man perched on the roof of his house which he is planning to rebuild after its destruction in an air raid. Chongqing was the most bombed capital city in World War II. (Source: Prints and Photographs Division, U.S. Library of Congress, LOT 11511-7, WAAMD #131)	208
Fig. 10.2	Resilient city in action. Female nurses are carrying an air raid victim through the streets of Chongqing. (Source: Prints and Photographs Division, U.S. Library of Congress, LOT 11511-7, WAAMD #123)	211
Fig. 14.1	Weimer Pursell's poster (1943) for the Office of War Information shows how crisis awareness and public power stimulated innovative thinking even in American car culture. (Source: National Archives (NWDNS-188-PP-42), via Wikimedia Commons)	285
Fig. 14.2	The power of persuasion. During world wars the best artists were engaged in national campaigns to reduce consumption, spare natural resources, and reuse and recycle for the public good. After the war, national medias were harnessed by private interests to increase consumption and use of natural resources. (Source: Poster designed by Vanderlaan. National Archives (NWDNS-79-WP-103), via Wikimedia Commons)	292
Fig. 14.3	The long shadows of war. A World War II veteran and his son looking at the Golden Gate Bridge in spring 1945. (Source: San Francisco History Center, San Francisco Public Library)	295

SECTION I

Introduction

CHAPTER 1

Environmental History, the Second World War, and Urban Resilience

Simo Laakkonen

A Shock City

Suddenly the whole class jumped: with a metallic clang, a siren right over our heads rang out, or more like howled and roared. Its plaintive wail almost split our eardrums as it plunged down to the pits of our stomachs. We schoolchildren looked at each other. We looked at our teacher. She stared back, eyes big as plates, face white as a sheet. It was the second class of the day. We knew right then and there that something terrible was happening. We couldn't immediately put it into words. But gradually, as the frenzied cry of the alarm continued to rise and fall, our class acknowledged reality. "What's that? What on earth is that?" our teacher asked so softly we could barely hear her. "It's an air raid siren! The Russkis are coming," our class yelled as one. "IT'S WAR!"[1]

For these Finnish schoolchildren, the war started as a surprise attack by the Soviet Union on November 30, 1939, when Soviet bombers appeared without warning in the skies over Helsinki and bombed Finland's capital.[2] The war that later came be known as World War II came as a shock to millions due to its unprecedented scale but also due to the introduced new technologies, strategies, and tactics. People were shaken by the surprisingly

S. Laakkonen (✉)
University of Turku, Turku, Finland

© The Author(s) 2019
S. Laakkonen et al. (eds.), *The Resilient City in World War II*,
Palgrave Studies in World Environmental History,
https://doi.org/10.1007/978-3-030-17439-2_1

quick advance of tanks and other land forces, the mass bombings of civilians, the annihilation of opponents by means of total war, and the merciless attitude toward the vanquished nations.[3] It was in absolute terms the world's most destructive war, claiming approximately 50–70 million human lives, depending on the method of calculation. In addition, the war injured millions of people and other living creatures. It dramatically weakened opportunities for action in nature conservation, animal protection, and environmental protection during or after the war.[4] Finally, World War II gave birth to the Cold War, which for half a century divided the world into competing socio-economic and military blocs that threatened to desolate planet earth with their arsenals of weapons of mass destruction.

However, from an urban point of view, World War II was and remains until today a paradox. In public imagination, this total global war was waged above all by omnipotent states and armies. Towns and cities were swept aside by military commanders into a marginal role; they were assigned the role of good servants, they were to be obedient and industrious, yet humble if not invisible. Even towns and cities where decisive operations or battles took place were generally regarded simply as battlegrounds, passive sites where external active forces clashed. And yet due to the industrial nature of modern warfare, state powers were completely dependent on the innovations, products, and services provided by the towns and cities. We argue that no other war in human history has been waged with such ferocity and devastation done *to cities, against cities, and in cities.*

Towns and cities were of crucial importance to warfare during World War II. The war was waged by the most urbanized and industrialized powers in the world, including the United States, Germany, United Kingdom, and Japan. University towns had a central role in research and development in all belligerent countries. War was waged in the air, on land, and on and under the seas by sophisticated machines fabricated by women and operated by men. Warfare between the major powers depended completely on the mass production of industrial products in towns and cities: Osaka, Detroit, and Essen are three of numerous examples. Consequently, the social and demographic impact of the war economy was significant in belligerent cities.

In the United States, for instance, the draft quickly depleted the American industrial labor force and provided for the first full employment after the Great Depression. At their peak, the armed services commanded 12,500,000 men. Industry needed to replace these employees and add another 6,000,000. In all, the war uprooted 15,000,000 male and female defense workers and 16,000,000 servicemen in the United States only. While the war created some boomtowns on the coasts, to staggering environmental consequences, many towns and cities in the interior faced decreasing populations and reducing environmental stress.[5] In the USSR, war devastated hundreds of towns and cities while evacuation of factories and millions of workers from the war zone created boom towns in the interior of the country, above all in the Ural region.[6] In addition, it may be claimed that frontlines and fortresses provided with developed infrastructure including medical, postal, and even cultural services, and accommodating millions of men and hundreds of thousands of animals, could be conceptualized as a new urban form stretching over continents.[7]

Civilians and towns have always suffered from war. Yet World War II was the first war in which military strategies systematically aimed at and succeeded in devastating towns and cities and killing civilian populations on a massive scale. The first signs of the new strategy became apparent in the German air raids conducted on British towns during World War I, which killed around six hundred civilians. German air forces tested this new strategy more broadly during the Spanish Civil War, killing some thousands of defenseless urban inhabitants, and then launched it at full scale during its World War II attacks on Polish and British urban centers. The Japanese air force also adopted a strategy of terror bombing in China in the late 1930s and 1940s. The heavy bombers of the Allied forces, which had been planned before the war, bombed systematically towns and cities in Germany and Japan, killing 1.5 million civilian residents and seriously injuring more than 2 million. Millions were evacuated and another 16 million were made homeless.[8]

When military created urban firestorms, they took the power of nature into their own hands. On the night of February 13, 1945, over 90 percent of Dresden's historic beautiful city center was destroyed and about 25,000 people were killed. Not only the inhabitants of the city suffered of the incendiary bombings but also Allied prisoners of war. London-born paratrooper Victor Gregg was taken as a prisoner and put to compulsory work. He was later caught sabotaging a factory and was sent for execution in Dresden on the very day that the air raid began: "I had been through six years of war," recollected the 95-year-old veteran when he was interviewed. "I've lost all but three of the 28 blokes who I joined up with in 1937. But nothing prepared me for seeing women and children alight and flying through the air. Nothing prepared me for that." Gregg said that it took him 40 years to get over that "evil" night.[9] The angel that miraculously remained standing on the roof of the city hall of Dresden became a black angel.

It was by no means an accident that the atom bombs were dropped on cities, too. Therefore, the home front—or should we rather say the new urban frontline—needs to be taken into account in order to understand the fundamentally urban nature of World War II. Needless to say, the planners of World War III (all of whom represented the winners of World War II) aimed at completing this twentieth-century strategy of urban terror by relying almost exclusively on air raids and the annihilation of major urban centers and civilian populations of the opposing military alliances. The drone war in the Middle East is just the most recent example of this strategic continuum.

To conclude, World War II (and other wars) could be described as a series of shocks consisting of the fear of war, the onset of war, acts of war, and also of the cessation of war, and then the unforeseeable post-war consequences. We will therefore use the concept of *shock city* to explore the multidimensional environmental crises that World War II signified for towns and cities, above all in Europe and Asia.[10] The concept of shock city enables us to assess and compare the different impacts of war on both urban societies and environments.

A Model City

In the end, World War II was not only a military shock but an economic, social, political, and cultural one as well. Urban populations, institutions, infrastructures, and environments were heavily modified by war. Yet, conceptualizing war as a destructive shock alone would generate a biased impression of the relationship between the urban and natural worlds. In addition to the wails of alarms, other voices could also be heard in towns and cities, especially at the end of war: "Streets like these; warehouses rising above endless rows of hideous houses, factories built over gardens, no space for playgrounds, churches tucked away behind railway arches – streets like these must have no place in the post-war Britain. Homes that were built without thought of consequences are even worse than firebombs or high explosives."[11]

These were the opening words of a propaganda film entitled *Model City* issued by the Ministry of Information in Britain at the end of World War II. The commentary repeatedly announced that profound reforms "must be realized" in post-war Britain in order to provide a healthy and pleasant city with sufficient open space, trees, gardens, and sunlight for everyone. The message of the film was explicit. The new model city was to be a just and democratic city for all the inhabitants—and it had to be realized promptly. The wartime coalition government established in Britain in 1940 had understood that in order to win the war against Germany, the socially deeply stratified British society had to be radically reformed in order to make it worth defending. Political democracy was not enough if it did not deliver well-being. Consequently, the socio-economic outlines of this better society were rapidly laid out and agreed upon during the Blitz. Hence, every German bomb dropped on British towns and cities was a vote for a profound political change that finally broke down the pre-war class barriers. By December 1942, a report commonly known as the "Beveridge Report" recommended that the government should provide adequate income, health care, education, housing, and employment for all—after the war. Because most Britons were urban residents, towns and cities had a central place in these plans for a new society. In brief, the planned model city was an expression of this politically radical version of a new model society: the welfare state. The future model city was to provide a concrete reward for the defenders of the isles, a better society in which to work and live in the brave new post-war world—something that differed completely from the pre-war society.

However, wartime planning of this model city was not solely limited to socio-economic reforms. In the summer of 1943, an extensive survey was completed of the city of Hull, a port city in Yorkshire, which was considered to be "an example of a blitzed town." In addition to socio-economic issues, this study included ample amounts of information on and maps of, for example, outdoor leisure facilities and sites of cultural and historical importance, classifications of landscape types, the availability of open spaces, soil conditions, sewers, and watercourses, energy infrastructure, and areas affected by smoke pollution. Even the main sources of noise pollution throughout Hull were located on the survey. The Hull Regional Survey was completed to provide "a local and national model" for planning post-war reconstruction of urban nature and landscapes and the protection of urban soil, waters, and air. This showed that not only the urban society and the built environment but also urban nature and environment had to be reconsidered in the future model city, the *City of Tomorrow*.[12]

Due to the new ethos of the public good and the increased powers of the public authorities, new plans to protect urban nature were launched during and after the war. Also new nature conservation organizations were established in towns and cities, and urban inhabitants joined them during the war more actively than during peacetime.[13] Hence, it is helpful to address the concept of a *model city* as well in order to understand the revolutionary nature of wartime political developments and related post-war urban reforms.

Resilient City

The conflicting concepts of shock city and model city provide a common yet ambiguous framework for exploring the multifaceted urban environmental history of World War II. These coupled concepts emphasize that, in addition to being a destructive process, war promoted genuine progress. Hence, the concept of model city provides an unexpected but fruitful angle for exploring the different dimensions of war as a socio-environmental urban agent. It offers fruitful insights for discussing the impacts of war on the concepts of shock city and model city because World Wars represent concrete examples of the "paradox of progress" that the urban-industrial era has signified in world history. If the term used by Karl Marx and Joseph Schumpeter is adopted, war could be conceptualized as a distinctive form of creative destruction that incessantly revolutionizes the social structure from within, incessantly destroying the old one, incessantly creating a new one.[14] Consequently, shock city and model city are best understood as complementary and not contradictory images of a complex and integrated process.

However, as a rule even the most hard hit towns and cities, including such extreme cases like Hiroshima, Chongqing, Stalingrad, and Dresden, survived wartime destruction, recovered gradually, and flourish today. Consequently, while the concepts of shock city and model city are used to make sense of the relationship between war, cities, and the urban environment, the key concept of this book is *resilient city*.[15] Urban resilience refers here to the capacity of towns and cities to function and provide realistic living opportunities to their inhabitants no matter what adversities they encounter. Resilience depends on the capacity of city inhabitants (both human and non-human), communities, institutions, and infrastructure to face man-made and/or natural stress or shocks.[16] This variation of urban socio-environmental resilience in wartime, place, and contents is the key concept of this volume.

What Is the Place of the City and War in Environmental History?

On an international level, the impacts of the war on the urban environment were massive and highly diverse, if not downright contradictory.[17] Roger W. Lotchin described the circumstances of American cities in his presidential address at a meeting of the Urban History Association: "Although we know from the experience of Los Angeles that the war actually created the smog problem there, few studies of war and the environment exist. Yet the war poured tons of pollutants into the atmosphere and into the waterways. By 1945, places like Pittsburgh and the other Ohio and Mississippi River cities had so much smoke in the air that they had to turn the streetlights on at noon, dust bowl style. At war's end, the L.A. area found its beaches polluted, as were those of the city of St. Francis. And the conflict exerted tremendous pressures on the wood and mineral products of the nation as the war's voracious appetite for housing, tanks, ships, and planes grew ever greater. The pressure on species and nonrenewable resources must have been enormous. Yet this process does not seem to interest environmental or any other historians."[18]

Such examples inspire the posing of broader questions. What were the actual impacts of World War II on the urban environment and related ideas and practices? What kinds of solutions were proposed and adopted to assuage the difficult wartime and post-war situations? What were the potential short-term and long-term, positive and negative impacts of the war on the wartime and post-war urban environments? And, from a contemporary socio-ecological point of view, could shock cities during World War II have become in some way model cities? Could these wartime towns have some implications for the peacetime urban worlds of today as we seek the sustainable model cities of the future?

The impact of World War II on the environment in general and the urban environment in particular has been rarely addressed in historical studies conducted to date. The main reason for this neglect may be found in the historiography of environmental history studies. When the field emerged in the late 1970s in the United States, most studies focused on the wilderness and the expansion of agriculture, nature protection movements, and national parks.[19] In wilderness-deficit Europe studies focused on the countryside. In France, studies addressed rural history and peasants. In the United Kingdom, studies focused on picturesque landscapes and gardens. In Germany, the cultural history of the landscape, *Landschaft*, was of great importance.[20] In the northern regions of the continent, Scandinavia and Russia, forests and related debates were addressed instead.[21] In the colonial world, forests and plantations were central themes of the first wave of environmental history studies.[22] All of these studies were of high importance during the early years of environmental history. Gradually, however, scholars also started to ask what the role of the city, of urban-industrial environmental problems, was in studies of environmental history.

Urban-industrial environmental histories developed as the second wave of environmental history studies. In the United States, pioneers like Martin Melosi, Christine Rosen, and Joel Tarr addressed the importance of cities in understanding the development of the modern environmental discourse and focused on urban infrastructures and the related pollution issues.[23] In addition to studies on specific themes, holistic studies on the environmental history of a single city have been conducted on several North American urban centers.[24] Peter Brimblecombe, Bill Luckin, and Christopher Hamlin initiated studies on infrastructure and pollution in European cities.[25] Yet the only non-North American city to date that has been the subject of an extensive city-specific study is Helsinki.[26] Studies on the history of urban nature, open spaces, and town planning have developed recently on both continents. William Cronon integrated cities and their hinterlands in his influential study of Chicago.[27] The publication of several edited books[28] and special issues[29] on urban environmental history show the importance of international cooperation in developing the field.

While environmental studies explored the wilderness, the countryside, and the cities, some scholars started to concentrate on contextual changes that affected all of these elements. This was partly because most studies on the history of nature conservation and environmental protection have focused on peacetime developments. Consequently, increasing attention has recently been paid to the role of war and mass violence in environmental changes.[30] The emphasis on the contextual impact of environmental changes could be defined as one novel element in the current third wave of environmental history studies.

Studies on the environmental history of warfare now make up a rapidly growing international field. But this new area of investigation seems to some extent mimic previous waves of environmental history studies. Above all, scholars have addressed the American Civil War, the Vietnam War, and the Gulf Wars,[31] while relatively little attention has been paid to the World Wars.[32] Also, most studies on the environmental history of war have explored nature and the countryside and neglected the urban environment. To conclude, World Wars and urban history[33] are the two Achilles' heels of studies on the environmental history of war.[34]

One obvious reason for the lack of studies on the urban environmental history of World War II is the complexity of the themes. Cities are the most complex social and technical mechanisms devised by man, and urbanization is a massive, unplanned experiment in global landscape change. Wars are cataclysmic events that sculpt the contours of international systems.[35] If we add these two components, war and cities, to the most complex system on the globe, that is, nature in general and the urban environment in particular, the result is a fascinating but ever mutating socio-environmental triangle ripe for exploration.[36] This volume aims at detecting emerging trends in scholarship even though only some major themes and angles can be addressed due to the complexity of the field and very limited supply of studies. We are still lacking the "big picture" of the urban environmental outcomes of this war. Therefore, we see the volume as a sample of initiatives and not as an exhaustive review of what the field will become in the future.

Is There Any Need to Explore the Urban Environmental History of World War II?

There is a definite need to explore urban environmental history because cities have been and remain at the center of modern environmental issues due to their dialectic role. On the one hand, contemporary urban-industrial environmental problems began in cities. On the other hand, cities have been at the forefront of debating, studying, and solving these problems. The adoption of public health ideology, related practices, and technical innovations in towns and cities at the end of the nineteenth century created the base for modern urban-industrial environmental protection. Due to rapid urbanization of the world over the past century, towns and cities have transformed from local caretakers of their environment to key actors in planetary stewardship.[37] More than half of mankind now live in cities. The study of urban environmental history is one key to understanding the development of global environmental problems and solutions.

This book will be the first to provide an in-depth investigation of the relationship between World War II, cities, and the environment. There is a definite need to explore these factors due to the significance of World

War II. In all, World War II directly or indirectly involved 61 countries and 1.7 billion people, that is, three-quarters of the world's population. The war was waged in the arctic, temperate, and tropical zones and spanned via practically all continents except Antarctica. World War II destroyed and/ or changed and transformed towns and cities and thereby urban environments more than any other previous or subsequent war. It is time to connect the urban world, war, and the environment.

We consider World War II as a "long war" along the thoughts of Winston Churchill who addressed the World Wars and inter-war period as "another Thirty Years' War."[38] This "cluster war" started much earlier than 1939 in Europe, Asia, and Africa and came to an end later than 1945. The book argues for the importance of understanding war as a distinctive force in urban and environmental change. As Richard Tucker has maintained, neither can be understood without the other.[39] War and peace were both integral parts of the social and environmental history of the twentieth century. War was an important catalyst for structural changes that emerged in peacetime, and the other way around.

This proposal builds bridges between disciplines (urban, environmental, and military history), short-term and long-term developments, rural and urban areas. The volume covers a notably large and diverse sociogeographical area as it connects three continents (Europe, America, and Asia). We hope that that this volume would inspire historians in different continents to dig more deeply into the unknown urban environmental history of war.

We believe that historical studies have something important to offer current and future cities as well. Since the beginning of urban-industrial development, human-induced crises such as water, air, and soil pollution have compelled cities to find solutions. Towns and cities are responsible for this task today and will continue to be in the future. For both good and bad, wartime cities could be seen as ahistorical models for the sustainable cities of the future.[40] But if we think that the experiences of World War II are so distant and that we have nothing to learn from them, we are being ignorant. If we think that the environmental awareness we have today did not exist in any form before the war, we are misguided. If we think that the implications of war will not or cannot be adopted today or in the future because World War II was such an exceptional time, we are being naïve.

Recurrent climate changes have caused the most massive upheavals that planet earth has experienced during its existence. Now we are about to face another climate change, which will be more rapid than most of the previous changes. Consequently, few of us believe that the future environmental changes that the world will face will take place without grave socio-economic crisis and political conflicts including mass violence and wars. However, urban environmental history of World War II shows how rapidly both crises and in response public power and urban citizens can change urban societies and environments. Therefore, we believe that World War II may have some important implications for the future, because it was the most profound crisis of the twentieth century, affecting populations and urban environments all over the world. It is time to listen to the plaintive wail of the air raid sirens, the acoustic icon of the twentieth-century development, and address related processes in different continents. It is time to take a look at the urban environmental history of World War II.

Main Elements of the City

In order to examine what urban resilience signified in practice during World War II, we have adapted in this book a holistic multilayer concept of the city that addresses its three elements, that is, physical environment (urban nature), technological system (urban infrastructure), and human beings (urban society). Together these three parts constitute in socio-ecological theory the main elements of every city. The main parts of the book address these three urban elements. Most essays focus on administrational areas of given cities, while a few also explore interaction between city and hinterland. This volume addresses towns and cities on three continents, in North America, Europe, and Asia. The selected essays explore urban issues in both Allied and Axis countries.

The predominant definition of World War II says that its battles started in 1939 and ended in 1945. In practice, these standard years provide only a general framework. The war was quite short for the United States while in China it started in 1937 and effectively ended in 1949. We have chosen the first chapter of the book to address the Spanish Civil War (1937–39), because it was in many ways already a part of the "long" World War II, which represented only one, though an important link in a chain of several wars reaching from the Franco-Prussian War (1870–71) to the World Wars, Cold War (1946–1991), Yugoslav Wars (1991–2001) and until today (e.g., Korean crisis). The chapters in this book focus on World War II in terms of different time periods depending on the chosen case studies. Authors may refer to pre-war or post-war periods but mainly they discuss wartime events. Environmental histories of both the inter-war period and post-war recovery deserve books of their own.

As said before, the key concept of the book is urban resilience. There are, however, no standard definitions or explanations for it, not at least in a historical framework. Nevertheless, both editors and authors of this book believe that an important pathway toward urban resilience is the exploration of past failures and successes in order to understand ongoing developments, and anticipate potential future challenges.[41] This book attempts to provide such a pathway. Despite the lack of previous studies on the urban environmental history of World War II, all chapters will address the question of urban resilience through detailed empirical case studies.

Notes

1. Aarni Krohn, "Kruununhaka – kotikaupunkini," in Paavo Haavikko, ed., *Helsinki, kaupunki graniittisilla juurilla, avaralla niemellä* (Helsinki: Art House, 2000), 203.
2. For Winter War, see Pasi Tuunainen, *Finnish Military Effectiveness in the Winter War, 1939–1940* (London: Palgrave Macmillan, 2016).
3. For a rare global approach to the history of World War II, see Evan Mawdsley, *World War II: A New History* (Cambridge: Cambridge University Press, 2009).
4. John Sheail, "'Never Again': Pollution and the Management of Watercourses in Post-war Britain," *Journal of Contemporary History* 33, no. 1 (January 1998): 117–133; Rauno Lahtinen and Timo Vuorisalo, "'It's war and everyone can do as they please!' An environmental history of a Finnish city in wartime," *Environmental History* 9, no. 4 (October 2004): 675–696.
5. Roger Lotchin, "Turning the Good War Bad?: Historians and the World War II Urban Homefront," *Journal of Urban History* 34, no. 1 (November 2007): 33, 174–175.
6. Donald Filtzer, *The Hazards of Urban Life in Late Stalinist Russia: Health, Hygiene, and Living Standards, 1943–1953* (Cambridge: Cambridge University Press, 2010).
7. Simo Laakkonen, "Warfare: An Ecological Alternative for Peacetime? The Indirect Impacts of the Second World War on the Finnish Environment," in Edmund Russell and Richard Tucker, eds., *Natural Enemy, Natural Ally: Historical Studies in War and the Environment* (Corvallis: Oregon State University Press, 2004), 175–194.

8. Kenneth Hewitt, "Place annihilation," *Annals of the Association of American Geographers* 73, no. 2 (June 1983): 257–284; Jörg Friedrich, *The Fire: The Bombing of Germany, 1940–1945*. Translated by Allison Brown (New York: Columbia University Press, 2008).
9. James Burke, "Former British POW Describes the Horror of the Dresden Fire-Bombing," *Vision Times*, March 26, 2015. Also US prisoners of war testified the bombing. Kurt Vonnegut's well-known anti-war novel *Slaughterhouse-Five, or The Children's Crusade: A Duty-Dance with Death* (1969) was based on his experiences of the same bombing that he survived in the slaughterhouse.
10. The concept adopted by Harold Platt (*Shock Cities: The Environmental Transformation and Reform of Manchester and Chicago* (Chicago: University of Chicago Press, 2005) is presented in a heavily modified format here.
11. Imperial War Museums, *Worker and War-Front Magazine*, Issue Number 12. Production date of the original news reel is 1945-01, http://film.iwmcollections.org.uk/record/index/5161 (accessed August 21, 2013).
12. Post-war planning and reconstruction in Britain: Hull Regional Survey, Ministry of Works, 1943. Imperial War Museum. For other reconstruction plans see Peter J. Larkham and Keith D. Lilley, *Planning the 'City of Tomorrow'. British reconstruction planning, 1939–1952: An annotated bibliography* (Pickering: Inch Books, 2001); Andrew Jenks, "Model City USA: The Environmental Cost of Victory in World War II and the Cold War," *Environmental History* 12, no. 3 (July 2007): 552–77; Richard Hornsey, "'Everything is made of atoms': the reprogramming of space and time in post-war London," *Journal of Historical Geography* 34, Issue 1 (January 2008): 94–117; Peter Shapely, "Governance in the Post-War City: Historical Reflections on Public-Private Partnerships in the UK," *International Journal of Urban & Regional Research* 37, no. 4 (July 2013): 1288–1304.
13. Raymond Dominick, *The Environmental Movement in Germany: Prophets and Pioneers, 1871–1971* (Bloomington: Indiana University Press, 1992), 84, 122; John Sheail, "War and the development of nature conservation in Britain," *Journal of Environmental Management* 44, no. 3 (July 1995): 267–283.
14. Joseph A. Schumpeter, *Capitalism, Socialism, and Democracy* (London, New York: Harper & Brothers, 1942), 83. Please note that in our text the term "social structure" is substituted for "economic structure."
15. The following books are the most important previous edited volumes on the history of the resilient city: Lawrence J. Vale and Thomas J. Campanella, eds., *The Resilient City. How Modern Cities Recover from Disaster* (New York: Oxford University Press, 2005). This book provides a rich historical

overview of cities and their inhabitants that have faced natural and manmade disasters over the nineteenth and twentieth centuries. It also discusses devastation wrought on Berlin, Warsaw, and Tokyo during World War II. The second book, Howard Chernick, ed., *The Resilient City: The Economic Impact of 9/11* (New York: Russell Sage Foundation, 2005), discusses more recent urban crisis. See also Michael Burayidi, ed., *City Resilience*, 4 vols. (London: Routledge, 2015).

16. For different definitions of resilience see Fikret Berkes, Johan Colding, Carl Folke, eds., *Navigating Social-ecological Systems: Building Resilience for Complexity and Change* (Cambridge: Cambridge University Press, 2003).

17. Philip Ziegler's popular account *London at War 1939–1945* (London: Pimlico, 2002) and Cynthia Simmons, Nina Perlina, *Writing the Siege of Leningrad: Women's Diaries, Memoirs and Documentary Prose* (Pittsburgh: University of Pittsburgh Press, 2002) focus on everyday wartime life. Otherwise similar studies seem to focus on the Great War. See e.g., Jay Winter, Jean-Louis Robert, eds., *Capital Cities at War: Paris, London, Berlin 1914–1919* (Cambridge: Cambridge University Press, 1997) and Maureen Healy, *Vienna and the Fall of the Habsburg Empire: Total War and Everyday Life in World War I* (Cambridge: Cambridge University Press, 2004).

18. Roger Lotchin, "Turning the Good War Bad? Historians and the World War II Urban Homefront," *Journal of Urban History* 33, no. 2 (January 2007): 176.

19. Donald Worster, ed., *The Ends of the Earth: Perspectives on Modern Environmental History* (New York: Cambridge University Press, 1988); Donald Worster, "Transformations of the Earth: Toward an Agroecological Perspective in History," *Journal of American History* 76, no. 4 (March 1990): 1087–1106.

20. W. G. Hoskins, *The Making of the English Landscape* (London: Hodder and Stoughton, 1955); Thomas Lekan, *Imagining the Nation in Nature: Landscape Preservation and German Identity, 1885–1945* (Cambridge, Mass.: Harvard University Press, 2004).

21. Douglas R. Weiner, *A Little Corner of Freedom: Russian Nature Protection from Stalin to Gorbachev* (Berkeley: University of California Press, 1999); Ronny Pettersson, ed., *Skogshistorisk forskning i Europa och Nordamerika. Vad är skogshistoria, hur har den skrivits och varför?* (Stockholm: Kungl. Skogs- och Lantbruksakademien, 1999).

22. For example, Richard P. Tucker and John F. Richards, eds., *World Deforestation in the Twentieth Century* (Durham, NC: Duke University Press, 1987).

23. James E. Krier and Edmund Ursin, *Pollution and Policy: A Case Essay on California and Federal Experience with Motor Vehicle Air Pollution, 1940–1975* (Berkeley: University of California Press, 1977); Martin V. Melosi, ed., *Pollution and Reform in American Cities, 1870–1930* (Austin and London: University of Texas Press, 1980); for discussion, see Martin V. Melosi, "The Place of the City in Environmental History," *Environmental History Review* 17, no. 1 (March 1993): 1–23; Christine Meisner Rosen and Joel A. Tarr, "The Importance of an Urban Perspective in Environmental History," *Journal of Urban History* 20, no. 3 (May 1994): 299–310.
24. Edited volumes of the environmental history of a single city have explored, for example, St. Louis, Pittsburgh, Houston, Boston, Phoenix, New Orleans, Seattle, Los Angeles, and Montreal.
25. Bill Luckin, *Pollution and Control: A Social History of the Thames in Nineteenth Century* (Bristol: Adam Hilger, 1986); Peter Brimblecombe, *The Big Smoke: A History of Air Pollution in London since Medieval Times* (London: Routledge, 1987); Chris Hamlin, *A Science of Impurity: Water Analysis in Nineteenth Century Britain* (Berkeley: California University Press, 1990); Alain Corbin, *The Foul and the Fragrant: Odor and the French Social Imagination*, (Cambridge, MA: Harvard University Press 1986); André E. Guillerme, *The Age of Water: The Urban Environment in the North of France, A.D. 300–1800* (College Station, TX: Texas A&M University Press, 1988); Franz-Josef Brüggemeier and Thomas Rommelspacher, *Blauer Himmel über der Ruhr: Geschichte der Umwelt im Ruhrgebiet, 1840–1990* (Essen: Klartext, 1992); Ian D. Biddle and Kisten Gibson, eds., *Noise, Audition, Aurality: Histories of the Sonic World(s) of Europe, 1500–1945* (Farnham: Ashgate, 2015).
26. Simo Laakkonen, Sari Laurila, Pekka Kansanen and Harry Schullman, eds., *Näkökulmia Helsingin ympäristöhistoriaan. Kaupunki ja sen ympäristö 1800- ja 1900-luvuilla* (Helsinki: Edita, 2001) (Approaches to the environmental history of Helsinki. The city and its environs in the nineteenth and twentieth century, in Finnish.)
27. Dorothee Brantz and Sonja Dümpelmann, eds., *Greening the City: Urban Landscapes in the Twentieth Century* (Charlottesville: University of Virginia Press, 2011); William Cronon, *Nature's Metropolis: Chicago and the Great West* (New York, London: W. W. Norton & Company, 1991).
28. See, e.g., Dieter Schott, ed., *Energy and the City in Europe: From Preindustrial Wood-Shortage to the Oil Crisis of the 1970s* (Stuttgart: Steiner, 1997); Christoph Bernhardt, ed., *Environmental Problems in European Cities in the 19th and 20th Century* (Münster: Waxmann, 2001); Dieter Schott, Bill Luckin, and Geneviève Massard-Guilbaud, eds., *Resources of the City: Contributions to an Environmental History of Modern Europe*

(Aldershot: Ashgate, 2005); Andrew C. Isenberg, ed., *The Nature of Cities: Culture, Landscape, and Urban Space* (Rochester: University of Rochester Press, 2006); Geneviève Massard-Guilbaud and Richard Rodger, eds., *Environmental and Social Justice in the City: Historical Perspectives* (Cambridge: The White Horse Press, 2011).
29. Simo Laakkonen and Sari Laurila, eds., "The Sea and the Cities," a special issue, *Ambio. A Journal on the Human Environment* 30, no. 4–5 (August 2001): 263–326; Stephen Bocking, ed., a special issue, *Urban History Review/Revue d'Histoire Urbaine* 34, no. 1 (Fall 2005 automne): 3–112; Geneviève Massard-Guilbaud, ed., "Ville et environnement," a special issue, *Histoire Urbaine* 1, no. 18 (avril 2007): 5–156; Geneviève Massard-Guilbaud and Peter Thorsheim, eds., "Cities, Environments, and European History," *Journal of Urban History* 33, no. 5 (July 2007): 691–847.
30. Arthur H. Westing, *Ecological Consequences of the Second Indochina War* (Stockholm: Almqvist & Wiksell, 1976); Susan Lanier-Graham, *The Ecology of War: Environmental Impacts of Weaponry and Warfare* (New York: Walker, 1993); A. Corvol and J. P. Amat, eds., *Forêt et guerre* (Paris: L'Harmattan, 1994); Günther Bächler et al. *Kriegursache Umweltzerstörung Vol. 1–III* (Zürich: Rügger, 1996).
31. Jay E. Austin and Carl E. Bruch, eds., *The Environmental Consequences of War: Legal, Economic and Scientific Perspectives* (Cambridge: Cambridge University Press, 2000).
32. As exceptions see Edmund Russell, *War and Nature: Fighting Humans and Insects with Chemicals from World War I to Silent Spring* (New York, Cambridge University Press, 2001); Joachim Radkau and Frank Uekötter, eds., *Naturschutz und Nationalsozialismus. Geschichte des Natur- und Umweltschutzes* (Frankfurt am Main: Campus Verlag, 2003); Franz-Josef Brüggemeier, Mark Cioc, and Thomas Zeller, eds., *How Green Were the Nazis? Nature, Environment, and Nation in the Third Reich* (Athens: Ohio University Press, 2005).
33. Rainer Hudeman and François Walter, eds., *Villes et guerres mondiales en Europe aux XXe siècle* (Paris and Montreal: L'Harmattan, 1997); Marcus Funck and Roger Chickering, eds., *Endangered Cities: Military Power and Urban Societies in the Era of the World Wars* (Boston: Brill, 2004); Stefan Goebel and Derek Keene, eds., *Cities into Battlefields: Metropolitan Scenarios, Experiences and Commemoration of Total War* (Farnham: Ashgate, 2011). The main approach of these studies is cultural history. For other approaches see e.g., Nico Wouters, "Municipal Government during the Occupation (1940–45): A Comparative Model of Belgium, the Netherlands and France," *European History Quarterly* 36, no. 2 (April 2006): 221–246.

34. Lahtinen and Vuorisalo, "'It's war and everyone can do as they please!'"; Matthew Evenden, "Lights Out: Conserving Electricity for War in the Canadian City, 1939–1945," *Urban History Review/Revue d'Histoire Urbaine* 34, no. 1 (Fall 2005): 88–99; Laakkonen, "Warfare: An Ecological Alternative for Peacetime?" in *Natural Enemy*, 175–194; Timothy Cooper, "Challenging the 'refuse revolution': War, waste and the rediscovery of recycling, 1900–1950," *Historical Research* 81, no. 214 (2008): 710–731; M. Riley, "From Salvage to Recycling: New Agendas or Same Old Rubbish?", *Area* 40, no. 1 (2008): 1–11.
35. Jari Niemelä et al. "Introduction," in Jari Niemelä et al., eds., *Urban Ecology: Patterns, Processes, and Applications* (Oxford: Oxford University Press, 2011), 2.
36. Dieter Schott, "Urban Environmental History: What Lessons Are There to Be Learnt?," *Boreal Environment Research* 9, no. 6 (2004): 519–28.
37. Sybil P. Seitzinger et al., "Planetary Stewardship in an Urbanizing World: Beyond City Limits," *Ambio* 41, no. 8 (December 2012): 787–94.
38. Winston Churchill, *The Second World War. Volume 1.* (London: The Educational Book Company Ltd., 1954), ix.
39. Richard P. Tucker, "The Impact of Warfare on the Natural World: A Historical Survey," in *Natural Enemy*, 37.
40. Friedrich Huchting, *Prüfung alter Verwertungstechnologien aus Mangel- und Kriegszeiten*. Umweltforschungsplan des Bundesministers des Innern. Abfallwirtschaft (Berlin: Umweltbundesamt, 1979); Johanna M. Jurkola, Treading lightly on the environment: Using Second World War fabric saving and clothing reuse techniques to inform contemporary women's clothing design, Thesis, University of Alberta (2003); Timothy Cooper, "War on waste? The politics of waste and recycling in post-war Britain, 1950–1975," *Capitalism Nature Socialism* 20, no. 4 (February 2009): 53–72.
41. See David D. Woods and Erik Hollnagel, "Prologue: Resilience Engineering Concepts," in Erik Hollnagel, David D. Woods, and Nancy Leveson, eds., *Resilience Engineering: Concepts and Precepts* (Boca Raton, FL: Taylor & Francis, 2006), 6.

SECTION II

Urban Environment

CHAPTER 2

Critical Networks

Urban Water Supply in Barcelona and Madrid During the Spanish Civil War

Santiago Gorostiza

INTRODUCTION

On June 18, 1940, Winston Churchill addressed the House of Commons in what has become one of his most famous speeches.[1] After the collapse of Poland and France, and after the battle of Dunkirk, Britain was bracing itself for the coming German attack. "I do not at all underrate the severity of the ordeal which lies before us", stated the British Prime Minister, announcing the beginning of the Battle of Britain, "but I believe our countrymen will show themselves capable of standing up to it, like the brave men of Barcelona".[2]

The Spanish Civil War (1936–1939) had finished on April 1, 1939, only five months before the beginning of World War II in Central Europe, but it made a lasting impression on contemporary observers. The bombing of Barcelona by Italian and German air forces during 1938 and 1939 heralded a new era of war where cities could no longer be considered part

S. Gorostiza (✉)
Institut de Ciència i Tecnologia Ambientals, Universitat Autònoma de Barcelona (ICTA-UAB), Barcelona, Spain

© The Author(s) 2019
S. Laakkonen et al. (eds.), *The Resilient City in World War II*, Palgrave Studies in World Environmental History, https://doi.org/10.1007/978-3-030-17439-2_2

of the rear-guard, and where the targeting of civilian environments from the air was to become paramount. Churchill was not mistaken—the tribulation which was looming over Britain and particularly its cities was probably even more severe than the one that struck Barcelona's citizens. The capacity of civil populations to cope with the large disturbances caused by protracted aerial bombings was to become one of the central struggles of World War II.[3]

Targeting cities and civilians also meant striking at the infrastructures that enable urban life—disrupting the flows of water, energy, and materials that ultimately make this life possible.[4] By exploring how the water supply systems of Madrid and Barcelona coped with the turmoil caused by war, in this chapter I examine the critical role of infrastructure and utility workers' knowledge in enhancing urban resilience under fire. In other words, I intend to unveil the role of human labour—through workers and their organization—in sustaining urban metabolism and making it resilient in the face of acute challenges. Episodes of urban stress provide an opportunity to analyse the performance of utility companies and can reveal historical, social, and political issues embodied in city infrastructures.[5]

The case studies developed in this chapter are relevant not only because they were the prologue of World War II, announcing a new era of total war, but particularly due to their political dimensions. Both in Barcelona and Madrid, workers played a key role in the management of their utility company. The comparison between the two cities is also significant because they represented—both in the 1930s and today—different models of water management. Since the mid-1800s, the water company of Madrid (Canal de Isabel II, which changed its name to Canales del Lozoya during the 1930s) had been the apple of the eye of the Spanish state, which had invested heavily to guarantee water supply to its capital, building dams and aqueducts that secured water in adequate quantity and quality. In contrast, Barcelona epitomizes private water management. Back in the 1930s, after failed attempts of municipalization, the Sociedad General de Aguas de Barcelona extracted water from the local aquifers but lacked the security guaranteed by reservoirs and aqueducts. In some neighbourhoods, water provision per capita averaged 30 litres per day, while in Madrid the average was ten times greater. Quality was also a problem in Barcelona, for not only did typhoid fever rates remain comparatively high, but waters also suffered from an increasing salinization caused by potash mining upstream on the Llobregat River. Neither the Barcelona private water company nor the state provided the funds needed to cope with these problems.[6]

This chapter explores how the water companies of Madrid and Barcelona performed during what is arguably the most stressful episode of their history: the Spanish Civil War. It expands on the previous work of the author about the environmental history and political ecology of water management in both cities.[7] In the case of Barcelona, the main primary sources are the records of the water company—Arxiu General d'Aigües de Barcelona (AGAB). In the case of Madrid, a detailed report located at the archive of Fundación Pablo Iglesias, originally censored by Governmental authorities, plus several published sources by Canales del Lozoya are combined with historical press and bulletins.[8] The chapter roughly follows a chronological narrative, starting with the defence of the reservoirs supplying water to Madrid in the first weeks of the war. The second section moves to Barcelona and introduces the reforms carried out by the anarcho-syndicalist management that took over the water utility company. The third section considers the key role of the water infrastructure in the urban combat that took place in Madrid. The fourth section provides a discussion of the water management in Barcelona and Madrid until the final defeat of the Republic at the hands of Franco's forces. Finally, in the conclusion I underline the historical and political significance of both cases together with the critical role of workers to enhance the resilience of water supply services in cities at war.

Defending Madrid's Waters at Its Sources

On July 17 and 18, 1936, after several months of preparations, the Spanish army launched a military coup against the young Second Spanish Republic, a democratic regime established in 1931. Carried out with the support of conservative groups, monarchists, and the fascist political party (Falange Española), the coup was designed to end all resistance swiftly. However, the failure of the rebels to take hold of the main Spanish cities, along with the divisions in military and police forces and the determined resistance of the leftist trade unions and parties, ignited a civil war that lasted almost three years.

The commitment of workers was key to thwarting the coup in Madrid and Barcelona. Rumours of the military action, which started in the Spanish Protectorate of Morocco on July 17, spread rapidly throughout the country. On July 19, before the insurgents were able to seize the streets of Madrid, a mix of unionists and loyal police and military forces surrounded several military barracks. Unable to force their way out, insurgent military troops fired on the masses encircling the main military garrison of Madrid. Alarmed by the growing extent of the coup, the Republican government ordered the distribution of weapons to the civilian population. Several workers of Madrid's water company, Canales del Lozoya, participated in the struggle against the military, cutting the water supply to the besieged barracks. One of Canales' most renowned engineers was killed in the action.[9]

By July 20, the military coup had been defeated in Madrid and in most major Spanish cities, with the exception of Seville, in the south. But the rebellious troops were consolidating their control of some regions, including Navarre and Castile, north of the capital. Several insurgent military units soon departed towards the mountain pass of Somosierra, which gave access to the Madrid region, and forced the Republican column assigned to defend it into retreat. By July 24, the way to Madrid from the north lay open, and the capital's water reservoirs and dams—Puentes Viejas and El Villar—stood in no man's land.[10]

As he arrived at the region accompanied by militiamen from the water company, the Government's Delegate in Canales del Lozoya found the Republican troops preparing fortifications south of the reservoirs, and burst out in anger. If the reservoirs were not protected, its occupation by the insurgent troops could decide the fate of Madrid. "What will we defend if the dam gates of Madrid's water are in the enemy's hands?", he wrote to his superiors.[11] Taking advantage of a halt in the advance of insurgent troops, which lacked ammunition, Republican militiamen—with workers of Canales del Lozoya among them—advanced to the reservoirs and took up positions around them. Soon Republican forces occupied the near town of Buitrago de Lozoya and established a defence line north of the reservoirs. This time, when the rebel military troops resumed their advance, the battlefront stabilized; and by the end of July, further reinforcements secured the region (see Fig. 2.1). Both sides soon fortified their defence lines. Although combat went on, the so-called Guadarrama battle had ended up with a Republican victory, and the battlefronts in this sector would experience little change until the end of the war (see Fig. 2.2).[12]

The reservoirs had been saved by a narrow margin, but the military forces encamped around them, together with the proximity of enemy forces, remained a concern for Republican military and sanitation authorities. Water transfers between Puentes Viejas—the reservoir closer to the battlefront—and El Villar—where the water aqueduct to Madrid started—

Fig. 2.1 "Madrid will not lack water." Guards, soldiers, militiamen, and women in front of El Villar dam. Published in *Ahora*, July 30, 1936, p. 22. (Source: Hemeroteca Digital de la Biblioteca Nacional de España. Photo by Contreras and Vilaseca)

were cancelled by the Republican military, due to fears of pollution. Tactical concerns also mattered in this decision, since lowering the levels of Puentes Viejas reservoir would also change the terrain under dispute.[13] Water quality analyses were increased and so was chlorination of Madrid waters. But with no transfers between Puentes Viejas and El Villar, Madrid would run out of water in a matter of weeks. Moreover, the city's consumption increased during August and early September.[14] To address this problem, Canales del Lozoya and the Madrid city council agreed to impose restrictions on the use of water, starting on September 15. Watering gardens and streets was prohibited and the company called for the collaboration of the population to find water leaks.[15]

At the same time, Canales' engineers worked against the clock to come up with alternative plans for the water supply of the city, in case the Lozoya's waters were lost. The distribution network of Canales del Lozoya was unified with the private supplier Hidráulica de Santillana, and plans to take water from the Manzanares and Jarama Rivers were drafted. However, water transfers between Puentes Viejas and El Villar resumed in early October, with a transfer of three million cubic metres that alleviated the supply shortage of the city, and a commission was established to monitor

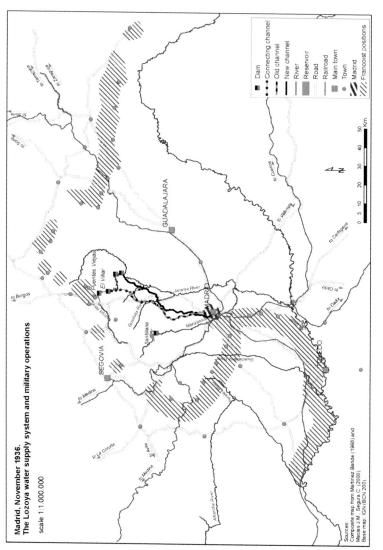

Fig. 2.2 The Lozoya water supply system and military operations, Madrid region, November 1936. (Source: Own elaboration based on J. M. Macías and C. Segura, eds., *Historia del abastecimiento y usos de agua en la Villa de Madrid* (Madrid: Confederación Hidrográfica del Tajo y Canal de Isabel II, 2000) and J. M. Martínez Bande, *La lucha en torno a Madrid en el invierno de 1936–1937* (Madrid: Librería Editorial San Martín, 1968))

transfers, with members of the army, a representative of the city council, and also a delegate of Canales del Lozoya. Moreover, the efforts to save water in the city came to fruition, with a 25% reduction in consumption in October from the previous year.[16]

Throughout these months, the hectic activity of Madrid's public water company also involved a significant internal restructuring. The status of the workers in the company—most of them connected to the socialist trade unions—was improved via a top-down reform which granted them participation in the company's board, but overall control remained firmly in the hands of the Republican government.[17] Meanwhile, more than 600 kilometres to the east, behind Republican lines and far from the battlefronts, the water supply of Barcelona was undergoing a major reorganization.

Revolutionizing Water Management in Barcelona

As in Madrid in July 1936, the joint action of workers and police forces in Barcelona defeated the military coup after a fierce fight in the streets. In Catalonia, however, the workers' victory went hand in hand with a social revolution in which the anarcho-syndicalist trade union Confederación Nacional del Trabajo (CNT) took the lead. In Barcelona, the summer of 1936 saw the unfolding of a deep transformation of economic and social life, which included the collectivization of services, industries, and agriculture. As pointed out by Myrna B. Breitbart, Spain was perhaps the only country in the twentieth century "where anarcho-communism and anarcho-syndicalism were adopted extensively as revolutionary theories and practices in urban and rural areas". This applies particularly to Catalonia, parts of Aragón and Andalusia.[18]

The workers of Barcelona's private water company took an active part in this process. As early as July 24, they seized the company and dismissed its managers, who fled to the regions controlled by the insurgents. One week later, a workers' assembly approved joining the CNT. The Sociedad General de Aguas de Barcelona thus became the Sindicato de Obreros de Aguas de Barcelona (Workers' Trade Union of Barcelona's Waters). A first round of reforms approved by the worker-controlled company increased salaries, reduced working hours, and extended pensions, while reorganizing the operation of the company. They also established a school for the children of the 800 workers of the company, located at the mansion and gardens previously housing the company's director. During the following weeks, workers debated how to address the precarious situation of Barcelona's water supply. Several reforms approved in agreement with the Catalan government were announced in September 1936, addressing what was probably the most important objective of the worker-controlled company: to improve access to water in sufficient quantity and quality.[19]

First, the reforms increased the water allowance of each apartment block in those cases where the previous laws concerning water supply had been ignored. This situation was common enough, since landlords had often resisted the application of the former law. Now, with the reforms of September 1936, landlords would pay for this minimum consumption, while tenants would pay any consumption in excess of that amount. Second, fixed payments for water meters were abolished and homeowners were forced to buy and install their own meters. Finally, the water price was unified into a single tariff for the whole city. Prior to this reform, prices varied according to the elevation of each neighbourhood above sea level. All in all, the rationale behind the reform was to expand water consumption in the city, expecting that the increase in water sold would compensate for the lower prices set.

Accordingly, starting in October 1936, the offices of the Barcelona water company engaged in hectic activity to revise all contracts to comply with the new directives.[20] Contracts in perpetuity, an old privilege enjoyed by certain homeowners, were swiftly terminated—a significant improvement in the efficiency of the supply system. Water provision was automatically revised and increased in many city blocks.[21] At the same time, piped water was installed in neighbourhoods that had previously been left aside by the private management of the company, while the new worker-controlled company increased its control of water sources by absorbing private vendors.[22] The records of the water company also show how water supply expanded in many factories and workshops, where showers and even swimming pools were installed, in accordance with the anarcho-syndicalists' own view of the importance of hygiene (Fig. 2.3).[23] This expansion of water facilities affected even churches, many of which had been devastated in the aftermath of the street combat of July 1936 and were converted to warehouses, community kitchens, or car parks, with water supply installed accordingly.[24]

Meanwhile, more and more workers had been joining militias and departing for the battlefronts in Aragón or Madrid. Completing a rapid advance from the south of Spain, in late October 1936 the army led by General Franco was about to set foot in the Republican capital.

Fig. 2.3 Swimming pool at the workshop n°33 of the Industry of Socialized Wood (Galileu street, Sants neighbourhood, Barcelona), a collectivization run by the CNT. Summer 1936. (Source: Merletti collection, Arxiu de l'Institut d'Estudis Fotogràfics de Catalunya, reference ACP-1-2799p)

Urban War and Infrastructure in the Siege of Madrid

Francisco Franco had been one of several military generals conspiring to organize a coup against the Spanish Republic, but circumstances during the summer of 1936 made him the leader the of the insurgents. First, General Franco led the only experienced troops of the Spanish army: the colonial units formed by the Spanish Foreign Legion (*Legionarios*) and the Protectorate Moroccan soldiers (*Regulares*), units particularly feared by the Spanish population. In the early moments of the war, German and Italian support allowed for an airlift of these units from North Africa to Seville, evading the Republican control of the Gibraltar strait. Starting in August 1936, Franco's troops started a swift and unstoppable advance towards the capital, leaving a trail of violence and repression. In October 1, 1936, Franco was proclaimed *Generalísimo* of the armies and Head of the Spanish State. The advance towards the capital, coming from the south, was completed by the end of the month, and attracted international attention, as the fate of the Spanish Republic seemed to be hanging in the balance.[25]

By the time the Francoist troops started the assault on Madrid, the city was flooded with refugees and the authority of the Republican government, which so far had been unable to stop the advance of the insurgent troops, was at a low ebb. The difficulties experienced by the city's water supply, nonetheless, seemed to have been overcome. With the northern battlefronts under control, water transfers resumed and chlorine supply secured, water restrictions in the city were lifted on October 30. On this very day, however, the increasing attacks on the city killed one of the employees of Canales del Lozoya while at work on the supply infrastructure of the city centre. As the aerial assault on the capital became systematic—making Madrid one of the first cities to experience sustained aerial bombing—the Republican government abandoned the city on November 6 and moved to València. The assault of the insurgents, however, coincided with the arrival in Madrid of a growing stream of Soviet military equipment, together with the International Brigades, a military unit formed by foreign volunteers, which marched in the streets of Madrid chanting "*¡No pasarán!*" ("They shall not pass!"). Boosted in their morale, the defenders offered a dogged resistance against the Francoist soldiers who were unaccustomed to combat in urban environments.[26]

The systematic bombing of the city caused grave damage to the water infrastructure (see Fig. 2.4). Starting in November, bomb impacts on key points of the network caused critical surges in water leaks, as well as more casualties among firefighters, Canales' workers, and militiamen. Several artillery shells hit the urban reservoirs, near the city centre, but they survived intact. Unable to penetrate the city from the south, the Francoist forces shifted their efforts to the west, and around November 15 were able to break the front, crossing the Manzanares River and entering Madrid's brand new University City. The efforts of the defenders of the sector were finally able to hold the battlefront, but the buildings captured and retained by the Francoist units stood only one kilometre away from the offices and urban reservoirs of Canales del Lozoya. During the following days, the involvement of the German Condor Legion and the use of fire-bombing techniques brought the water supply system to the brink of collapse, with water leaks reaching a peak around November 18, coinciding with the largest German air raid until that date. On this very date, Germany and

Fig. 2.4 Bombing damages in Carrera de San Jerónimo, Madrid city center. (Source: Ministerio de Cultura y Deporte. Archivo General de la Administración, Fondo "Archivo Fotográfico de la Delegación de Propaganda y Prensa de Madrid durante la Guerra Civil", F-04042-53501-001r, ANTIFAFOT)

Italy gave official diplomatic recognition to Franco, thus committing themselves to the insurgent forces.[27]

During these critical days, the experience, knowledge, and organization of the workers of Canales del Lozoya proved the significance of water infrastructure for the defence of the city. As leaks became widespread, disrupting water pressure and hindering the work of firefighters, repairs by the Canales employees also multiplied, regaining control of the network and cutting losses after a few days. The company's engineers delivered detailed maps of the underground infrastructure to the Republican military command, warning about its importance. Using the subterranean networks, unfamiliar to Franco's soldiers, company workers systematically cut water supply from the buildings occupied by the insurgent troops, making the lack of water an issue of concern for the rebels. One of Canales' engineers, Federico Molero, led the formation of the so-called subterranean battalion. Joined by miners, city council underground workers and Canales' employees, this unit exploited the unpreparedness of Francoist troops for subterranean warfare by sneaking into their territory via underground infrastructure such as sewers or distribution networks, and then blowing up enemy buildings in the University City. These actions contributed to offset the military superiority of the Francoist troops and caused a certain psychosis among the insurgents.[28]

The Francoist offensive had entered the city and almost reached its heart. But by November 22, 1936, lacking supplies and fresh troops, it had to be cancelled. Newly arrived Soviet planes were contesting control of the skies over Madrid with the German Condor Legion, and the gritty determination of Madrid's defenders brought the battle to a stalemate, which was in fact a major propagandistic victory for the Republic. The long-lived myth of the International Brigades was born in Madrid, which in the Republican imaginary had become "The Tomb of Fascism". Despite renewed efforts to encircle Madrid during the following months, the capital would remain in the hands of the Republic almost until the very end of the war. Until then, Francoist troops would remain only one kilometre from the city's urban reservoirs, but with no connection to the water supply. Aside from the damage caused by bombing, the service of Canales del Lozoya to the rest of the city continued without major problems.[29]

Living Under Fire: Water Supply in Barcelona and Madrid Until the End of the Spanish Civil War

The siege of Madrid mobilized the support of the rest of the Spanish Republic. Several thousand militiamen from Barcelona and the rest of Catalonia joined the defence of the city, and as early as October 1936, the Catalan government had established committees in charge of coordinating the material and human support to the capital. By the end of 1936, the

Fig. 2.5 A present of the city of Barcelona to Madrid: Six water-tank trucks to keep supply going in the bomb-torn neighbourhoods. December 20, 1936 (Source: Pérez de Rozas, Arxiu Fotogràfic de Barcelona)

city council of Barcelona gave Madrid six water-tank trucks to keep the water supply going in those neighbourhoods where the bombing had destroyed the infrastructure (Fig. 2.5). When the evacuation of the capital was announced in early 1937, hundreds of refugees from Madrid found shelter in Barcelona, where several public events were organized in support of the capital, under the slogan "To defend Madrid is to defend Catalonia". Later that year, an expedition of Catalan firefighters travelled to Madrid to support the work of their counterparts in the capital.[30]

Following the Collectivizations' Decree approved by the Catalan government, the transformation of the Barcelona water company into a collectivized entity was approved in July 1937, resulting in the creation of Aigües de Barcelona Empresa Col·lectivitzada (Collectivized Water Company of Barcelona, ABEC). Managed by their own elected workers, ABEC carried out several reforms in the city's water supply during 1937, while facing serious difficulties. The city's water consumption had increased significantly since the beginning of the war, but non-payments hindered the company's revenues. Nonetheless, the workers completed improvements at their wells and pumps that allowed for continued increases in the water supply, and pursued the purification of the River Llobregat's waters, salinized by upriver potash mining companies—a problem that had been repeatedly denounced by the anarcho-syndicalists in previous years. In contrast to the collectivist management of Barcelona, Canales del Lozoya in Madrid remained public, even if workers were given representation in the board. During 1937, once the worst moments of the siege ended, the company implemented reforms similar to the ones approved in Barcelona the previous year. Water contracts granting water in perpetuity were cancelled and payment for the renting of water meters was suppressed.[31]

In the face of the food shortages, water became a fundamental resource in Barcelona and its surroundings, which increasingly turned to urban and peri-urban gardening. Neighbourhood committees occupied monasteries and their lands, putting them to cultivation. By the end of 1936, hundreds of gardeners from different neighbourhoods, connected to the anarcho-syndicalist CNT, joined to form the Colectividad Agrícola de Barcelona y su radio (Agricultural Collective of Barcelona). With more than two thousand workers and circa 850 hectares of land, this collective is one more example of the extent of the anarcho-syndicalist reorganization of Barcelona's economy, society, and environment. In war-torn Madrid, urban gardening was also practiced, but the proximity of the battlefronts made its organization difficult. Vegetable gardens, in any case, required abundant water and became a headache for the collectivized company. While ABEC negotiated special contracts for Barcelona gardeners, in many cases they simply connected to its network without asking for permission. The extent of these practices forced the company to make public calls for residents to stop tapping its distribution systems.[32]

The collectivized water company also got involved in passive defence. After Barcelona was first bombed in 1937, a network of air-raid shelters was built throughout the city's neighbourhoods. ABEC built air-raid shelters in its facilities and provided a fundamental support to those citizens and associations excavating shelters and asking for pipe diversions or installation of electrical equipment. Starting in 1938, the intensification of Italian air raids coming from the island of Mallorca put the city's water supply to the test. Bombing in January 1938 caused more death than in all 1937, but the worst was yet to come.[33]

In March 1938, while a Francoist offensive pushed the Republican lines back towards the Mediterranean, a three-day long Italian air raid on Barcelona left almost a thousand dead and many more injured—among them several ABEC workers. As in Madrid, bombing caused a string of water leaks and dropping water pressure that disrupted supply and degraded the network. By mid-April, Francoist troops had reached the Mediterranean and cut off Catalonia from the rest of Republican Spain. Most critically, they occupied the Pyrenees' dams that supplied electricity to Barcelona's industry. The loss of this strategic energy infrastructure had direct consequences for the water supply of Barcelona. ABEC was forced to reopen its old steam engines, powered by coal, but the high prices of fuel caused the company's expenses to rise. Moreover, despite the efforts of the company, water pressure remained irregular, hindering the supply to higher neighbourhoods and causing the concern of the Fire Department and the complaints of several citizens. On the base of the higher costs faced by ABEC, the Catalan government authorized the company to increase the price of water, which doubled the tariffs established in September 1936. Additionally, a ban was imposed on superfluous water uses.[34]

By August 1938, about 318,000 refugees were in Barcelona. Under recurrent air raids and suffering from lack of food and supplies, the city's conditions were rapidly deteriorating. A few weeks after defeating the Republican army in the Battle of the Ebro, Francoist troops marched towards Catalonia. Barcelona was occupied on January 26, 1939, almost without resistance, as hundreds of thousands fled the city towards the border with France. The former owners of Sociedad General de Aguas de Barcelona rapidly recovered control of the company and showed their gratitude to the Francoist army, paying for a brand new military residence, while many of the workers in positions of responsibility followed the path of exile. Despite the fact that the returning managers publicly despised the worker-controlled management of the war years, the standardization rates, the increase in water supply, and particularly the elimination of perpetual contracts eventually benefited the private management of the company. Nonetheless, Barcelona finished the war with the highest rates of deaths from typhoid fever since 1914. Madrid, in contrast, was the only Spanish province with a mortality rate for typhoid below 5 cases per 100,000 people. The capital resisted until late March 1939, but was occupied with no opposition right before the end of the war, on March 28, 1939. Canales del Lozoya was soon renamed as Canal de Isabel II—the name it had had until 1931 and the same it still holds today. The new management, initially in the hands of the military, found a water supply system that had fulfilled its functions during the war and carried out significant improvements, such as the interconnection between the networks of Canales del Lozoya and Hidráulica de Santillana, which increased the overall security of public water service.[35]

Conclusions

Starting in 2012, archaeological works in the north of Madrid unearthed several fortifications constructed around the dams of Puentes Viejas and El Villar during the Spanish Civil War. Under the initiative of the regional government, a marked path and several information signs have been set up to explain what is now called the "water battlefront", in reference to the strategic importance of this sector during the first weeks of the war. These fortifications are not the only traces of the importance of water supply during the Spanish Civil War that have been unearthed in recent years. In the summer of 2010, maintenance works uncovered an artillery shell embedded at the top of one the city's urban reservoirs. This unexploded projectile on the top of an intact, almost hundred-year-old urban reservoir constitutes a telling symbol of the extraordinary resilience shown by the city of Madrid in the face of military aggression between 1936 and 1939, as well as of the critical importance of urban water infrastructure in cities at war.[36]

However, the symbolism of the artillery shell found in Madrid's urban reservoir must not obscure those who made the city's infrastructure work: the citizens and employees of the water companies. By organizing effectively and resisting the assaults on their cities, citizens, and water company workers from Barcelona and Madrid played a critical role in enhancing the resilience of urban water supply services, and sustaining the urban metabolism of cities themselves. The heart of Madrid's water supply system stood only one kilometre from the battlefronts for most of the conflict, and yet remained working, despite the attempts of the insurgents to disrupt it. What is more, Canales del Lozoya workers turned their infrastructural knowledge into a tactical advantage used against the invaders. In Barcelona, a water company reorganized and managed under anarcho-syndicalist principles expanded supply and maintained service under recurrent bombing raids, while providing at the same time key technical support for the extension of air-raid shelters throughout the city.

The history of each water company helps explain the different outcomes in the two cities. Starting in the mid-1800s, the Spanish state had taken the reins of Madrid's water supply, devoting significant resources to build a modern infrastructure, including dams and aqueducts, in order to guarantee the supply of water to its capital city. In contrast, the extension of water supply in Barcelona had been left to the private sphere, after failed attempts of municipalization in the early 1900s. Not only was water consumption much lower than in Madrid, but the system also depended on extraction from aquifers and suffered from several quality issues. When the military coup brought war to both cities, nonetheless, the companies succeeded in maintaining a continuous supply of water—the lifeblood that underpinned urban resilience. Water was intimately intertwined with food supply, as proven by the extension of urban gardening in Barcelona; with health, as revealed by the (differing) rates of typhoid fever in both cities; and eventually dependent on energy, as shown by the problems experienced in Barcelona's water supply after the hydropower dams of the Pyrenees fell into Franco's hands. In other words, as depicted by the Republican artist José Bardasano, water was a critical resource (Fig. 2.6).

Fig. 2.6 Poster by José Bardasano: "Water in bad conditions causes more casualties than shrapnel", 1937. (Source: Pavelló de la República CRAI Library)

The scale of destruction visited upon cities during World War II dwarfs the experience of Spanish cities during that country's Civil War. The siege of Leningrad, the German attack on Coventry, or the Allied fire-bombing of Tokyo may make the cases of Barcelona and Madrid look rather small. But the German Condor attacks during the battle of Madrid, the three-day Italian bombing of Barcelona, or the destruction of the Basque town of Guernica—which inspired Picasso's most well-known painting—announced a new era of total war where cities became crucial battlefronts subject to aerial assault. The targeting of civilian objectives during the Spanish Civil War emerges arguably as one of the clearest precedents of war crimes of World War II, and one where urban environmental history can provide valuable insights.

In 1940, the will and organization of the citizens of Barcelona in the face of aggression was the example chosen by Winston Churchill to prepare British citizens for the Battle of Britain. The following year, the epic of the battle of Madrid and the cry "¡No Pasarán!" was in the mouths of many during the siege of Leningrad and in 1942 at the battle of Stalingrad. The tragic efforts of the workers of the Leningrad waterworks to maintain water supply are today memorialized at the water museum of St. Petersburg. In cities under fire, to keep the water running was to keep urban metabolism alive. The defence of Madrid proved that functioning infrastructure was a critical factor for the resilience of war-torn cities which would have been impossible to achieve without the workers' dedication, skills, knowledge, and organization. The motto chosen by anarcho-syndicalist workers of Barcelona's collectivized water company illustrates well the importance of people in the maintenance of urban technical networks: "Water running through your pipes is like blood running through our veins: Life".[37]

Notes

1. I am grateful to David Saurí and Hug March for their support throughout our research on water supply in Madrid and Barcelona. This chapter is partly based on two articles I co-authored with them: Santiago Gorostiza, Hug March, and David Sauri, "Servicing Customers in Revolutionary Times: The Experience of the Collectivized Barcelona Water Company during the Spanish Civil War," *Antipode* 45, no. 4 (September 2013): 908–25; and Santiago Gorostiza, Hug March, and David Sauri, "'Urban Ecology under Fire': Water Supply in Madrid during the Spanish Civil War (1936–1939)," *Antipode* 47, no. 2 (March 2015): 360–79. I also thank

the publishers for giving me permission to use here the map of Madrid region (Fig. 2.2). I am thankful to Alejandro Pérez-Olivares, Ekaterina Chertkovskaya, Simo Laakkonen and John McNeill for their comments on earlier versions of this chapter. I am indebted to José Manuel Naredo, who kindly granted access to Federico Molero's letters and patiently replied to all my questions about him. I am grateful to all the archives that granted permission to publish their images, and particularly to Carolina Peña Bardasano and to the team at the Pavelló de la República CRAI Library (Barcelona). Finally, I acknowledge financial support from the Spanish Ministry of Science, Innovation and Universities, through the "María de Maeztu" program for Units of Excellence (MDM-2015-0552).
2. Hansard (1940). War situation (House of Commons debate), 18 June. https://api.parliament.uk/historic-hansard/commons/1940/jun/18/war-situation, last accessed May 3, 2018.
3. On the proportion between civil and military casualties between World War I and World War II, see Javier Rodrigo, "Presentación. Retaguardia: Un Espacio de Transformación," *Ayer* 76, no. 4 (2009): 13–36 (20); Alan Kramer, *Dynamic of Destruction: Culture and Mass Killing in the First World War* (Oxford: Oxford University Press, 2007), 334. On the importance of urban water infrastructure during World War II, see J. Bowyer Bell, *Besieged: Seven Cities Under Siege* (New Brunswick and London: Transaction Publishers, 2006), particularly the cases of Warsaw, Singapore and Leningrad.
4. Urban bombardment and infrastructure destruction do not belong to the past. In September 2016, battles in Aleppo, Syria left 1.75 million people without water supply. Several other crisis in besieged cities such as Sarajevo, Yugoslavia, or Grozny, Chechenya in the 1990s, illustrate the critical importance of water. For Aleppo, see *The Guardian*, 24 September 2016, "Syria bombings leave 1.75 million without running water in Aleppo."
5. On infrastructure and urban flows, see Maria Kaika, *City of Flows: Modernity, Nature, and the City* (New York: Routledge, 2005); Stephen Graham, "When Infrastructure Fails," in Stephen Graham, ed., *Disrupted Cities. When Infrastructure Fails* (New York & London: Routledge, 2009), 1–26. On aerial bombardment and "urban ecologies under fire" see Kenneth Hewitt, "Place Annihilation: Area Bombing and the Fate of Urban Places," *Annals of the Association of American Geographers* 73, no. 2 (June 1983): 257–84. On the environmental history of work, see Stefania Barca, "Laboring the Earth: Transnational Reflections on the Environmental History of Work," *Environmental History* 19, no. 1 (January 2014): 3–27.
6. On the water supply of Barcelona and Madrid, see Enric Tello and Joan Ramon Ostos, "Water Consumption in Barcelona and its Regional Environmental Imprint: A Long-Term History (1717–2008)," *Regional*

Environmental Change 12, no. 2 (2012): 347–61; Hug March, "Taming, Controlling and Metabolizing Flows: Water and the Urbanization Process of Barcelona and Madrid (1850–2012)," *European Urban and Regional Studies* 22, no. 4 (2015): 350–67; Manel Martín Pascual, *Aigües de Barcelona. 150 Anys al Servei de la Ciutat (1867–2017)* (Barcelona: Fundació Agbar, 2017); Rosario Martínez Vázquez de Parga, *Historia del Canal de Isabel II* (Madrid : Fundación Canal de Isabel II, 2001).

7. Santiago Gorostiza, Hug March, and David Sauri, "Servicing Customers in Revolutionary Times: The Experience of the Collectivized Barcelona Water Company during the Spanish Civil War," *Antipode* 45, no. 4 (September 2013): 908–25; Santiago Gorostiza, Hug March, and David Sauri, "'Urban Ecology under Fire': Water Supply in Madrid during the Spanish Civil War (1936–1939)," *Antipode* 47, no. 2 (March 2015): 360–79; Santiago Gorostiza, "Potash Extraction and Historical Environmental Conflict in the Bages Region (Spain)," *Investigaciones Geográficas*, no. 61 (2014): 5–16; Santiago Gorostiza and David Saurí, "Salvaguardar un Recurso Precioso: La Gestión del agua en Madrid durante la Guerra Civil Española (1936–1939)," *Scripta Nova* 17, no. 457 (2013).
8. The censored text is "Ante la sublevación. Capítulo III," AAVV-AMTC-149-56. Fundación Pablo Iglesias archive. It was originally a part of Manuel Torres Campañá, *Gestión de la Delegación del Gobierno de la República durante 1936* (Canales del Lozoya: Madrid, 1937), 445–449.
9. On the beginning of the coup, see Paul Preston, *The Spanish Civil War: Reaction, Revolution and Revenge* (London: Harper Perennial, 2006), 99–114. On Canales del Lozoya, see Julián Diamante, *De Madrid al Ebro. Mis Recuerdos de la Guerra Civil Española* (Madrid: Fundación Ingeniería y Sociedad, 2011), 55–57; and José María Sanz García, *Las Aguas de Madrid, en Paz y en Guerra, de la Segunda República* (Madrid: Ayuntamiento de Madrid, 2000), 31.
10. Ramón Salas Larrazábal, *Historia del Ejército Popular de la República, Volumen I* (Madrid: La Esfera de los Libros, 2006), 326–336; Sanz García, *Las Aguas de Madrid*, 26–27.
11. "Ante la sublevación. Capítulo III," AAVV-AMTC-149-56, Fundación Pablo Iglesias archive, 13.
12. Vicente Rojo, *Historia de la Guerra Civil Española* (Barcelona: RBA, 2010), 233–237.
13. Sanz García, *Las Aguas de Madrid*, 31–33.
14. "Ante la sublevación. Capítulo III," AAVV-AMTC-149-56, Fundación Pablo Iglesias archive, 22.
15. See for instance *ABC*, September 10, 1936, 13; *ABC*, October 1, 1936, 14; Federico Bravo Morata, *Historia de Madrid. Volumen XI Extra* (Madrid: Fenicia, 1985), 20; Sanz García, *Las Aguas de Madrid*, 31–33.

16. "Ante la sublevación. Capítulo III," AAVV-AMTC-149-56, Fundación Pablo Iglesias archive, 32–43; letter signed by Federico Molero, engineer of Canales del Lozoya and member of the Communist Party, Moscow, December 1, 1966, personal collection of José Manuel Naredo; Bravo Morata, *Historia de Madrid*, 106–107; Canales del Lozoya, *Obras Públicas: Aportación a la Guerra* (Madrid: Talleres Espasa, 1937), 45; Sanz García, *Las Aguas de Madrid*, 31–33.
17. *ABC*, September 27, 1936, 9; *Boletín Oficial de los Canales del Lozoya*, 1936, n°690, 31 December 1936.
18. Myrna Margulies Breitbart, "Spanish Anarchism: An Introductory Essay," *Antipode* 10–11, no. 3–1 (December 1978): 60–70. For an account of the role of CNT in the Spanish Civil War, see José Peirats and Chris Ealham, *The CNT in the Spanish Revolution* (Hastings: The Cañada Blanch Centre for Contemporary Spanish Studies, 2001). There is a wide literature on the collectivizations but it has often focused on regional or local experiences, particularly agrarian. For a recent review, see Assumpta Castillo, "Anarchism and the Countryside: Old and New Stumbling Blocks in the Study of Rural Collectivization during the Spanish Civil War," *International Journal of Iberian Studies* 29, no. 3 (September 2016): 225–39. For the case of industry and services in Barcelona, see Antoni Castells Durán, *Les Col·lectivitzacions a Barcelona, 1936–1939* (Barcelona: Editorial Hacer, 1993).
19. Regional Government to SGAB workers committee, 24 September 1936, box 7373, AGAB. See also *La Vanguardia*, September 30, 1936, 8. On the school, named after pedagogue Francesc Ferrer i Guàrdia, see *Luz y Fuerza*, 5, April 1937, 2.
20. Service note n°17, October 3, 1936, box 10,998, AGAB.
21. On perpetual contracts, see Manel Martín Pascual, "Aigua i Societat a Barcelona entre les Dues Exposicions (1888–1929)" (unpublished PhD thesis, Universitat Autònoma de Barcelona, 2007), 240–249.
22. See for instance letter from L'Hospitalet city council, 22 October 1936, box 7371, AGAB; see also letter from Subcomité de Defensa de Barriada "19 de Julio" to SGAB workers committee, February 8, 1937, box 7375, AGAB.
23. On the relations between anarchism and human ecology in the Iberian Peninsula see Eduard Masjuan, *La Ecología Humana en el Anarquismo Ibérico* (Barcelona: Icaria, 2000). Regarding the installation of water-using facilities in factories during the war, see for instance Anglo-Española de Electricidad E.C. to Aguas de Barcelona E.C., February 2, 1938, box 7377, AGAB.
24. See for instance Regional Government to Aguas de Barcelona E.C., February 2, 1938, box 7377.

25. On the airlift and the advance of the Francoist troops towards Madrid, see Preston, *The Spanish Civil War*, 116–120. On international correspondents in the Spanish Civil war, see Paul Preston, *We Saw Spain Die: Foreign Correspondents in the Spanish Civil War* (London: Constable, 2008).
26. On the battle of Madrid, see Preston, *The Spanish Civil War*, 163–198. On the aerial bombings, see Josep Maria Solé i Sabaté and Joan Villarroya i Font, *España en Llamas: La Guerra Civil desde el Aire* (Madrid: Ediciones Temas de Hoy, 2003), 45–60. On the death of Canales' employee, see *Boletín Oficial de los Canales del Lozoya*, n°690, 31 December 1936. The Francoist plans for the occupation of Madrid took water supply into consideration, see Alejandro Pérez-Olivares, "La Victoria Bajo Control. Ocupación, Orden Público y Orden Social del Madrid Franquista (1936–1948)," (PhD thesis, Universidad Complutense de Madrid, 2017), 83–95, available at http://eprints.ucm.es/45481/
27. On the impacts of bombing and the efforts of Canales del Lozoya to control the leaks, see Bowyer Bell, *Besieged*, 22–26; Canales del Lozoya, *Obras Públicas*, 45; Diamante, *De Madrid al Ebro*, 79–80, 96, 99–100. On the casualties of firefighters in Madrid, see Juan Carlos Barragán Sanz and Pablo Trujillano Blasco, *Historia del Cuerpo de Bomberos de Madrid. De los Matafuegos al Windsor (1577–2005)* (Madrid: La Librería Ediciones, 2006), 258–259. On the Francoist offensive in the University City, see Fernando Calvo González-Regueral, *La Guerra Civil en la Ciudad Universitaria* (Madrid: La Librería Ediciones, 2012), 59–61; and Preston, *The Spanish Civil War*, 189.
28. On the war efforts of Madrid water company, see Canales del Lozoya, *Obras Públicas*, 36–41. On mapping the city's underground infrastructures, see "Ante la sublevación. Capítulo III", AAVV-AMTC-149-56, Fundación Pablo Iglesias archive, 10–11. The role of Canales' engineer Federico Molero has been referred to by Diamante, *De Madrid al Ebro*, 65 and 74; Enrique Lister, *Nuestra Guerra. Memorias de un Luchador* ([Zaragoza]: Silente, 2007), 143 and 147; and Pedro Montoliú, *Madrid en la Guerra Civil* (Madrid: Sílex, 1999), 264–267, among others, but can be read in Molero's own words in letter signed in Moscow, December 1, 1966, personal collection of José Manuel Naredo. On the subterranean war from the Francoist perspective, see Servicio Histórico Militar, *Guerra de Minas en España: 1936–1939: Contribución al Estudio de esta Modalidad de Nuestra Guerra de Liberación* (Madrid: Imprenta del Servicio Geográficos del Ejército, 1948), 59 and 69.
29. Lack of water pressure remained a problem in Madrid and Francoist newspapers announced that the capital was falling short of water. However, this problem existed prior to the war. See Montoliú, *Madrid en la Guerra Civil*, 27, 40 and 496–497; Sanz García, *Las Aguas de Madrid*, 25.

30. On the campaign "To defend Madrid is to defend Catalonia", see Antoni Segura and Andreu Mayayo, *Defensar Madrid és Defensar Catalunya: Solidaritat en Temps de Guerra (1936–1939)* (Barcelona: Ajuntament de Barcelona. Serveis Editorials Municipals, 2009). On the water-tank trucks given by the Barcelona city council to Madrid, see *La Vanguardia*, December 22, 1936, 2. On the Catalan firefighters expedition to Madrid, see Marc Ferrer i Murillo, "Bombers i Defensa Passiva al Vallès durant la Guerra Civil", in *El Vallès: Segona República, Guerra Civil i Postguerra (1931–1945)*, 2017, 91–108 (100).
31. On the water consumed in Barcelona and the non-payments, see *Solidaridad Obrera*, January 14, 1937, 4, and Josep Ferret i Pujol, *L'Aprofitament de les Aigües Subterrànies del Delta del Llobregat: 1933–1983* ([El Prat de Llobregat]: Comunitat d'Usuaris d'Aigües de l'Àrea Oriental del Delta del Riu Llobregat, 1985), 27. On the problem of salinity in Barcelona's waters see Santiago Gorostiza, Jordi Honey-Rosés, and Roger Lloret, *Rius de Sal: Una Visió Històrica de la Salinització dels Rius Llobregat i Cardener durant el segle XX* (Sant Feliu de Llobregat: Edicions del Llobregat: Centre d'Estudis Comarcals del Baix Llobregat, 2015); Santiago Gorostiza and David Sauri, "Dangerous Assemblages: Salts, Trihalomethanes and Endocrine Disruptors in the Water Palimpsest of the Llobregat River, Catalonia," *Geoforum* 81 (May 2017): 153–62. Regarding the reforms in Madrid, see Martínez Vázquez de Parga, *Historia del Canal de Isabel II*, 230–239.
32. On the agricultural collective of Barcelona, see Clara Garcia, Joan Milà, and Aurora Rius, "La Colectividad Agrícola de Barcelona. Una Aportació al Fet Col·lectivista a Catalunya," *Quaderns d'Història Contemporània. Revolució i Guerra Civil. Recerques a l'Arxiu Històric Nacional de Salamanca (SGC)*, 1983, 131–43. On the complaints about water used for gardening, see several letters received by ABEC during 1938, boxes 7377 and 7378, AGAB. ABEC also published a warning note in the press, *La Vanguardia*, August 19, 1937, 2. On urban gardening in Madrid, see *Campo Libre*, August 20, 1938, 12–13 and *La Vanguardia*, October 7, 1938, 5.
33. On the petitions of citizen committees to ABEC, see for instance City Council to ABEC, August 13 and 17, 1938, box 7377, AGAB. On aerial bombings in Barcelona, see Santiago Albertí and Elisenda Albertí, *Perill de Bombardeig!: Barcelona Sota les Bombes, 1936–1939* (Barcelona: Albertí Editor, 2004).
34. On the impact of the loss of the Pyrenees' dams on the Catalan industry, see Josep Maria Bricall, *Política Econòmica de la Generalitat* (Barcelona: Edicions 62, 1970), 47–55. On the water ban published by the Catalan government and the permission to increase water price, see *Diari Oficial*

de la Generalitat de Catalunya, April 28, 1938, n°118, 371; and *Diari Oficial de la Generalitat de Catalunya*, June 23, 1938, n°174, 1035. On the complaints and concerns about water pressure, key for the work of firefighters, see, among other, Barcelona Fire Brigade to ABEC, January 18, 1938, box 7377, AGAB; ABEC to Collectivised Tram Company of Barcelona, July 22, 1938, box 7378, AGAB.

35. On the refugees in Barcelona, see Joan Serrallonga i Urquidi, *Refugiats i Desplaçats dins la Catalunya en Guerra, 1936–1939* (Barcelona: Base, 2004), 190. On the impact of typhoid fever in Barcelona in 1939, see Pere Conillera i Vives, *L'Aigua de Montcada: L'abastament Municipal d'aigua de Barcelona: Mil Anys d'història* (Barcelona: Institut d'Ecologia Urbana de Barcelona, 1991), 111. On the military residence, see Santiago Gorostiza, "Diagonal 666. Un Monument a l'ocupació de l'exèrcit Franquista," *L'Avenç*, 389 (April 2013): 42–50. On typhoid fever rates in Madrid, see Pedro Matos Massieu, *Canal de Isabel II, Memoria 1939–1945* (Madrid: Ministerio de Obras Públicas, 1947), Matos Massieu, 1947, 129, 135, 139, 298–299 and annex 14. On the occupation of Madrid, see Alejandro Pérez-Olivares, "Objetivo Madrid: planes de ocupación y concepción del orden público durante la Guerra Civil española," *Culture & History Digital Journal* 4, no. 2 (December 2015): 1–13, available at http://cultureandhistory.revistas.csic.es/index.php/cultureandhistory/article/viewArticle/84/276 and Daniel Oviedo-Silva and Alejandro Pérez-Olivares (Coords.), *Madrid, una ciudad en guerra (1936–1948)* (Madrid: Los Libros de la Catarata, 2016).

36. *El País*, October 11, 2013, "Aflora el 'Frente del Agua,'" https://elpais.com/ccaa/2013/10/10/madrid/1381413281_187426.html Last accessed May 2, 2018; *ABC*, July 29, 2010, "Una bomba de la Guerra en el Canal," http://www.abc.es/20100724/madrid/bomba-guerra-canal-20100724.html Last accessed May 3, 2018.

37. Advertisement published in *Luz y Fuerza*, 7, July 1937. On the museum of Saint Petersburg Vodokanal, see http://www.vodokanal-museum.ru/en/muzejnyj_kompleks/kratkaya_informaciya/

CHAPTER 3

Fortress City

The Militarized Landscape of Seattle

Katherine Macica

INTRODUCTION

Seattle played an important role in the waging of World War II. Although relatively isolated in the far northwestern corner of the continental United States, weapons and materiel flowed from the city and onto battlefields around the world. Trucks and trains brought supplies from across the country to the army's and navy's supply depots in Seattle to be deployed to the Pacific Theater. At the peak of production, sixteen B-17 "Flying Fortress" and six B-29 "Superfortress" heavy bomber aircraft rolled out of Boeing's two Seattle-area factories every day, destined for bases in Europe, North Africa, and the Pacific. Thousands of ships of all sizes and functions slid off the shipways into Puget Sound during the war, steaming to every corner of the globe.[1]

The transformation of Seattle from a seeming colonial backwater into a major weapon in America's arsenal of democracy did not happen overnight or by chance. Instead, local and federal planners and policymakers, along with leaders in business and industry, sought to use the war as an opportunity to remake the city and the region. The efforts of planners and

K. Macica (✉)
Loyola University Chicago, Chicago, IL, USA

© The Author(s) 2019
S. Laakkonen et al. (eds.), *The Resilient City in World War II*,
Palgrave Studies in World Environmental History,
https://doi.org/10.1007/978-3-030-17439-2_3

the demands of the military, along with a new understanding of the region's resources, coalesced to bring war industries to the Pacific Northwest and Seattle. The growth of manufacturing industries during the war broke the virtual monopoly that extractive industries had on the region's economy and altered Seattle's land use patterns. World War II transformed Seattle by generating new networks of capital and commodities and by creating a lasting military-industrial complex in the city. The changes wrought by war permanently altered the economy of Seattle and engendered closer ties between the city, region, and federal government.

This essay examines these changes from the point of view of landscape history. The aim is not to explore urban landscape in its totality but to focus on militarized landscapes, that is, landscapes that were to varying extents mobilized to achieve military aims in wartime Seattle.[2]

THE PREWAR ECONOMY AND ENVIRONMENT

Nestled among lush evergreen forests, between the Cascade Mountains to the east and Puget Sound to the west, Seattle seemed an unlikely major contributor to the allied war effort. Large cities in the Midwest and East coast, like Chicago, Pittsburgh, and New York, had served as industrial centers for decades, while Seattle grew to become a trade and processing center for the resources of the Northwest's hinterlands. Policymakers, planners, and business leaders in the years prior to the war found it difficult to imagine how such a geographically isolated city with a relatively small population and industrial base could play an important role in the mobilization of the nation's economy.[3] However, over the first decades of the twentieth century, environmental and economic changes in Seattle set the stage for its ascendency as a fortress city during World War II.[4]

By the early twentieth century, Seattle had established itself as the commercial hub for processing and exporting the region's vast natural resources across the country and around the world. The city's and region's economies relied primarily on extractive industries like mining, fishing, agriculture, and most importantly, logging.[5] Seattle enjoyed a geographically advantageous position on Puget Sound that facilitated the transportation of goods via land and over water. Two transcontinental rail lines, the Northern Pacific Railway and Great Northern Railway, connected Seattle with the mining and agricultural areas in its hinterlands and with markets in the east.[6] Likewise, Puget Sound served as an important means of transport for timber cut from forests surrounding the Sound, and provided a protected, deep water port for ocean-going vessels to ship Northwest products around the world. Although well-positioned for trade and exploitation of natural resources, Seattle's distance from population centers and markets prevented its economic base from moving beyond extractive industries in the early twentieth century (Fig. 3.1).

Seattle experienced significant environmental changes and indirect impacts of war in the first decades of the twentieth century, laying the foundation for the more dramatic upheaval wrought by World War II. For centuries, Native Americans and later, Euro-American settlers had been remaking the city's landscape to meet their changing needs.[7] Alterations to Seattle's physical landscape encouraged the growth of industry in the city, and at the same time, new land uses led to environmental change. Several large-scale reclamation projects undertaken primarily by the Army Corps of Engineers between the turn of the century and the 1920s reconfigured Seattle's waterways for the purpose of facilitating commerce and industry. The Lake Washington Ship Canal, completed in 1916 by the Army Corps of Engineers, connected Lake Washington to Puget Sound via Lake Union just to the north of the city center. The Lake Washington Ship Canal provided a waterborne transportation route between logging and mining operations east of Lake Washington and Seattle's sawmills and port facilities. The canal proved a boon to sawmills and other materials processing industries around Lake Union, improving networks between sites of extraction to the east of Seattle and sites of production and distribution in the city.[8]

At the same time, city planners turned their attention to the area south of the city center as a possible new industrial zone. The Duwamish Valley, south of downtown Seattle, seemed an ideal location for industry, with its relatively flat and open landscape contrasting with the hilly city to the north. The Duwamish Valley was also proximal to several modes of transportation, including both transcontinental railways and the Duwamish River, which flowed west from the Cascade Mountains and emptied into Elliott Bay just south of downtown Seattle. However, the meandering Duwamish River had to be tamed in order to create a rational space for industry. Beginning in 1895, engineers worked to remove the many twists and turns of the river's channel. The bulk of the work was completed between 1911 and 1924, resulting in a ready-made industrial waterway that was deeper, wider, and straighter than the original river, and the perfect location for new industries to grow. The construction of the Duwamish Waterway and regrading of the city's hills produced millions cubic yards of

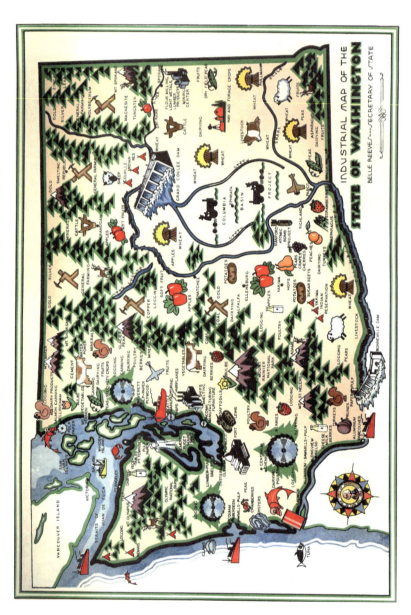

Fig. 3.1 This resource map, produced by the State of Washington for school children in 1945, illustrates the diversification of the state's economy as a result of the war. (Source: Office of the Secretary of State of Washington, Washington State Archives)

soil that engineers used to create Harbor Island, a 350-acre artificial island, at the mouth of the Duwamish River in Elliott Bay.[9] The construction of the Lake Washington Ship Canal, Duwamish Waterway, and Harbor Island reoriented the location of industry in the city and represented the most significant environmental changes in Seattle prior to World War II.

Although Seattle and the Northwest were not highly industrialized by the 1910s, the region's natural resources were in high demand. With the outbreak of World War I, the need for lumber for ships, aircraft, and other necessities of war proved a boon to Seattle's nascent industries. The Northwest lumber industry boomed during the war, with local species like Sitka spruce and Douglas fir being in high demand for aircraft construction. While much of this lumber made its way across the Atlantic Ocean to aircraft manufacturers in England, France, and Italy, beginning in 1917, some of it remained in the Northwest and was used by the recently incorporated Boeing Airplane Company for seaplanes built for the US Navy.[10] Boeing joined the growing ranks of manufacturers sprouting up in Seattle to produce weapons and material for the military. The lumber, shipbuilding, and aircraft industries brought prosperity to the region during World War I and planted the seeds for a more diversified economy. The industrial boom and attendant prosperity proved temporary however, as defense-related demand dried up after 1918 and Seattle returned to processing and marketing raw materials in the interwar years.[11] While the reclamation projects of the 1900s and 1910s and the growth of industry during World War I extensively altered the landscape of Seattle, the changes wrought by World War II proved more revolutionary in their extent and scale.

Federal Investment in the Northwest

The transformation of Seattle into a fortress city during World War II required massive amounts of capital. Although Seattle served as a manufacturing and commercial hub for the region since the turn of the century, prior to World War II, capital generally remained in the pockets of large corporations in the East. The region's and city's economy relied upon extraction and processing, depleting the region's natural resource without significant monetary gain at the local level. Indeed, just prior to the war, more than 75 percent of the region's workers relied on extractive industries for their livelihood.[12] As Secretary of the Interior Harold Ickes explained in 1941, the problem was that the East saw the Northwest as "a nut to be cracked, the succulent meat of which was to fatten Wall Street.

The empty shell was left for the farmers and the tradesmen of the West to gnaw upon until they could hopefully produce another nut to meet the fate as of its predecessor."[13] Federal agencies established during the New Deal, particularly the National Resources Planning Board, worked with local planning organizations, like the Washington State Planning Council, to research the economic and social problems of the region and provide recommendations for policymakers. President Franklin Roosevelt and Secretary of Interior, Harold Ickes, were keenly aware of both the problems and potential of the Northwest and sought to adopt measures that would encourage the diversification of the region's economy. The federal funds that poured into the region as a result of New Deal programs and, to a greater extent, in the form of defense contracts, created new flows of capital between the two Washingtons and contributed to the development of Seattle as a Fortress City.

During the 1930s, the federal government undertook a variety of measures to promote development and industrialization in the region. Between 1933 and 1939, the federal government spent more than $580 million dollars in the state of Washington, the 9th highest rate of expenditure per capita among the 48 states.[14] This federal funding went toward building roads, maintaining public lands, rural electrification, public housing, and reclamation. The most important projects for the region and the city of Seattle were the construction of two massive dams on the Columbia River, designed to improve navigation, provide irrigation water, and generate electricity. Bonneville Dam and Grand Coulee Dam, completed in 1938 and 1942, respectively, brought millions of dollars into the region during their construction and created the means for the region to generate its own power and income.[15] The federally funded New Deal projects built and modernized the region's infrastructure, providing a strong foundation for the growth of the defense industry starting in the late 1930s (Fig. 3.2).

As the nation mobilized for war, local boosters and the federal government saw the opportunity to further the gains made during the New Deal era by encouraging the growth of industry in the Pacific Northwest through defense contracts. Seattle proved a suitable location for war industry, according to federal policy. When awarding defense contracts, the government looked for locations that were strategically situated to facilitate rapid production and distribution, relatively safe from enemy attack, and dispersed across the country to take advantage of all available manpower.[16] With its existing aircraft and shipbuilding industries, connec-

Fig. 3.2 Grand Coulee dam began powering Northwest war industries in 1941, facilitating the growth of shipbuilding, aircraft manufacturing, and light metals processing in the region. (Source: Library of Congress, Prints & Photographs Division, FSA/OWI Collection, LC-USW33–035035-C)

tion to transcontinental rail lines, proximity to several major military installations, deep water harbor with access to the Pacific Ocean, and plentiful inexpensive electricity provided by the federal dams, Seattle clearly fit the bill. Federal capital poured into the region to militarize the landscape and economy, by financing the construction of production facilities and through defense contracts.

Boeing and the shipyards of the Todd Shipyards Corporation emerged as Seattle's two largest government contractors during the war. The development of their physical plants to meet the needs of war production required the investment of millions of dollars by the federal government. Between 1940 and 1945, Boeing received more than $16 million in federal funding to accommodate the production of B-17s. In the summer of 1940, Boeing advised the War Department that the company would need to expand Plant 2 in order to meet the increased production goals for its

B-17E aircraft. The War Department subsequently approved a $7.3 million appropriation to enlarge the facility, more than doubling the size of the plant, and to purchase new equipment. Over the course of the war, the Army Air Forces provided an additional $9 million for improvements to and equipment for Plant 2 to expedite the production of B-17s.[17] In addition to the money spent on Plant 2, the Army Air Forces invested a total of $8.8 million during the war to upgrade King County Airport, from which they took delivery of completed B-17s.[18] Likewise, the federal government spent a considerable amount of money in Seattle to facilitate the construction and repair of naval vessels. The Todd Shipyards Corporation operated two shipyards in Seattle, both located on Harbor Island. The Seattle-Tacoma Shipbuilding Corporation yard, designed specifically to construct destroyers, cost the navy $8.5 million to build in 1940. The Todd-Seattle Drydocks Corporation yard, a ship repair yard, required an additional $5.5 million to expand and modernize to meet the needs of the navy. The money spent by the federal government in Seattle for these and numerous other production facilities represented a massive influx of capital that was not forthcoming from the private sector. Indeed, Boeing had sunk most of its funds into the development of the B-17 in the mid-1930s and could not have expanded without government funding.[19] The investment created jobs in the short-term and grew the city's industrial infrastructure in the long-term.

In addition to financing capital improvements, the federal government awarded a total of $5.6 billion in defense contracts to Seattle firms during the war, making Seattle one of the top three cities in the nation in terms of per capita defense orders.[20] These contracts paid for raw materials, parts, equipment, and wages, drawing more products and people to the region. Clearly, the federal government channeled an unprecedented amount of money into Seattle and the Pacific Northwest because of the war. These funds built the military installations and factories that militarized Seattle's landscape and economy and contributed to the city's transformation into a fortress city.

Hydropower and Aluminum

As a fortress city, Seattle became a hub for processing, manufacturing, and transporting commodities and products essential for the nation's defense. Seattle's leaders and businesses had many decades of experience with this type of work, but the war altered the types of commodities flowing through the city, as well as the pace and scale of movement. Electricity and aluminum emerged as two commodities central to the city's industries. The war generated new inter- and intra-regional networks of these commodities, connecting Seattle with its hinterlands and beyond in new ways. Electricity undergirded the majority of industry in Seattle, and while the city had enjoyed electricity since the late nineteenth century, the quantities required by war industries necessitated the creation of new networks. To facilitate the transfer of electric power between different power sources and customers in order to even out the supply across the region, the War Production Board organized the Northwest Power Pool in May 1942. Seattle City Light owned its own power sources and had not previously tapped into the electricity generated by Bonneville and Grand Coulee dams. However, as industry boomed, so did the demand for electricity, especially from industries like electrometallurgy and shipbuilding.[21] The networks established through the Northwest Power Pool connected Seattle and its hinterlands to the Columbia River in new ways, transmitting electricity to factories and homes hundreds of miles away from the power source.

Electricity also played an important role in the emergence of a commodity entirely novel to the Pacific Northwest: aluminum. One of the major criticisms of the massive hydroelectric dam projects in the region was that the dams would produce much more electricity than the region could ever use. But federal planners saw the dams as a powerful driving force for change in the region, anticipating that the growth of population and industry would increase demand for electricity over time. Many officials also called for a surplus generating capacity as an important element of the nation's defense so as not to be caught off guard should the nation become involved in the looming war.[22] To take advantage of the inexpensive electricity generated by Bonneville and Grand Coulee dams, and to promote the growth of industry in the region, especially defense-related industry, the federal government recruited aluminum manufacturers to the Pacific Northwest. Aluminum smelting consumed a great deal of electricity, but at the beginning of the war there was not sufficient generating capacity in traditional industrial areas, so the industry had to move to where electricity was available.[23] The first aluminum plant in the region, and indeed the first ever on the Pacific coast, opened in 1940 in Vancouver, Washington. Over the course of the war, five additional aluminum plants were constructed in the Pacific Northwest, turning out nearly one third of the nation's aluminum.[24] The electricity generated by Bonneville and

Grand Coulee dams was essential to the success of the region's aluminum industry. Planners saw the establishment of the aluminum industry as an important step toward diversification of the region's economy. When the first plant was announced in late 1939, Secretary Ickes proclaimed that "the way is open for new industrial expansion in the Pacific Northwest... This is only the beginning."[25]

The production of aluminum in the Northwest led to the creation of new connections between sites of extraction and sites of production. Bauxite ores from Arkansas, Georgia, and as far away as Dutch Guiana (Suriname), traveled to the plants in Washington and Oregon to be converted into aluminum. The aluminum sheets, ingots, and alloys fabricated in these plants were used in a variety of defense industries across the country.[26] The aircraft industry was the largest consumer of aluminum during the war, using as much as 68 percent of the nation's production.[27] Due to its proximity to Northwest aluminum processing plants, Boeing enjoyed an advantage over other airframe manufacturers, as its two Seattle-area aircraft plants could rely on faster delivery and lower shipping costs than its competitors in California and on the East Coast.[28] The development of the aluminum industry and other new commodity chains that emerged as a result of war mobilization contributed to the diversification of Seattle's and the larger region's economy and provisioned the city's defense industries.

Aircraft Manufacture and Shipbuilding

Federal investment in Pacific Northwest during the New Deal era began a flow of capital and commodities that, spurred on by the war, enabled industry to boom in Seattle, and led to the establishment of a lasting military-industrial complex in the city. Seattle became a center for aircraft manufacture and shipbuilding. The aircraft industry in Seattle was based entirely upon the Boeing Aircraft Company. Boeing had been among the first industries to locate in the new industrial corridor along the Duwamish Waterway when it moved its headquarters and production facilities there in 1917. However, with the development of the B-17 in 1935, Boeing needed a larger plant to accommodate the construction of the heavy bombers. Boeing's Plant 2, located south of Plant 1 and adjacent to King County Airport, was built in 1936.[29] Over the course of the war, each subsequent order for aircraft from Boeing required the expansion of the company's physical plant. Plant 2 became a sprawling campus complete with a main assembly building, engineering building, paint and camouflage buildings, warehouses, and an aerodynamics lab and wind tunnel.[30] Even with all the expansions, Boeing could not produce B-17s quickly enough for the Army Air Forces, so the company developed a "feeder plant" system of seven smaller plants in the Seattle metropolitan area that would produce specific subassemblies of the aircraft, then ship those parts to Plant 2 for final assembly.[31] Because of the importance of the B-17 to allied strategy in Europe, Boeing could spare no space in Plant 2 for the production of its new B-29 bomber. In January 1942, the Army Air Forces ordered five-hundred B-29s, which were to be produced in Boeing's Wichita, Kansas facility[32] (Figs. 3.3 and 3.4).

However, as the war in the Pacific accelerated, the demand for B-29s outstripped the space available in Boeing and its contractors' facilities. To speed the production, the Army Air Force and Boeing negotiated with the navy to take over its newly built aircraft plant in Renton, just south of Seattle, where production of B-29s began in 1943. Situated on a 107-acre site at the southern tip of Lake Washington, the Renton facility was built in 1941–1942 and included a large main assembly building encompassing more than one million square feet of floor space, and a new runway.[33] Between 1935 and 1945, Boeing assembled 6981 B-17s and 1119 B-29s at the company's Seattle and Renton plants, respectively, totaling around half of all of those aircraft models produced.[34] The growth of Boeing's Plant 2, the construction of the Renton plant, and organization of the feeder plant system created new networks of production within the Seattle metropolitan area and changed the physical landscape of the city by creating a new and lasting aircraft production center on the city's south side.

Seattle's shipbuilding industry made a significant contribution to the nation's defense as well, while at the same time reshaping Seattle's waterfront. The Seattle-Tacoma Shipbuilding Corporation and Todd-Seattle Drydocks Corporation shipyards on Harbor Island in Seattle completely transformed the island into an industrial landscape. Although built more than 20 years earlier, Harbor Island remained largely undeveloped prior to the construction and expansion of the Todd yards beginning in 1940. The Seattle-Tacoma Shipbuilding Corporation yard encompassed more than 30 acres of land and included five slipways and a 1750-foot-long outfitting pier.[35] The yard was expanded in 1941, adding a 24-acre site just to the south to outfit the ships constructed at the main shipyard.[36] In addition to

Fig. 3.3 Workers assemble B-17Gs inside Boeing Plant 2. Subassemblies and component parts, such as wings, tail assemblies, and engines, arrived in Seattle via railroad from around the Puget Sound industrial region and across the country. (Source: U.S. Air Force photo, National Museum of the United States Air Force, VIRIN 060517-F-1234S-027)

Fig. 3.4 The Army Corps of Engineers' Passive Defense Division fabricated an artificial town with houses and streets on the roof of Boeing Plant 2 to protect the factory from Japanese aerial attack. The camouflage was made of chicken wire, spun glass, painted canvas, burlap, feathers, and local lumber. Newly assembled bombers are visible on the right side of the plant. (Source: Records of the Office of the Assistant Chief of the Air Staff, A-4, R & D Branch, Case Histories, 1941–1946, Box 31, Entry 22, Record Group 18, National Archives at College Park)

the Todd yards, 27 other shipyards in Seattle and dozens more in the metropolitan area built and repaired military and merchant vessels of all varieties for the United States and allied nations.[37]

The process of shipbuilding and repair necessitated the use of various commodities and materials on an unprecedented scale, creating new networks of raw materials and value-added products into and out of Seattle. Millions of board feet of local lumber went into the construction of shipyard buildings, piers, scaffolding, jigs, and templates.[38] Local lumber mills provided wooden keels for wood-hulled vessels constructed locally and across the country, as well as lumber and plywood for decking, framing, furniture, and other purposes for steel ships. Most of the steel used in the construction of ships in Seattle had to be brought in from the Midwest and East coast, however.[39] Seattle's shipyards expanded the industrial facilities of the city's waterfront, creating lasting improvements to the city's industrial infrastructure. Although aircraft and shipbuilding were the largest industries in Seattle, other processing and manufacturing operations, including foundries, food packers, and lumber mills, expanded in response to the war.[40]

THE CHANGING URBAN ENVIRONMENT

The growth of war industries in Seattle also sparked ancillary changes to the city's land use patterns and environment. The expansion of aircraft manufacturing and shipbuilding required the recruitment of thousands of additional employees to Boeing and the city's shipyards, as well as to the smaller firms that supported these and other war industries. The influx of workers to the Northwest led to a massive population boom in Seattle and its metropolitan area. The population of Seattle jumped from approximately 368,000 in 1940 to 480,000 in 1943, an increase of 30 percent.[41] As early as 1939, policymakers and business leaders in Seattle recognized that the city had inadequate housing to accommodate the expected growth in population. In 1940, the newly established Seattle Housing Authority began working with the federal government to finance and construct housing projects for low-income residents and military personnel. Seattle's first defense-related housing project was Sand Point Homes, built in 1940 to house married enlisted personnel stationed at Sand Point Naval Air Station, northeast of downtown Seattle along Lake Washington.[42] Over the next five years, the Seattle Housing Authority constructed around 6000 housing units for people working in war industries. These units took the form of permanent housing projects for families, temporary family housing, and temporary apartments and dormitories. The housing projects built by the Seattle Housing Authority were strategically located near the city's largest employers. Six were located south of downtown in close proximity to Boeing's Plant 2. Four were located just east of downtown with easy access to the shipyards on Harbor Island and Seattle's waterfront.[43] Other agencies, such as the King County Housing Authority, and private individuals also added to the housing stock of Seattle during the war. The new housing built to accommodate Seattle's booming population transformed acres of "blighted" areas and undeveloped land into new residential neighborhoods, and provided housing for Seattle residents for decades into the postwar era.[44]

Despite their genesis as wartime measures, the production activities in Seattle brought about long-lasting changes to the city's environment and economy. The industrialization of the Duwamish waterway, begun during World War I, was complete by World War II. What had once been a prime location for truck farms became the center of the city's industry, and remains so today.[45] What also endures today are a wide variety of industrial pollutants found in the soil and water around Seattle's industrial areas. During the war, Americans were generally more concerned about achieving victory than preserving the environment. Groups that had a vested interest in environmental protection, like the Department of the Interior, urged citizens to use natural resources wisely and to "not be misled by attempts to <u>improperly exploit</u> our oil, our fish, our timber, our grazing lands under the guise of necessity."[46] However, such caution did not necessarily extend to protecting the latent natural resources in urban areas from pollution. Dangerous chemicals, materials, and sewage entered the water and soil around the Duwamish waterway and Elliott Bay during World War II. Sewage systems designed for Seattle's prewar population and level of industrial activity became overburdened with wastes, while chemicals such as lead, arsenic, and mercury were released directly into the water and soil through handling and disposal procedures that were unsafe[47] (Fig. 3.5).

While the practices of companies like Boeing and Todd Pacific Shipyards polluted the environment, they also brought prosperity to Seattle and the Pacific Northwest, diversifying the economic base of the region. The militarization of the nation's economy ultimately led to recovery from the Great Depression.[48] Federal investment during the war made it possible for private companies to prosper in the postwar era. Many of the manufacturing firms maintained their wartime employment levels after the war. State-

Fig. 3.5 The war militarized Seattle's seascape as well as landscape. Warships became a common sight in Puget Sound as they deployed overseas or returned from battle for repairs. Here the destroyer USS *Little* (DD-803), built by the Seattle-Tacoma Shipbuilding Corporation, departs Seattle for Pearl Harbor in November 1944. *Little* supported the invasions of Iwo Jima and Okinawa before being sunk by kamikazes in May 1945. (Source: Naval History & Heritage Command, photo NH 107134)

wide employment in the transportation equipment manufacturing industry, dominated by shipbuilding and aircraft manufacture, grew 515 percent between 1939 and 1953, demonstrating the longevity of the war's changes to the region's economy.[49] The experience of Boeing offers the prime example of the ways in which patterns established during World War II continued to impact Seattle's economy in the decades after. The knowledge and experience with military aircraft design and contracting that Boeing gained from the war enabled the company to maintain its position as a top producer of military aircraft. In addition, the facilities constructed and expanded with federal funds during the war provided the infrastructure needed to keep up production of both military and commercial aircraft after the war. Indeed, Boeing designed and assembled the first 737, the top selling commercial passenger jet in aviation history, at its Seattle plant, originally built to produce B-17s. Production of 737s continues to this day at the company's nearby Renton plant, first used for the assembly of B-29s.[50]

Conclusions

Although fought thousands of miles away, World War II engendered material changes to Seattle's environment and economy. Seattle emerged from the war as both a shock city and a model city. The war transformed Seattle by generating new networks of capital and commodities and by creating a lasting military-industrial complex in the city. The influx of federal funds into the Northwest beginning with the New Deal and accelerating during the war led to a rapid shift of the region's economic base. Seattle had been a hub for processing and distributing the region's raw materials, but the war precipitated a boom in manufacturing. Large shipbuilding, aircraft, and other manufacturing facilities emerged virtually overnight on what was once open land, farms, and older infrastructure. The boom in war industries led to a massive migration of workers to the city, necessitating further changes to Seattle's infrastructure, including expanded transportation networks, and thousands of new housing units. The rapid growth of industry and population, along with the sense of urgency to produce war materiel as quickly as possible, resulted in more extensive use of the region's resources and pollution of the urban environment.

While the fast pace and scale of change made Seattle a shock city, Seattle also represents a model city. Federal and local planners intentionally used the war as a means of development to reshape and modernize the region. The work of planners during the New Deal to bring federal funding to the Northwest to spur the growth of industry in the region continued throughout the war. The national mobilization effort during World War II spawned an enormous demand for raw materials and manufactured products, creating the market and available capital needed for industry in the Northwest to flourish. Seattle businesses took advantage of this opportunity and ultimately achieved the planners' goal of a diversified economy in the region. Federal funding built the infrastructure that modernized Seattle and the Northwest. New multipurpose dams provided a surplus of inexpensive electricity for the city and region. That electricity powered new manufacturing facilities built for wartime production that became the basis of the city's postwar industrial infrastructure.

War mobilization militarized Seattle's environment and created a lasting military-industrial complex in the city. Seattle became an important part of the "gunbelt" of American cities whose growth was based on the development of the military-industrial complex during World War II. Cities along the west coast, including Seattle, Portland, San Francisco, Los Angeles, and

San Diego each attracted major war industries.[51] While the weapons manufactured in the United States wrought destruction abroad, they served as a force of creation for new urban industrial landscapes in the cities in which they were produced. War industries forged new militarized landscapes in Seattle. Boeing's three production campuses in Seattle and Renton, its system of warehouses and feeder plants throughout the Puget Sound region, and King County Airport became the starting point for the air war in Europe and the Pacific. The shipyards that dotted Puget Sound and Lake Washington likewise became militarized as production shifted from civilian to military craft and as battle-damaged vessels entered the ports for repair. These facilities represented the emergence of a resilient military-industrial complex that would become a cornerstone of the region's economy for decades to come.

Notes

1. Boeing reached its peak production of B-17s in 1943 and B-29s in 1945. Richard C. Berner, *Seattle Transformed: World War II to Cold War*, vol. 3 of *Seattle in the 20th Century* (Seattle: Charles Press, 1999), 59–60; Carlos Schwantes, *The Pacific Northwest: An Interpretive History*, rev. ed. (Lincoln: University of Nebraska Press, 1996), 411; 1. Mrs. Sutton Gustison, "The Boeing Story," *The Pacific Northwest Quarterly* 45:2 (1954): 44; *History of the U.S. Naval Station, Seattle, Washington*, RG 181, 13ND-5: Wartime Unit Histories, Box 4, Folder: "U.S. Naval Station – Seattle, Washington," and *Wartime History of the Supervisor of Shipbuilding, United States Navy, Seattle, Washington*, RG 181, 13ND-5: Wartime Unit Histories, Box 6, Folder 2, National Archives and Records Administration, Seattle, WA (NARA-Seattle).
2. For an extensive discussion of the concept see, Chris Pearson, "Researching Militarized Landscapes: A Literature Review on War and the Militarization of the Environment," *Landscape Research* 37, no. 1 (February 2012): 115–133; For a longer historical perspective, see Chris Pearson, Peter Coates and Tim Cole, eds., *Militarized Landscapes: From Gettysburg to Salisbury Plain* (London: Continuum, 2010); For postwar changes, see Simo Laakkonen, ed., "Militarized Landscapes: Environmental Histories of the Cold War," a special issue in *Cold War History* 16, no. 4 (November 2017): 377–481.
3. Seattle's population in 1940 numbered 368,302, compared to more than 3 million residents in Chicago and 1.5 million in Los Angeles. US Bureau of the Census, *Sixteenth Census of the United States: 1940, Population, Vol. 1 Number of Inhabitants* (Washington, DC: US Government Printing Office, 1942), United States Summary, table 12.

4. The use of the term "fortress city" is borrowed from Roger Lotchin's concept of the "metropolitan-military complex" as explained in *Fortress California, 1910–1961: From Warfare to Welfare* (New York: Oxford University Press, 1992), and is a play on the importance of Boeing and its "Flying Fortress" and "Superfortress" aircraft to Seattle.
5. National Resources Planning Board, Region 9, *Pacific Northwest Region: Industrial Development* (Washington, DC: US Government Printing Office, 1942), 2.
6. For more on the impact of the railroads on the Pacific Northwest see Richard White, *Railroaded: The Transcontinentals and the Making of Modern America* (New York: W.W. Norton, 2011); Bruce Cumings, *Dominion from Sea to Sea: Pacific Ascendancy and American Power* (New Haven: Yale University Press, 2009); and Schwantes, *The Pacific Northwest: An Interpretive History*.
7. See Matthew Klingle, *Emerald City: An Environmental History of Seattle* (New Haven: Yale University Press, 2007) for an overview of how Seattle's landscape changed from the eighteenth century to the present.
8. Robert Ficken, "Seattle's 'Ditch': The Corps of Engineers and the Lake Washington Ship Canal," *Pacific Northwest Quarterly* 77, no. 1 (January 1986): 11.
9. Lucile Carlson, "Duwamish River: Its Place in the Seattle Industrial Plan," *Economic Geography* 26:2 (April 1950), 147; Klingle, *Emerald City: An Environmental History of Seattle*, 75–76. The regrading of Seattle's hills during the first decades of the twentieth century also resulted in extensive changes to the city's environment, but did not significantly alter land use patterns. For more on the regrades see David B. Williams, *Too High and Too Steep: Reshaping Seattle's Topography* (Seattle: University of Washington Press, 2017).
10. Gerald Williams, "The Spruce Production Division," *Forest History Today* (Spring 1999): 3; Robert J. Sterling, *Legend and Legacy: The Story of Boeing and Its People* (New York: St. Martin's Press, 1992), 2–3.
11. National Resources Planning Board, *Pacific Northwest Region: Industrial Development*, 20–21; Klingle, *Emerald City: An Environmental History of Seattle*, 185–186.
12. National Resources Planning Board, *Development of Resources and of Economic Opportunity in the Pacific Northwest* (Washington, DC: US Government Printing Office, 1942), 12.
13. Harold Ickes, speech at the Chamber of Commerce, Spokane, WA, 19 August 1941, Box 341, Harold L. Ickes Papers, Library of Congress.
14. Leonard Arrington, "The New Deal in the West: A Preliminary Statistical Inquiry," *Pacific Historical Review* 38, no. 3 (August 1969): 314–315.

15. Paul Hirt, *Wired Northwest: The History of Electric Power, 1870s–1970s* (Lawrence: University Press of Kansas, 2012), 268–269.
16. National Resources Planning Board, *Industrial Location and National Resources* (Washington, DC: US Government Printing Office, 1943), 1; John B. Appleton, *Pacific Northwest Resources* (Portland, OR: Northwest Regional Council, 1943), 112.
17. "Facility Expansion of Boeing Aircraft Company, Seattle, Washington," and "Summary of Case History of Boeing Aircraft Company, Seattle, Washington," p. 1–3, RG18 Army Air Forces, Office of the Assistant Chief of Air Staff Materiel and Services (A-4), Research and Development Branch Case Histories, Box 31, "Boeing" folder, National Archives and Records Administration, College Park, MD.
18. "Summary of Case History of Boeing Aircraft Company, Seattle, Washington," 10–11.
19. "A Report of Accomplishment in Engineering, Production and Cost Reduction by Boeing Aircraft Company and Boeing Airplane Company for the year 1943," Boeing Archives, Bellevue, WA.
20. Schwantes, *The Pacific Northwest: An Interpretive History*, 411.
21. Hirt, *Wired Northwest*, 308, 312.
22. Hirt, *Wired Northwest*, 306; J.D. Ross, *First Annual Report of the Bonneville Administrator* (Washington, DC: US Government Printing Office, 1939), 20.
23. Carleton Green, *The Impact of the Aluminum Industry on the Economy of the Pacific Northwest* (Vancouver, WA: Aluminum Company of America, June 1954), 3–4, Bonneville Power Administration Library, Portland, OR (BPA Library).
24. Bonneville Power Administration Information Service, press release September 1, 1940, Box 7, Samuel Moment Collection, BPA Library; National Resources Planning Board, *Pacific Northwest Region: Industrial Development*, 29.
25. "Ickes Says Aluminum Firm Proves Bonneville Worth," *The Oregon Daily Journal*, December 25, 1939.
26. National Resources Planning Board, *Pacific Northwest Region: Industrial Development*, 29.
27. War Production Board, "Materials Handbook, Aluminum," (War Production Board, Statistical Services Division, issued July 6, 1945): 3-X, Samuel Moment Collection Box 244, Folder: "I WPB, All Products, Supply and Consumption Materials handbook, confidential," BPA Library.
28. *Aluminum at Work In the Air, On Land and Sea, In Space: The Story of Aluminum Products From the Boeing Company* (Portland, OR: Western Aluminum Producers, 1979), 3.

29. Robert J. Serling, *Legend and Legacy: The Story of Boeing and Its People* (New York: St. Martin's Press, 1992), 32–33; Boeing Aircraft Company Seattle, Washington "B-17 Production and Construction Analysis," Prepared by: Air Materiel Command Headquarters, Los Angeles AAF Procurement Field Office, Industrial Planning Section (29 May 1946): 17, Boeing Archives.
30. Boeing Aircraft Company, *Master Manual Plant 2, Boeing Aircraft Company* (April, 1943), Boeing Archives.
31. "B-17 Production and Construction Analysis," 32–34.
32. Peter Bowers, *Boeing Aircraft since 1916*, 2nd ed. (Annapolis: Naval Institute Press, 1989), 319; Boeing Aircraft Company Seattle, Washington "B-29 Production and Construction Analysis," Prepared by: Air Materiel Command, Western District Headquarters T-5 Research Office (21 January 1946): iii, 4, Boeing Archives.
33. "Summary of Case History of Boeing Aircraft Company, Renton, Washington," 1–4, RG18 Army Air Forces, Office of the Assistant Chief of Air Staff Materiel and Services (A-4), Research and Development Branch Case Histories, Box 32, "Boeing Airc. Co. Renton, Wash." folder, NARA-College Park. "B-29 Production and Construction Analysis," vi, 12–14.
34. Bowers, *Boeing Aircraft since 1916*, 286, 328.
35. "Work Rushed to Prepare For Building 20 Destroyers," *Seattle Daily Times*, November 10, 1940.
36. "An Appraisal of U.S. Naval Industrial Reserve Shipyard Plant 'B' Seattle, Washington, For General Services Administration Contract No. 60,602,253 August 1960 By Harry R. Fenton, M.A.I., and Jerrold F. Ballaine, M.A.I." p. 12. RG 291 Box 103, NARA-Seattle.
37. Art Ritchie, ed., *The Pacific Northwest Goes to War (State of Washington)* (Seattle: Associated Editors, 1944), 19, 34.
38. "Improved Timber Construction Speeds Expansion of the Shipyards," *Marine Engineering and Shipping Review* (July 1941), 82–84; Timber Engineering Company, *The Forest Fights!* (Washington, DC: Timber Engineering Company, 1942), Box 4, MSS2547 Oregon Shipbuilding Corporation, Oregon Historical Society, Portland, OR.
39. "Defense! Lumber's Part," *The Timberman* (July, 1941): 10–11; "Use of Wood in Ships," *The Timberman* (August, 1942): 85; National Resources Planning Board, *Pacific Northwest Region: Industrial Development*, 30.
40. Ritchie, *The Pacific Northwest Goes to War*, 53, 79, 150.
41. State of Washington, *Report of the Washington State Census Board for the Years 1943 and 1944* (Olympia: State of Washington, 1944), 14.
42. Housing Authority of the City of Seattle, *First Annual Report of the Housing Authority of the City of Seattle* (January 1, 1941).

43. Housing Authority of the City of Seattle, *Housing the People: Sixth Annual Report of the Housing Authority of the City of Seattle* (February 1, 1946), 8–11.
44. All of Seattle's permanent World War II-era housing projects have only recently been redeveloped. For example, the Rainier Vista project, completed in 1942, was redeveloped beginning in 1999. Seattle Housing Authority, "Rainier Vista Redevelopment," https://www.seattlehousing.org/about-us/redevelopment/rainier-vista-redevelopment, accessed December 9, 2017.
45. Carlson, "Duwamish River: Its Place in the Seattle Industrial Plan," 148–149; City of Seattle Department of Planning & Development, "Duwamish M/IC Policy and Land Use Study Draft Recommendations" (November 2013), 5, 9, http://www.seattle.gov/dpd/cs/groups/pan/@pan/documents/web_informational/p1903847.pdf, accessed December 1, 2017.
46. *Conservation: The Resources We Guard* (Washington, DC: US Department of the Interior, 1940), 29; emphasis in the original.
47. William McNamara, "Wartime Operation Problems at Seattle," *Sewage Works Journal* 16, no. 6 (November 1944): 1244–1245; US Environmental Protection Agency, Region 10, *Record of Decision: Lower Duwamish Waterway Superfund Site* (November 2014), 2, 4, https://semspub.epa.gov/work/10/715975.pdf, accessed December 1, 2017; US Environmental Protection Agency, Region 10, *Five-Year Review Report: Harbor Island Superfund Site, Seattle, King County, Washington* (September 2015), 9–10, https://semspub.epa.gov/work/10/100014200.pdf, accessed December 1, 2017.
48. Gerald D. Nash, *World War II and the West: Reshaping the Economy* (Lincoln: University of Nebraska Press, 1990), 4.
49. University of Washington Bureau of Business Research, *An Analysis of Manufacturing in the Puget Sound Area* (Seattle: University of Washington, July 1955), 110.
50. Dominick Gates, "Wrecking Ball Looms for Historic Boeing Plant 2," *Seattle Times*, January 13, 2010, http://www.seattletimes.com/business/boeing-aerospace/wrecking-ball-looms-for-historic-boeing-plant-2/, accessed 30 November 2017; The Boeing Company, "Renton Production Facility," http://www.boeing.com/company/about-bca/renton-production-facility.page, accessed November 30, 2017.
51. Ann R. Markusen, et al., eds., *The Rise of the Gunbelt: The Military Remapping of Industrial America* (New York: Oxford University Press, 1991), 4; 1. Roger W. Lotchin, *Fortress California at War: San Francisco, Los Angeles, Oakland, and San Diego, 1941–1945* (Berkeley: University of California Press, 1994).

CHAPTER 4

War and Urban-Industrial Air Pollution in the UK and the US

Peter Brimblecombe

INTRODUCTION

War represents a time of profound change. Such times reveal a narrowing of activities within a society, as it focuses around those factors relevant to victory. Thus, it is clearly a time of restriction often accompanied by a reduction in the democratic process. High ideals about an improved environment, open education and greater citizen participation may be lost. Despite this, the transitions imposed by periods of war have relevance for the environment with both positive and negative outcomes the result of focused activities. Scholarship in the area of environment and war seems to be preoccupied with recent environmental impact of war, most notably during the 1991 Gulf War, when the Kuwaiti oil fires released large amounts of smoke into the air.[1]

This chapter combines national and local approaches. The aim is to explore the interaction between war and urban-industrial air pollution in the US and the UK, and also to highlight air quality above all in two cities, London and Los Angeles. They represent archetypes of two key forms of

P. Brimblecombe (✉)
School of Energy and Environment, City University of Hong Kong, Kowloon, Hong Kong

© The Author(s) 2019
S. Laakkonen et al. (eds.), *The Resilient City in World War II*, Palgrave Studies in World Environmental History, https://doi.org/10.1007/978-3-030-17439-2_4

air pollution: (1) classical smog that derives directly from combustion products emitted from chimneys or exhausts, and is often produced by the combustion of solid fuels much of which derive from large fixed stationary sources (i.e. primary pollution), and (2) photochemical smog, the product of reactions of precursor compounds such as volatile organic materials and nitrogen oxides in sun-lit atmospheres, typically arising from liquid fuels burnt in mobile sources (i.e. secondary pollution).

The Second World War can be seen as a turning point in our understanding of air pollution and its regulation. This view frequently derives from the observation that post-war growth stimulated a huge increase in the emission of pollutants and gave rise to more air pollution in cities—especially increasing car ownership and resultant emissions of volatile organic matter. However, a merely quantitative view hardly captures the enormous paradigm shift which arose with changes from pollutants that derived from solid fuel combustion in stationary sources and liquid fuels consumed largely by the transport fleet.[2] Scientists and policymakers soon began to understand that air pollution no longer simply mirrored its sources, but rather it had become the product of chemical reactions in a sun-lit atmosphere.

Interwar Years

There was an increasing societal interest in smoke abatement in the UK and the US from the late nineteenth century. In the UK, where regulations developed earliest, a formalised arrangement was reached that promoted a professionalised role for sanitary or nuisance inspectors, who with the passing decades became public health and ultimately environmental health officers, along with a degree of specialism in the area of smoke abatement.[3] They were increasingly informed by a widening number of influential books, from both sides of the Atlantic, on air pollution and its effects which include the following: W. E. Gibbs' *Clouds and Smokes* (1924), Shaw and Owens' *The Smoke Problem of Great Cities* (1925), Cohen and Ruston's *Smoke, a Study of Town Air* (1925), Henderson and Haggard's *Noxious Gases* (1927), Henry Obermeyer's *Stop That Smoke!* (1933), R. Whytlaw-Gray's *Smoke: A Study of Aerial Disperse Systems* (1932), R. J. Schaffer's *The Weathering of Natural Building Stones* (1932) and Raistrick and Marshall's *The Nature and Origin of Coal and Coal Seams* (1939). These added to regular annual reports such as *The Investigation of Atmospheric Pollution* from the Department of Scientific and Industrial Research formed in the UK during the Great War, in 1915.

The interwar years saw increasing moves towards smoke prevention by local governments. Nevertheless, wartime frequently saw such implementation overturned or at least reduced. A number of environmental issues that worried public health officials in the interwar years are worth examining. In the UK, there was a long debate about burning bings.[4] These were piles of waste coal which caught fire and led to the spread of smoke and sulphur dioxide over wide areas. County medical officers raised the issue with the Ministry of Health in the 1930s, and although the grievance was found legitimate, it was difficult to get the Government to intervene. However, the Alkali Inspectorate took over under W. A. Damon (Chief Inspector from 1929 to 1955) and provided technical assistance in the best practicable means of extinguishing fires. By 1935, the Minister of Health promoted discussion of how industry could be encouraged to take action and develop a better understanding of coal waste fires. However, there was no effective way to coerce mine owners to act, as the Health Departments were in a weak statutory position and needed to work with the Mines Department. This required a level of interdepartmental collaboration that has emerged as a key requirement in achieving solutions to many contemporary environmental issues. John Sheail[4] emphasised the role of individual local incidents in driving the issue forward. In the end the application of what had been learnt in technical studies was only effective when fears of enemy bomber activity led to regulatory change. This change came as part of the support for the Air Raid Precautions Department of the Home Office: "many in calling attention to the burning refuse-heaps which, in the event of war, would serve as beacons, enabling enemy aircraft to inflict considerable damage to property and possibly serious loss of life."[4]

Policy decisions by local governments prompted the construction of electricity generating stations, many of which that have lasted to this day. There was an increasing pressure to build very large power stations within cities, which meant new power stations such as London's Battersea faced considerable opposition. There was not just a problem with siting these, but also concern that they caused pollution, stimulating considerable public and professional agitation.[5]

A further issue that preoccupied experts through the 1930s was the severe pollution episode in the heavily industrialised Meuse Valley near Liége, Belgium, in 1930, which resulted in more than 60 deaths. This disaster led to the first scientific proof of the potential for atmospheric *pollution* to cause deaths and disease, and it clearly identified the most likely causes.[6] The Meuse "Death Valley" tragedy had considerable impact in the

UK, with the famous Scottish physiologist J. S. Haldane writing about it and several articles appearing in key medical journals.[7] In the US, it may be that the economic crash of the late 1920s had muted the impact of the Meuse Valley incident on research. Nevertheless, regulations were evolving in the US in the interwar years with improvements in the amount of smoke control. The Smoke Prevention Association published the *Manual of Ordinances and Requirements* guide to regulation in 1938. The Association also analysed local ordinances from 80 cities (3 of which were in Canada) with populations above 37,000. These regulations were variable and doubtless applied locally with different degrees of enthusiasm. The *Manual* also raised concern that 45 of the 80 cities had no full-time employees devoted to smoke abatement.[8]

It is notable that St. Louis, Missouri, was surprisingly active in pollution regulation during the interwar period. This can be attributed to the foresight of its Mayor Bernard Dickmann and his work in 1934 with Raymond Tucker, professor of mechanical engineering at Washington University, to spearhead the campaign to abate smoke. Their work received support from a very severe air pollution episode in November 1939. This smog, the infamous Black Tuesday, was widely discussed at the time.[9] The civic authorities convened a smoke elimination committee, which worked hard to change the quality of the coal used in the city. In April the following year the council passed a law requiring the use of low volatility fuels. Black Tuesday and the subsequent success of fuel restrictions had a strong influence on Raymond Tucker, who became the St. Louis City Smoke Commissioner and later its mayor. He was notable as a key adviser on pollution problems in post-war Los Angeles.[10] He argued: "The fundamental fallacy of all smoke programs is the belief of the citizens of the community that air can be purified by legislation. The mere passage of an ordinance and the appointment of a smoke commissioner does not endow him with divine power." It is also noted in line with a long-held European view on the importance of stoking[11] from the late nineteenth century, that "installation of mechanical-firing equipment for the burning of solid fuels, such as exist in Illinois, eliminates to a great extent the inherent handicaps that are attendant to hand-firing."

Another key interwar issue concerned the reduction of smoke from trains, which had long been a problem in the UK, through the electrification of the railways.[12] The precedent was set by New York City, which in 1903 passed a law that essentially forced the New York Central and New Haven railroads to electrify their commuter lines serving New York by banning the use of steam locomotives south of the Harlem River after mid-1908. The decision followed a deadly 1902 train accident that occurred on the smoky, tunnelled approach to Grand Central Terminal. Chicago passed a similar ordinance, forcing Illinois Central to electrify its busy lakefront commuter line by 1927. The debate that this stimulated might be typified by that which occurred in the newspapers with the opening of the Illinois Central station in 1926. The adoption of electrified suburban railways seemed symbolic, and Chicago newspapers of the time renewed their campaign for wider electrification. Smoke abatement reasons prompted the construction of electrified approaches to Cleveland Union Terminal from 1930 to 1953. Enthusiasm for electrification continued in New York and Philadelphia as lines there were converted. In the end such plans were much delayed by the war. However, the move away from coal power was often driven by other factors, such as the need for long tunnels where smoke accumulated, or due to congestion, or in passenger terminals where electrification allowed for more movements each day.[13]

The War Years

As the wartime economy often requires higher output from heavy industry, there is clearly the potential for higher levels of pollution. There is also a sense of patriotism that can lead to a loosening of restrictions, such as those from smoke abatement legislation, that are normally placed on industries. Limitations on output could be seen as showing a lack of commitment to the cause. There is also a restriction on the range of activities that are less directly related to the war effort. Thus, the rate of academic publication may be reduced because of a shortage of paper or printing factories, conferences cancelled or limited by the number of experts available to attend, or restrictions on travel.

The UK Department of Scientific and Industrial Research reported that the smokiness of many cities in Great Britain had by 1937 increased, after a notable reduction during the depression. The arrival of the war was accompanied by views that air quality was very much worse than before.[14] The *Manchester Guardian* of 3 April 1940 covered a meeting of the National Smoke Abatement Society, which remained very active during the war,[15] under an article headlined "More Smoke in War Time," reporting that smoke had increased because of the extra work caused by war. Robert Veitch Clark, medical officer of health for the City of Manchester from 1922, pressed to maintain the activities of the Society even though there was a war. It was accepted that enforcement had become more difficult because of the shortage of staff and the poor quality of fuel. The view that air pollution was worse during the war could have been true at specific locations, but as seen in Fig. 4.1, the general trends towards improved air quality continued, the notable exception being insoluble ash deposit at Attercliffe near Sheffield.

In 1943, although in the middle of a demanding war, the National Smoke Abatement Society was able to hold a conference to consider the claims of clean air in the post-war reconstruction of British cities (National

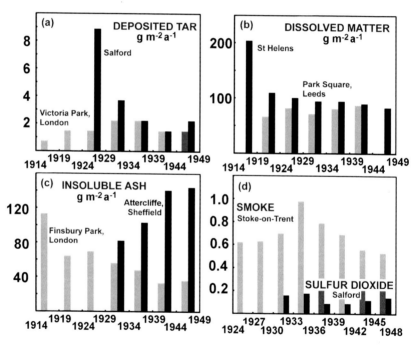

Fig. 4.1 Average annual deposit of (**a**) tar, (**b**) dissolved matter, (**c**) insoluble ash and (**d**) the concentrations of sulphur dioxide and smoke at sites from England in the first half of the twentieth century. (Note: Victoria Park, Park Square, Finsbury Park and smoke at Stoke-on-Trent are shown as light bars. The units for sulphur dioxide are parts per million and smoke mg m^{-3}. Source: Data from A. R. Meetham, *Atmospheric Pollution: Its Origins and Prevention* (London: Pergamon Press, 1952))

Smoke Abatement Society, 1943). In 1945 the Department of Scientific and Industrial Research's report on the "Leicester Survey," whose publication had been delayed by the war, appeared and drew attention to the important part played by turbulence in the dispersion of pollutants from the air above a city and there were also detailed surveys of the deposit of airborne material, such as that in Bilston in the winter of 1943.[14] Evidence of continued smoke abatement activities during the war, though perhaps at a lesser rate, can readily be found. For instance, a letter to *The Guardian* of 23 November 1943 from the famous reformer Arnold Marsh (General Secretary to the National Smoke Abatement Society) complained that new industrial areas were being developed and these were promoting an oversimplified view that prevailing wind would offer sufficient protection. He felt it was actually a lack of determination that caused manufacturers to continue emitting smoke when modern technology could make industries smokeless.

In the US, wartime activities of the Air Pollution Control Association have been reviewed by William G. Christy.[16] The Smoke Prevention Association of America held its 1941 meeting in Atlanta, Georgia, 3–6 June, just two weeks before German forces began Operation Barbarossa and invaded the Soviet Union. Nevertheless, the minutes of the 1941 meeting show an organisation still preoccupied with pre-war issues. We read that a Ladies Auxiliary (formed the year before) was in operation and there was a detailed report on "Abating Locomotive Smoke in Chicago" by I. A. Deutch and Sam Radner of the Chicago Smoke Department. Deutch was later Director of the Los Angeles County Office of Air Pollution Control and played a significant role in addressing the complexity of the city's problems.[17]

However, the 1941 meeting was before the American entry into the war that came after the 7 December Japanese attack on Pearl Harbour. The following day President Roosevelt asked Congress to declare war on Japan, so the US became a combatant in the Second World War. It meant that wartime restrictions soon limited travel, reducing the number of attendees to the Association's meetings from 1942 to 1945.[16] However, as during the First World War, the Association played a critical role in helping the federal government develop coal conservation policies. In the First World War such activities in the smoke abatement area were seen by Ann Marie Todd to be part of environmental patriotism,[18] although she viewed the Second World War more in terms of saving energy and materials, with a focus on recycling endeavours.

At the June 1942 Association meeting in Cleveland, issues relevant to the war were addressed more directly, notably with a focus on efforts to conserve coal. Fuel economy was critical, and the seriousness of the effort to conserve coal was emphasised during this meeting, as it took place just as American forces were engaged in the intense air and sea battle with the Japanese fleet off Midway Island.[19] Thus the war had moved fuel efficiency to the forefront of concern in the conference presentations. Here T. J. Thomas, of the Office of Fuel Coordination Washington, talked about "Coal—Its Relation to the War," stressing the role of pollution control in fuel conservation. Professor Sumner B. Ely gave a paper on "Smoke Abatement in Wartime." He had retired from his academic post shortly before to head Pittsburgh's newly restructured Smoke Bureau in 1941. The conference was also addressed by J. F. Barkley on "How Smoke Regulation Departments Can Assist in the War Project of Control, Allotment and Conservation of Coal." At the 1943 meeting in Pittsburgh, there were 13 papers on general subjects and three on railroads. The 1944 meeting in Detroit was marked with the presentation of 26 papers, 21 on general subjects and 5 concerned with railroads.

During the war there was increased interest in meteorological factors affecting pollution concentration,[20] much as wind was an important consideration raised by Arnold Marsh in the UK. There was a suggestion that the US Weather Bureau statistics would be of assistance to authorities in US cities with studies on meteorology and smoke abatement.[21] In 1945 the Los Angeles Air Pollution Control Office, which had been working on smog control since 1942, issued a paper on the various sources of sulphur dioxide, which was then thought to be the main cause of eye irritation,[22] although as we know now this view was far too simple, and in reality the photochemical smog had emerged.[23]

The 1945 annual meeting of the Air Pollution Control Association in Columbus, Ohio, ended in being delayed until 16 October because of the Government's ban on conventions during the war. The Association Secretary Chambers read a resolution of warm appreciation from the National Smoke Abatement Society of Great Britain. He explained that this had come in answer to the resolution of the previous year in which the Association had congratulated the Society for carrying on its work without interruption during the trying conditions of the war. In the ensuing discussion it was decided to invite Arnold Marsh, executive director of the British Society, to participate in future meetings of the American Association.[19]

There continued to be much concern through the war that "smoke needed to be eliminated, not merely abated" and "wartime air quality conditions were very bad, reinforcing the need for strict enforcement once the war ended."[24] Pittsburgh as a steel city took special efforts and in February 1941, the mayor set up a Commission for the Elimination of Smoke that aimed to represent the community in general. Mayor Cornelius D. Scully declared that "Pittsburgh must, in the interest of its economy, its reputation and the health of its citizens, curb [smoke]." The city's smoke control ordinance was implemented in October 1947 after a delay caused first by the war and then by a shortage of smokeless fuels. In the UK the big achievement was in the reduction of smoke from coal waste or bings as a result of concerns that these fires would serve as beacons and make nearby coal mines targets for enemy bombers. The ease with which the fires associated with waste coals were put out after being a problem for so many years was a testament to the importance of the research that had been undertaken before the war.

The biggest transition was the development of a new type of smog in Los Angeles. There were eye stinging episodes in 1942 and 1943. Quite quickly it was evident that this was somewhat different from traditional pollution and some began to think that it might have a different source. Initially, officials suspected that it arose in the production of artificial rubber, which used butadiene—increasingly used because rubber was much in demand for the war. However, such a suggestion may not have been welcome because the production was critical to the war effort. Even though butadiene is a highly reactive hydrocarbon and can cause smog, the main culprit in 1940s Los Angeles may well have been the volatile organic compounds from evaporating fuel, which served as an important smog precursor. Although the chemistry of smog formation took until the 1950s to be understood, perceptive policymakers in Los Angeles became aware of the dangers of attributing the smog to a single source. The lack of an obvious source detached the pollutant production from the resulting pollution and added significantly to the problem of regulation. The novel air pollution problems that had emerged in Los Angeles took half a century to resolve using formalised techniques such as air quality management. Parallel to these changes was a growing understanding of the socio-political context of air pollution and the way both politicians and the public needed to engage with the problem, particularly as it began to have consequences for the choice of fuel, transport, urban bonfires and so on. In more recent years it is apparent that the changes originating during the Second World

War shifted health impacts from respiratory outcomes to include cardiovascular effects driven by the decreasing size of airborne particles that allowed them to penetrate deep into the lung.

IMMEDIATELY POST-WAR

The end of the war in 1945 allowed the implementation of many pollution policies that had been on hold. It saw the publication of a number of important books: the more popular work Arnold Marsh's *Smoke*, (1947) and A. R. Meetham's *Atmospheric Pollution: Its Origins and Prevention* (1952). Nevertheless, there was a disastrous smog in London in December 1952, which killed many thousands. The Great Smog is partly blamed on post-war fuel shortages that obliged many householders to burn a low-grade fuel called *nutty slack*. The episode in 1952 led to the iconic Clean Air Act of 1956.[25,26] Between 1945 and 1960, the North American railroad industry switched from coal to diesel, and railroad coal consumption fell dramatically.[19] Pittsburgh saw its fuels regulated under a smoke control ordinance implemented in October 1947 that had lain dormant through the war. The most critical question in terms of implementation continued to be enforcement against domestic consumers. The Bureau of Smoke Control solved the enforcement problem, as had St. Louis, by focusing on the coal distribution yards (approximately 30) and the coal truckers.[24] The Bureau's new ordinance forbade yards to sell, and truckers to deliver, high volatility coal.[27]

There are interesting differences between the US and UK abatement practices as Halliday notes[27]: "It is probably worthwhile to consider briefly the factors which have caused the Americans to apply city control and the British to apply national control. In both cases the clue to the solution is rationalised fuel supply. In the US unorganised processes, such as the development of waste natural gas supplies and of oil burners for house heating, at a price competitive with coal, caused householders and others to turn to the right sort of fuel for the purpose of reducing smoke emission. In addition, the US being a federation and extremely large in area, with an expanding economy and fiscal funds in plenty, the conditions were right for city control of smoke emission, aided by state legislation to give the cities the necessary powers. In Great Britain the question of fuel supply was difficult under the constraints of post-war conditions."

While the war may have caused a lull in the development of new pollution control activities, it also meant a flurry of activity in the years that followed. This led to the emergence of important legislation in terms of

Clean Air Acts in both the UK and the US, in 1956 and 1963. However, ultimately it was the recognition of the relevance of evaporating fuel from automobiles and its importance as a precursor to photochemical smog that has had the most lasting influence. It was this that forever changed the way in which we think about air pollution, its chemistry and regulation.

Notes

1. Timothy O'Riordan, *Environmentalism* (London: Pion limited, 1981).
2. Peter Brimblecombe, "Deciphering the chemistry of Los Angeles smog, 1945–1995," in James Rodger Fleming and Ann Johnson, eds., *Toxic Airs: Body, Place, Planet in Historical Perspective* (Pittsburgh: University of Pittsburgh Press, 2014), 95–110.
3. Peter Brimblecombe, "Historical perspectives on health: The emergence of the Sanitary Inspector in Victorian Britain," *Perspectives in Public Health* 123, no. 2 (2003): 124–131; Peter Brimblecombe, "Origins of smoke inspection in Britain (circa 1900)," *Applied Environmental Science & Public Health* 1 (2003): 55–62.
4. John Sheail, "'Burning bings': a study of pollution management in mid-twentieth century Britain," *Journal of Historical Geography* 31, no. 1 (2005): 134–148.
5. Catherine Bowler and Peter Brimblecombe, "Battersea Power Station and environmental issues 1929–1989," *Atmospheric Environment. Part B. Urban Atmosphere* 25, no. 1 (1991): 143–151.
6. Alexis Zimmer and Benoit Nemery, "The Meuse Valley (1930): Just fog or industrial pollution," *Air Pollution Episodes* 6 (2017): 27–42; Benoit Nemery, Peter H. M. Hoet, and Abderrahim Nemmar, "The Meuse Valley fog of 1930: an air pollution disaster," *The Lancet* 357, Issue 9257 (March 2001): 704.
7. J. S. Haldane, "Atmospheric pollution and fogs," *British Medical Journal* 1, no. 3660 (1931): 366.
8. .H. B. Meller, "Practical Procedures and Limitations in Present-Day Smoke Abatement," *American Journal of Public Health and the Nations Health* 29, no. 6 (1939): 645–650.
9. Raymond R. Tucker, "Smoke Prevention in St. Louis," *Industrial & Engineering Chemistry* 33, no. 7 (1941): 836–839.
10. Brimblecombe, "Deciphering the chemistry of Los Angeles smog, 1945–1995."
11. Peter Brimblecombe, "London 1952: An enduring legacy," in Peter Brimblecombe, ed., *Air Pollution Episodes* 3 (London: World Scientific Europe Ltd., 2017), 57–72.

12. Catherine Bowler and Peter Brimblecombe, "The difficulties of abating smoke in late Victorian York," *Atmospheric Environment. Part B. Urban Atmosphere* 24, no. 1 (1990): 49–55.
13. Scott Lothes, "A brief look at the eras in American railroad electrification," September 25, 2009 (http://trn.trains.com/railroads/railroad-history/2009/09/the-stages-of-us-railroad-electrification).
14. E. C. Halliday, "A historical review of atmospheric pollution," *Air Pollution* (Geneva: World Health Organization, 1961): 9–38. A. R. Meetham, *Atmospheric Pollution: Its Origins and Prevention* (Oxford: Pergamon Press, 1952): 163.
15. Gary Willis, How did British voluntary organisations concerned with the environment balance their commitment to protect it with supporting the Second World War effort. MRes diss., University of London, 2017.
16. William G. Christy, "History of the air pollution control association," *Journal of the Air Pollution Control Association* 10, no. 2 (1960): 126–174.
17. Joshua Dunsby, "Localizing Smog," in E. Melanie DuPuis, ed., *Smoke and Mirrors: The Politics and Culture of Air Pollution* (New York: New York University Press, 2004), 170.
18. Anne Marie Todd, *Communicating Environmental Patriotism: A Rhetorical History of the American Environmental Movement* (Abingdon, Oxon: Routledge, 2013).
19. Bill Beck, "The First Half-Century: Pre- and Post-World War II Growth," *Journal of the Air & Waste Management Association* 57, no. 5 (2007): 636–639.
20. R. D. Fletcher and D. E. Smith, "Meteorological factors affecting smoke pollution," in Proceedings of the Smoke Prevention Association of America; 38th Annual Conference Pittsburgh, (1944), p. 123.
21. R. D. Fletcher, "Meteorology and smoke abatement," in *Proceedings of the Smoke Prevention Association of America; 39th Annual Conference* (1945), p. 31.
22. H. O. Swartout, and I. A. Deutch, "The 'Smog' Problem," *Los Angeles County Office of Air Pollution Control* (1945).
23. Peter Brimblecombe, "Deciphering the chemistry of Los Angeles smog, 1945–1995."
24. Joel A. Tarr, "Changing fuel use behavior: The Pittsburgh smoke control movement, 1940–1950," *Technological Forecasting and Social Change* 20, no. 4 (December 1981): 331–346.
25. Brimblecombe, "London 1952: An enduring legacy."
26. Ralph H. German, "Regulation of Smoke and Air Pollution in Pennsylvania," *University of Pittsburgh Law Review* 10 (1948–1949): 493.
27. Halliday, "A historical review of atmospheric pollution."

CHAPTER 5

Imagined Resilience

U.S. Conservation Campaigns and Fat Salvage

Sarah Frohardt-Lane

INTRODUCTION

In recent years, environmental activists seeking to invigorate Americans' efforts to combat climate change have suggested that the country draw on its World War II experiences as a model for action.[1] U.S. entry into World War II brought a surge in demand for resources at the same time that the war interrupted Americans' access to many materials on which they heavily depended. As a result, the U.S. government prioritized scarce materials for military purposes. It rationed rubber, sugar, and other commodities for the American public. Government agencies and private industries encouraged civilians to reduce their use of scarce resources and to modify their habits. Although the depression years had fostered conservation due to financial necessity, in wartime home-front messages stressed a patriotic responsibility not to consume more than necessary.

As the United States' home-front story of World War II is often told, Americans reduced their consumption for the sake of the war effort and salvaged scarce materials to be repurposed into weapons. At home and in

S. Frohardt-Lane (✉)
Department of History, Ripon College, Ripon, WI, USA

© The Author(s) 2019
S. Laakkonen et al. (eds.), *The Resilient City in World War II*,
Palgrave Studies in World Environmental History,
https://doi.org/10.1007/978-3-030-17439-2_5

their communities, men, women, and children were resourceful at making do and doing without, as a popular saying of the war had it. From this self-congratulatory national narrative, it would seem logical that Americans' mindsets and actions in World War II should inspire the types of behaviors and attitudes necessary to confront the "World War III" of climate change.[2]

This chapter explores the question of whether U.S. conservation campaigns in World War II should be considered models for future action. Using a case study of the fat salvage campaign, I examine the effort to collect household fats that began in Chicago, Illinois, and became a national fat salvage effort in mid-1942. I pay particular attention to the discrepancy between the way that the waste fats program was sold to the public (and has been remembered in popular memory ever since) and what the program actually entailed. As I argue, the salvage drive was a program to avoid more fundamental changes to American consumption patterns that would have required adaptation to scarce resources. The fat salvage campaign encouraged Americans to imagine themselves as resilient in the face of wartime, getting by with less and getting creative about repurposing resources. Instead of providing a historical example of Americans' resilience, however, the case of waste fats offers a cautionary tale about the ability of private industry to control conservation campaigns to serve their own ends.

Chicago's Fat Salvage Initiative

After the United States entered World War II, fats were among the many resources that became short in supply and high in demand. Fats and oils were needed for industrial soap, lubrication, textiles, paint, and countless other industrial and military purposes, as well as to meet civilian consumers' demands for a variety of products.[3] According to one estimate, in order "to meet the requirements for military production and to maintain a reasonable level for civilian use," the United States needed to fill a "deficit of 2.3 billion pounds of hard fat" when it entered World War II.[4] Most glamorously, salvaged fat that was no longer usable in the home kitchen when rendered produced glycerine, and this glycerine could make explosives and medicines for wounded soldiers on the frontlines. Fat renderers separated out glycerine as they turned waste fats into tallow in the process of making soap. This idea of housewives providing bullets to soldiers through saved kitchen grease was what advertising and news stories emphasized about fat salvage throughout the war and what became a part of popular accounts of the home-front.

Cities were the primary locales for waste fat collection and for fat salvage publicity. Because urban residents, butchers, and renderers were in relatively close proximity, it did not require a lot of extra resources (such as gasoline) to run a program in which individual households regularly brought their fats to the butcher shop and renderers in turn purchased the containers of waste fats that the butcher had amassed. It was also easier for fat salvage messages to reach Americans living in cities, whether in car cards on streetcars and buses, posters in downtown businesses, storefront window displays, daily newspapers, or radio announcements.

From the outset of the fat collection drive, officials expected urban residents to contribute more fat per person than rural residents. In fall 1942, for instance, the War Production Board (WPB) estimated that rural residents in Illinois should save on average 2.8 pounds of fat while city dwellers should save on average 4.5 pounds of fat that year. It was not until spring 1945—after the War Food Administration had taken control of the fat salvage program from the WPB—that the fat salvage campaign started focusing on ensuring rural cooperation in the program.[5] For the bulk of the period that waste fats were collected, the focus was on convincing Americans living in cities to turn in used household fats.

The fat salvage drive began as a local effort in Chicago, Illinois in late January 1942, less than two months after the United States entered the war.[6] Targeting "local housewives" with a message about the necessity of saving waste fats, fat renderers asked women to save fats from their cooking and then, once they could no longer use the fat any more, to collect it in a container and bring it to their butcher. Initially, in the Chicago program, the butcher paid the housewife 5 cents per pound of fat that she brought in and then sold the fat to a renderer for 6 cents per pound.[7] Renderers turned the salvaged fat into components for use by industries and the military.

Contributing waste fat to the salvage campaign required a housewife to take multiple steps, adding an additional burden to women who experienced a number of disruptions to running their households as a result of the war. Women were asked to strain hot grease (e.g. what was left in the frying pan after cooking bacon) into a collection can through cheesecloth and to collect hardened fat (e.g. what was left on a plate) and reheat it to remove food particles. Ads repeatedly emphasized how much fat the country could collect if each housewife saved a tablespoon every day.

Although the focus of saving waste fats was on convincing each housewife to cultivate a habit of saving kitchen fat every day, in Chicago community groups became involved in the campaign. Beginning in the spring of 1942, the Chicago Commission of Civilian Defense organized communities within the city into groups of volunteers on a variety of homefront committees, including a conservation committee that oversaw salvage drives.[8] The fat salvage program in Chicago had female volunteers publicize the need for saving waste fats and collect the fats that individual households set aside. Children became active in fat salvage through boy scout and girl scout troops. Scouts distributed posters to butchers to publicize the campaign and went door-to-door collecting cans of waste fats and turning them over to butchers, thus encouraging greater fat salvage by eliminating the housewife's need to bring in cans of fat herself. In this way, the campaign provided concrete actions for housewives and children to contribute to the war effort, and in Chicago the fat salvage program was explicitly touted as a solution to the problem of individuals needing to feel useful while far removed from the theaters of war.[9]

When war broke out, there was tremendous enthusiasm on the U.S. home-front to make a positive contribution to the war effort. As the *Chicago Daily Tribune* reported just before the Chicago fat salvage program became a national one, "Leaders of the drive offer the collection of fats and greases as an answer to the question frequently asked by Chicago housewives, 'What can I do to help win the war?'" Presenting the idea of a housewife's used fats traveling "from containers to cannon," the campaign came to feature "a special emblem showing a frying-pan pouring grease into the breach of an anti-aircraft gun being fired at the enemy" (Fig. 5.1).[10]

While the overall civilian defense program was dominated by men in industry, the fat salvage campaign itself in Chicago was directed by Ethel Kelly, invariably identified in news stories as Mrs. Walter J. Kelly, who was an "editor of a neighborhood newspaper" and active volunteer for the Chicago Metropolitan branch of the Office of Civilian Defense and many other community organizations.[11] As chairperson for fat salvage in Chicago, Kelly regularly provided information to the press and to volunteers to encourage participation in fat salvage collection. She stressed the contribution that each housewife made when she saved waste fats. As Kelly described, it was important to use "dynamic, colorful words in the whole campaign." Calling housewives who saved fats

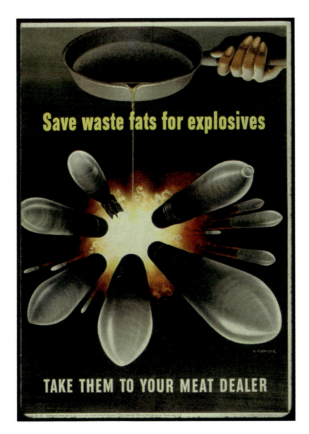

Fig. 5.1 This poster by Henry Koerner (1943) printed for the Office of War Information illustrates the message that household waste fat helped the war effort. (Source: National Archives poster (NWDNS-44-PA-380), via Wikimedia Commons)

"1942 Molly Pitchers," Kelly encouraged women to picture themselves as "loading guns from their kitchens."[12]

By late June 1942, Chicagoans had turned in 1.5 million pounds of household fats.[13] News stories lauded "Mrs. Chicago's" willingness to do her part to support the war by saving fats that could directly benefit the military. These accounts of Chicago housewives' fat salvage efforts praised their contributions to the war and their willingness to do whatever was necessary on the home-front to help achieve military success.[14]

NATIONWIDE CAMPAIGN

After months of organizing and collecting fat salvage as part of civilian defense work in Chicago, the city became a "guinea pig" for a two-week, highly publicized fat salvage drive in late June to test out what became a nationwide waste fat collection program in July.[15] The national fat salvage campaign followed many of the themes of the Chicago model. At the direction of the federal government's War Production Board, "members of the soap, glycerin, and rendering industries" created an entity to publicize the need for fat salvage and coordinate the collection process in early 1942. The Colgate-Palmolive-Peet Company, Lever Brothers Company, and Procter & Gamble Company were the three main corporations that formed the American Fat Salvage Committee (AFSC) and funded a massive, multimillion-dollar publicity campaign that began in July.[16] The AFSC employed Kelly, head of the Chicago fat salvage program, to speak to women in other cities about how to properly salvage fat and run a fat salvage campaign throughout urban America.[17] As the *New Yorker* assured readers who were skeptical that "New York can learn anything from Chicago," Kelly "is *the* authority on the mass collection of kitchen fat." Kelly explained how Chicago's fat salvage program had been organized and facilitated replication of the "Chicago plan," though she discouraged other cities from involving scout troops in the fat salvage efforts, in part because of the concern that it added an unnecessary layer to the process.[18]

AFSC promotional materials emphasized the need for saving waste fats and sought to motivate housewives to collect fats in the kitchen. The AFSC bought advertisements in city newspapers across the country. It also sent suggested news stories to newspapers and magazines throughout the United States and promoted radio announcements, comics, posters, and other media that stressed the urgency of fat salvage. For much of the war, this publicity was usually in the form of appeals to patriotism and home-front duty. Ads praised women for their efforts to help out and presented fat salvage as a way to contribute to the war effort. In what the AFSC deemed a "typical exampl[e] of magazine cooperation" with fat salvage efforts, a magazine story asked: "A Vital War Job for Every Kitchen: Is it Being Done In YOUR Home?" The piece began by detailing the daily routine of a housewife to whom the reader was intended to relate. "At breakfast time today Mrs. John

Dorr fried some bacon for her husband and some munitions for her son in North Africa." Walking through "how she did it," the article described how Mrs. Dorr set aside "leftover grease from her frying pan" for a month, then took the pound of fat to the butcher shop to be manufactured into glycerine. Lest the process seem too abstract, the article delineated: "Three weeks from now [the fat that Mrs. Dorr collected] will turn into explosives in a munitions factory. In two months those kitchen drippings may save the life of the Dorrs' boy in North Africa. Or your boy's life." Following the common pattern of beginning as a seeming public interest story, the text then shifted to providing information about the fat salvage campaign and how to participate.[19] Anything but subtle, fat salvage ads offered a specific action that women could do to help their country and by extension help their family members serving in the military.[20]

Fat salvage propaganda targeted women and was, in the words of one historian, "the one real 'women's campaign' for scrap."[21] Publicity implied it was women's role to conserve resources for the fighting men who deserved them. Advertisements and news stories specifically asked women (identified as housewives) to save fat and portrayed conservation as women's wartime job. At times such appeals were framed explicitly as giving resources to the soldiers who deserved them. As one ad explained: "Listen, lady! Used fats are so urgently needed for munitions, medicines and other vital war materials that every drop—every single drop—is precious!" … "Our fighting men need this help from you—they have a right to expect it!"[22] This advertisement, which closely mirrored propaganda aimed at convincing women to conserve gasoline, had clear messages about who needed to sacrifice and change behaviors, and who had the right to these resources.[23] Rather than encouraging a sense of equal sacrifice, this type of publicity placed the burden to conserve on housewives (Fig. 5.2).

Beyond the nature of the waste fat promotional materials, some faulted fat salvage procedures for being too onerous because they required women to strain out food particles before collecting waste fats. As one Office of Price Administration (OPA) official noted, it was "just asking a bit too much to have the housewife go to all the trouble and inconvenience" of straining fats before taking them in, when it would have been less burdensome and more efficient for renderers to strain large quantities after the point of collection.[24]

Fig. 5.2 Myriad materials were collected in salvage drives all over the world. Women had a crucial role in recycling due to their increased importance in both households and working places. (Source: San Francisco History Center, San Francisco Public Library)

Points-for-Fats Program

As the war progressed, it was clear that far from all households were saving waste fats, despite "campaign publicity [that] constantly emphasize[d] the drastic need."[25] Extremely bold claims about the impact that this salvage would have on the U.S. military's ability to successfully prosecute the war aside, several months into the campaign most housewives had not taken any waste fats to the butcher. A January 1943 survey indicated that while 92% of housewives "knew about the fat salvage campaign," 86% "were convinced there was a real need to save fats," and 85% "knew their used fats were needed for gunpowder," only 59% "were saving their used cooking fats," and a mere 31% "had turned in fats to meat dealers."[26] In public pronouncements, the AFSC declared a goal of collecting 275 million pounds of household fat in 1943.[27] But it actually collected not quite one-third of this target amount, at just under 90 million pounds.[28]

Claims of direct necessity for the war effort had not produced the level of fat salvage that the American Fat Salvage Committee sought. Their promotional materials evolved into appeals to self-interest, not just patriotic duty. With meat rationed, household shoppers had to use red points to purchase most meat during the war.[29] The AFSC convinced the OPA, the government entity that oversaw rationing, to award two red points for every pound of salvaged household fat that individuals sold to their butcher. The more fat a household salvaged, the more meat they could buy (and the more fat they could then salvage to redeem for more points). The implementation of the "points-for-fats" program on December 13, 1943, had an immediate impact on the amount of fat salvage collected. From approximately 9 million pounds a month collected in December, total salvaged household fat in the United States shot up to nearly 15 million pounds in January. In Chicago, according to a contemporary newspaper story, "salvage of grease tripled" in one week.[30]

But introducing points for fat had some unintended consequences: by using points to encourage fat salvage, the points-for-fats program motivated some people to game the system and try to get as many points as possible. Soon industrial fats—for which no points were awarded—started showing up in large quantities in the collection of household fats for which salvagers were reimbursed. OPA staff had to spend time distributing red points, seeking to prevent fraud, and attempting to ensure that the points were fairly allocated.[31] As fat salvage totals rose after the announcement that the government would give two red points per pound of fat, OPA personnel cautioned "taking with a grain of salt the figures indicating heavy increases in the collection of salvage fats" because of the likelihood that much of the new fat collected was disguised industrial fat.[32]

Giving red points for waste fats also complicated the involvement of community organizations in the fat salvage campaign. According to the discussion at one OPA meeting in March 1944, paying points for fats made it more difficult for community groups to assist with fat salvage collection. Girl scouts who had been accustomed to taking fat from housewives to the butcher now were supposed to obtain the appropriate number of points for each housewife and were not allowed to have possession of these points unless their group was registered as a renderer in their city.[33]

This seemed to be a problem especially in Chicago, where scouts had become involved in waste fat collection. Because of the various difficulties with the campaign and strong suspicions that the fat salvage was not working as it had been intended, some administrators advocated unsuccessfully for a "30-day suspension" until they could work through the inconsistencies and graft that had accompanied the introduction of red ration points in exchange for salvaged household fat.[34]

Total pounds of fat collected stayed higher after points for fats was introduced in late 1943 though they dwindled as the war progressed. But giving points did not stimulate the widespread participation rates that the AFSC had desired, and it is unclear how much of the increase in collection was actually from household fats.

Purpose of the Fat Salvage Campaign

It appears that the greatest discrepancy between supply and demand of fats was at the very end of the war, when American housewives had become disillusioned with the fat salvage campaign.[35] Despite the 4 cents and two red points issued per pound of fat collected, participation in the fat salvage had declined. As Allied victory appeared only a matter of time, it became harder for the American Fat Salvage Committee to convincingly portray saving waste fats as essential for victory. Previous appeals to collect fat for the sake of the military seemed nearly irrelevant. As Judith Russell, OPA's Special Assistant to the Director of the Food Rationing Division, explained in her 1947 account of the fat salvage campaign, at the end of the war "there was perhaps an element of boomerang in appeals so long addressed to patriotic motives primarily."[36] In the war's final months the chickens came home to roost: having heard from AFSC publicity for the last two years that the collection of waste fats was for explosives to win the war, housewives turned in decreasing amounts of waste fat as the war neared completion.

As V-E Day approached, the campaign stressed that "it's the Japs who have our fats.... So keep saving used fats until V-J day the day when we can celebrate final victory over our last and toughest enemy—Japan."[37] Advertisements also became much more direct in their appeals to self-interest rather than a sense of patriotic duty or willingness to help out: in the month following V-E Day, the *Chicago Daily Tribune* promoted fat salvage under the heading REWARD in large letters. Those who read the fine print discovered that the "reward" was the "2 red points" per pound of fat they sold to their meat dealer.[38] But these appeals had limited impacts. Despite catering to self-interest and expanding the message of types of needs for waste fats, as one headline noted, "Fats Salvage Drop[ped] Sharply After V-E Day."[39]

The AFSC lobbied OPA to allocate additional points for fat salvage beyond the two points per pound that had been in effect since December 1943. In August 1945, when the war ended and fat salvage continued to decline, an OPA administrator explained that the AFSC believed "What is needed now is a 'stepping up' of the selfish motive—a third red point."[40] Beginning in October 1945, OPA began to give four points per pound of fat the housewife sold to the butcher.[41]

Why did the government increase the number of points it issued for household salvage fat after the war was over? As much of the public had begun to suspect before the war had ended, the fat salvage campaign was not primarily conducted in order to turn waste fats into bullets, bombs, or balms for the military. Instead, much of the salvaged fat was used to make soap.[42] Concerned that a drastic reduction in soap supplies would necessitate soap rationing, and that rationing would mean that Americans would truly learn to "do without" as much soap as they desired, the fat salvage committee hit upon a way to prevent Americans from learning to adapt to the resource shortages of wartime by getting by with less.[43] Inundating the public with a message of the wartime necessity of saving waste fats, the soap industry amassed fats to make more soap supplies available. That was why Procter & Gamble, Colgate-Palmolive-Peet, and Lever Brothers underwrote the fat salvage campaign and carried out such an intensive publicity effort to make it successful.

As early as May 1942, the Department of Justice conducted a "soap investigation" to determine whether allocating "large quantities of fats and oils" to soap producers was a wise use of limited wartime supplies. When Assistant Attorney General Thurman Arnold wrote to Thomas Blaisdell, a member of the War Production Board's Planning Committee, to get to the bottom of the question of how essential glycerine really was, Blaisdell replied that "the direct use for military purposes by the Army and Navy is small" and that increasing soap production in order to make more glycerine "would seem to be wasteful production." In fact, Blaisdell wrote, "the pressure of the War Production Board has not been for an expanded production of glycerine. On the contrary, the WPB has been convinced that it would be entirely possible to reduce the production of glycerine without harm to the military effort and without serious inconvenience to the civilian population."[44] Despite these claims from the WPB before the fat salvage campaign began, the program's public justification was firmly rooted in assertions of glycerine shortages.

Some OPA employees raised concerns over the claims that the AFSC propagated and exerted pressure to phase out the most erroneous propaganda.[45] Critics pushed the AFSC to appeal more to housewives' self-interest—saving fats not to save the country but to be able to buy more meat for their families because they would earn more red points.[46] Others, including OPA head Chester Bowles, wanted the fat salvage campaign to be transparent that it was about generating fat to make more soap. Bowles berated soap executives for their "over-dramatization" of the fat campaign. "As I understand it," Bowles stated, "90 percent of all fats collected through the red-points program go to making soap. Certainly no one would suspect this from looking at the advertisements which have been sent to us. Although soap is mentioned, it is brought in in the most inconspicuous way, well muffled with medicines, gunpowder, etc." While Bowles noted the burden these claims caused for OPA staff tasked with trying to prevent misinformation from going out under its authority, the AFSC touted the fat salvage program as "one of the most successful joint Government-industry public relations jobs ever undertaken."[47]

It is striking that the points-for-fats program essentially began at the moment that restrictions on glycerine ended.[48] At the same time that it became clear that waste fats were not needed for glycerine to help military campaigns, the OPA allocated ration points to stimulate waste fat collection. From that point forward, the OPA was involved in the fat salvage campaign through implementing the red point system in exchange for fats, and throughout the war signs for waste fat collection in butcher shops stated that they were approved by the War Production Board, which oversaw the salvage campaign.[49]

Although some pushed back against fat salvage advertising, other government employees voiced concern that a more transparent publicity campaign would "precipitate the need for [soap] rationing immediately."[50] If the public became certain that waste fats were being used to make soap because soap was in short supply, the reasoning went, consumers would panic and try to buy more soap than they needed just to stockpile it for the future. As a result, the OPA would have to commence soap rationing, with all the corresponding headaches of setting up another rationing program to oversee.[51] Therefore, despite some government queasiness with the "alternate facts" of private industry's household fats promotion campaign, advertisers were essentially allowed to continue misleading the public about the purpose of the conservation campaign, its successes, and the direct relevance of civilians' actions to the war effort.

The fat salvage campaign was one of the most heavily advertised of the many joint government-industry efforts to convince the American public to alter their behaviors during World War II.[52] In the decades since, popular histories and personal accounts have emphasized the "can-do" attitude, unity of purpose, and willingness to sacrifice among the American public during World War II. In this narrative—part of the larger narrative of the "greatest generation" that fought and won World War II—the country rose to wartime challenges in ways that subsequent Americans have not been called to do. Fat salvage stands as one memorable example of housewives' contributions on the home-front to help end the war sooner.

More scholarly accounts of the salvage campaigns in the war have challenged the material benefits of these attempts to reuse and conserve resources. Instead, they have emphasized the utility of scrap drives to build public support and create a sense of helping the war effort, which was important to home-front morale. As others have suggested, the collection of war materials was not always a necessity but a convenient way to make Americans on the home-front feel that they were contributing in some way.[53] Beyond elevating civilians' sense of rising to the occasion, or collecting some usable materials, fat salvage and other similar conservation campaigns instilled a sense of pride in the sacrifice of doing without resources, even if there was very little adaptation that actually occurred.[54] Encouraging Americans to see relatively minor alterations to daily life as temporary wartime sacrifices, and lauding them for these actions, conservation campaigns gave the public a false sense of accomplishment in having substantively altered their lives to meet changing environmental realities. In this wartime strategy, one sees the seeds of "greenwashing" that have characterized much of the private sector's response to environmental concerns in recent decades.

To some extent, however, the fat salvage campaign was more insidious than merely assuaging Americans' collective sense of obligation without making significant adaptations to resource shortages. For the fat salvage campaign, in particular, collecting waste fats was intended, in part, to prevent conservation. As the war progressed, this anti-conservation motivation became more pronounced. Americans did not reduce their use of soap during World War II. Instead, demand for soap rose and soap production increased during the war compared to before U.S. entry.[55] According to Russell, "It was unprecedented demand rather than curtailed production that brought about the [soap] shortage."[56] Indeed, as Russell has noted, during the wartime years of fat salvage, some were critical of the campaign and "thought that it would have been in the public interest to have curtailed soap production in order to help safeguard world supplies of fats and oils, to have rationed limited supplies of soap, and to have taught consumers how to conserve it" (Fig. 5.3).[57]

The transition in the fat salvage campaign's messages about homemade soap illustrates how soap companies concerned with maintaining demand for their products undermined conservation impulses and promoted wartime use of their products. Early on, some AFSC ads and planted news stories that emphasized the dire shortage of fats and oils told consumers it was fine to conserve soap. A mid-1942 quiz distributed to newspapers for

Fig. 5.3 "DOWN WITH HITLER!" San Francisco school children dramatize the impact of their scrap paper drive on the war's outcome. (Source: San Francisco History Center, San Francisco Public Library)

publication included among its questions the statement, "We have been saving our kitchen greases for years and use them for soap or give them to other people who we know are making practical use of them." It supplied the unambiguously positive reply: "Any one doing this is to be congratulated for the conservation and use of these fats and we would suggest that you continue to use them as you have in the past."[58] Such a response suggested that "making do" and "doing without" store-bought soap made sense given the shortage of fats. But this conservation message was hardly compatible with fat salvage campaign themes that became more prominent in ensuing months.

Some fat salvage publicity directly contradicted the soap conservation message and became more pronounced by the middle of the war. Instead of congratulating women who made their own soap on reducing their use of commercial soap that was in short supply, "experts" warned about the dangers of home soap-making and advised American women not to try to get by with less commercial soap even though there was a shortage of fats to produce it. A mid-1943 column by Orson Munn, editor of *Scientific American*, and part of the AFSC's publicity team, excoriated home soap-makers.[59] Explaining that "homemade soap is false economy" because "commercial soap is inexpensive and generally excellent in quality" Munn admonished that "No woman who values good looking hands should subject them to the ravages of home-made soap." The travesties caused by making and using homemade soap (instead of buying soap during the war) were more than just the individual "assault upon self-esteem" and the "damage done to bodies and to clothing"; women who made soap were also guilty of unpatriotic behavior. According to Munn:

> Home soap-making operates against our cause in two evil ways. It takes waste fats directly out of the Government's fat salvage campaign, and it also decreases the manufacture of commercial soap, which is the largest single source of the nation's glycerine supply. Therefore, it is not an exaggeration to say that the home-making of soap tends to sabotage our war effort. At this time, there is no patriotic ground upon which the practice can be defended, nor is there any basis of common sense for it.[60]

In Munn's account, it was "evil" to address the wartime shortage of soap by making one's own soap and women who did so were hurting the country. His column was among the promotional materials AFSC sent to numerous newspapers and magazines.[61]

The soap industry prepared a broader defense of commercial soap in the event that soap was rationed during the war. Rationing risked the possibility of learning to make do with less, and thus the planned soap rationing program would have stressed the importance of soap to civilian life, even as it ostensibly prepared civilians to have limited access to soap during the war. In internal drafts of soap rationing "copy policy," Kenyon and Eckhardt—the advertising firm that oversaw fat salvage—explained that soap "almost like the air we breathe and the water we drink, has become so 'taken for granted' that we are apt to forget how vitally important it is to health, morale and even mortality."[62] Soap was not only essential for civilians but had dire importance for the war effort overseas: it reduced risk of diseases, was crucial for diplomacy, and was "an excellent propaganda weapon" in areas overtaken by the Allies, making these civilians "easier to govern" and prisoners "more manageable."[63] Ostensibly offering ways for housewives to conserve soap (with a 26-point list, including "mak[ing] soap jelly by dissolving scraps in boiling water"), these documents simultaneously sought to elevate the importance of soap in the public imagination.

Because soap rationing did not ultimately go into effect, however, testimonials about the benefits of soap largely lost out to military necessity explanations for fat salvage until the war was nearly won. Astute observers likely noticed changing justifications from the start of gathering waste fat in Chicago when the United States entered World War II to the ongoing pleas for more fats over a year after the war ended, and after wartime rationing programs had ceased. What had been framed as a way to help win the war by conserving resources became explicitly framed as a way to increase personal consumption. No longer claiming leftover home fats were needed for "making medicines and munitions"[64] postwar explanations pointed to a dearth of soap. As one ad put it in early January 1946, "Where there's fat, there's soap." The ad encouraged housewives to save fat so that it would "bring back more soap to your dealer's shelves sooner."[65] Into late 1946, the *Chicago Daily Tribune* ran columns and advertisements that urged women to keep saving fat "if we are to have more soap."[66] At war's end, associating fat salvage with its true purpose of keeping up soap supplies could appeal to Americans as consumers who did not want to do without.

Conclusions

The history of World War II fat salvage in the United States illustrates several points. Most significantly, as I have argued, those who promoted and organized fat salvage did so in order to keep soap consumption at high rates and prevent consumers from learning to make do with less. Instead of rationing soap or learning to conserve, demand and production increased. Secondly, the AFSC's attempts to produce fat salvage compliance relied upon misleading information at times, if not outright lies, that were approved by government agencies. Third, even these claims of dire necessity were insufficient to motivate a large majority of American housewives to partake in the fat salvage program, whether that was because they saw the daily fat saving as too burdensome or because they were skeptical that doing so would make more bullets. Finally, in the process of setting aside grease to cool and separate, skimming off oils from leftover foods, and straining fats to take to the butcher, housewives bore an extra burden that principally served to advance commercial soap interests. Although this may have "boosted morale," it was a burden on women's daily life that had little tangible positive good for the war effort. The fat salvage campaign made claims on the time of American women while it mainly helped corporate interests in maintaining consumer demand.

In popular culture, wartime waste fat collection is still remembered as a campaign that helped change behaviors. Housewives saving bacon fat for bombs evokes a time in which Americans were more willing to sacrifice for the greater good. At first blush, conservation and salvage drives that cities embraced during World War II might seem to offer evidence of Americans' resilience in difficult times and a model for collective efforts in the face of more trying circumstances that climate change will produce. But as the case of the fat salvage campaign demonstrates, most Americans did not seem motivated by a sense of obligation to set aside waste fats, nor did appeals to their own self-interest substantially change their behavior for the duration of the war. Rather than providing crucial wartime resources, conserving fats produced imagined resilience, providing the American public with a sense of accomplishment for solving a dire glycerine shortage that did not exist. More fundamentally, to the extent that women and children did participate in collecting waste fats, their efforts served a program that helped prevent behavior modification—in the form of getting by with less soap—and ensured increased soap production and consumption.

In August 1942, right after the Chicago fat salvage initiative became a nationwide program, Chicago's fat salvage coordinator Ethel Kelly extolled fat salvage as helpful in the "establishment of a habit of national thrift." Picking up on Kelly's point, Mary Webster White, the head of conservation programs in Washington, D.C., predicted that conservation campaigns like fat salvage would result in positive cultural changes. "As a nation we haven't been over-thrifty," White said. "I don't think we'll ever be quite so wasteful again."[67] Of course, the postwar years demonstrated that the opposite was true. But the seeds of overconsumption were sprouting during the war itself, packaged by self-serving industries in an ethos of conservation and military necessity, and misremembered ever since as Americans' willingness to sacrifice and ability to adapt to difficult circumstances.

Notes

1. Bill McKibben, "A World at War," *The New Republic*, August 15, 2016, and Naomi Klein, *This Changes Everything: Capitalism vs. The Climate* (New York: Simon and Schuster, 2014).
2. The term is McKibben's. For more on wartime salvage campaigns in the United States, see Hugh Rockoff, "Keep on Scrapping: The Salvage Drives of World War II," NBER Working Paper No. 13418, (September 2007), 14, http://www.nber.org/papers/w13418; Hugh Rockoff, "The U.S. Economy in WWII as a Model for Coping with Climate Change" NBER Working Paper No. 22590 (September 2016), http://www.nber.org/papers/w22590.pdf; James J. Kimble, *Prairie Forge: The Extraordinary Story of the Nebraska Scrap Metal Drive in World War II* (University of Nebraska Press, 2014); Robert William Kirk, "Getting in the Scrap: The Mobilization of American Children in World War II," *The Journal of Popular Culture* 29, no. 1 (Summer 1995): 223–233.
3. "Suggested OWI Release on Fat Salvage Program," File General- Fat Salvage, Box 808, Entry R 112 Rationing Department, National Office, Food Rationing Division, Meat Rationing Branch, Papers Concerning Fat Salvage, 1943–45, Record Group 188, Records of the Office of Price Administration, National Archives at College Park, College Park, MD.
4. Judith Russell, "The Fat Salvage Campaign," in *Studies in Food Rationing, General Publication No. 13. Historical Reports on War Administration: Office of Price Administration.* (Washington, DC: U.S. Government Printing Office, 1947): 233.
5. "Ask Chicago Women for 1,274,000 Pounds of Greases a Month," *Chicago Daily Tribune*, September 5, 1942; Wilder Breckenridge, American Fat Salvage Committee, Inc., to weekly newspaper publishers,

March 15, 1945, File Fat Salvage Campaign, Box 808, Record Group 188, Records of the Office of Price Administration, Entry R 112 Rationing Department, National Office, Food Rationing Division, Meat Rationing Branch, Papers Concerning Fat Salvage, 1943–45, National Archives at College Park, College Park, MD.

6. "Growth and Progress of the Household Salvage Fats Program," undated, File General- Fat Salvage, Box 808, Entry R 112 Rationing Department, National Office, Food Rationing Division, Meat Rationing Branch, Papers Concerning Fat Salvage, 1943–45, Record Group 188, Records of the Office of Price Administration, National Archives at College Park, College Park, MD.

7. "Copy Policy Fats Salvage Program," undated, File Salvage Programs, Tin Cans, Scrap Metal- General File, Box 583A, Entry 88, Record Group 208, Records of the Office of War Information, National Archives at College Park, College Park, MD. When this program extended to the rest of the country, the rate was 4 cents per pound of waste fat.

8. "300,000 To Serve Civilian Defense," *Chicago Daily Tribune*, March 8, 1942.

9. "City Will Serve as Guinea Pig in Grease Salvage," *Chicago Daily Tribune*, June 23, 1942; "Household Salvage Fat Program" meeting notes, March 7, 1944, 6, File General- Fat Salvage, Box 808, Entry R 112 Rationing Department, National Office, Food Rationing Division, Meat Rationing Branch, Papers Concerning Fat Salvage, 1943–45, Record Group 188, Records of the Office of Price, National Archives at College Park, College Park, MD.

10. "City Will Serve as Guinea Pig in Grease Salvage"; "Advertising Will Spur Big Campaign to Save Kitchen Fats for War Use," *New York Times*, June 24, 1942.

11. "Works at Many Jobs, But Still Saves Grease" *Chicago Daily Tribune*, April 26, 1942, N5; "Conservation is Theme of Defense Rally," *Homewood-Flossmoor Star*, 30 October 1942, 1.

12. "Greases Wheels of Fats Campaign," *New York Times*, July 21, 1942.

13. "City Will Serve as Guinea Pig in Grease Salvage."

14. John Wilhelm, "Your Grease- From Skillet to Firing Line," *Chicago Daily Tribune*, July 5, 1942.

15. "City Will Serve as Guinea Pig in Grease Salvage."

16. Russell, 234; Progress Report No. 1 Fat Salvage Campaign 1943, pages 1 and 10, File Progress Report No. 1 Fat Salvage Campaign 1943, Box 218, Entry 68, Records Concerning the Waste Fats Program, compiled 1943–1944, Record Group 208, Records of the Office of War Information, National Archives at College Park, College Park, MD.

17. Eugene Kinkead, "Chicago Molly Pitcher," *New Yorker*, August 29, 1942, 11.

18. Ibid., 12.

19. "A Vital War Job for Every Kitchen," in Progress Report No. 1 Fat Salvage Campaign 1943, File Progress Report No. 1 Fat Salvage Campaign 1943, Box 218, Entry 68, Records Concerning the Waste Fats Program, compiled 1943–1944, Record Group 208, Records of the Office of War Information, National Archives at College Park, College Park, MD.
20. For more on the conflation of wartime and family goals, see Robert B. Westbrook, "Fighting for the American Family: Private Commitments and Political Obligation in World War II," in Richard Wightman Fox and T. J. Jackson Lears, eds., *The Power of Culture: Critical Essays in American History* (Chicago: The University of Chicago Press, 1993).
21. Susan Strasser, *Waste and Want: A Social History of Trash* (New York: Henry Holt and Co., 1999), 253.
22. "Can a Woman Who Cooks for Two Save Used Fats?," *Chicago Daily Tribune*, June 8, 1944.
23. Sarah Frohardt-Lane, "Essential Driving and Vital Cars: American Automobile Culture in World War II" in Ross Barrett and Daniel Worden, eds., *Oil Culture* (Minneapolis: University of Minnesota Press, 2014), 102.
24. Chester Bowles, Acting Administrator, to W. F. Straub, Director, Food Rationing Division, October 27, 1943, reprinted in Russell, "The Fat Salvage Campaign," 261.
25. Oregon State Salvage Committee, "Status of Glycerine," 4, http://sos.oregon.gov/archives/exhibits/ww2/Documents/services-salv8.pdf
26. "1943 Advertising and Promotion Plan," in Progress Report No. 1 Fat Salvage Campaign 1943, File Progress Report No. 1 Fat Salvage Campaign 1943, Box 218, Entry 68, Records Concerning the Waste Fats Program, compiled 1943–1944, Record Group 208, Records of the Office of War Information, National Archives at College Park, College Park, MD.
27. "A Vital Job in Every Kitchen," File Progress Report No. 1 Fat Salvage Campaign 1943, Box 218, Entry 68, Records Concerning the Waste Fats Program, compiled 1943–1944, Record Group 208, Records of the Office of War Information, National Archives at College Park, College Park, MD.
28. Russell, "The Fat Salvage Campaign," 262.
29. "Points-for-Fats-Program," File General- Fat Salvage, Box 808, Entry R 112 Rationing Department, National Office, Food Rationing Division, Meat Rationing Branch, Papers Concerning Fat Salvage, 1943–45, Record Group 188, Records of the Office of Price Administration, National Archives at College Park, College Park, MD.
30. Kate Massee, "Women in War Work," *Chicago Daily Tribune*, December 28, 1943.
31. "Recommendations to Improve our Present Fat Salvage Amendment," in File General- Fat Salvage, Box 808, Entry R 112 Rationing Department,

National Office, Food Rationing Division, Meat Rationing Branch, Papers Concerning Fat Salvage, 1943–45, Record Group 188, Records of the Office of Price, National Archives at College Park, College Park, MD.
32. John J. Madigan to Walter F. Straub, January 17, 1944; File Salvage, Box 597, Record Group 188, Records of the Office of Price Administration, Entry R59 Rationing Department, Nat'l Office, Food Rationing Division, Office of the Director, General Correspondence, 1942–45, National Archives at College Park, College Park, MD. The OPA did not give points for fat salvaged from industry because industries typically already were salvaging fat before the points for household fats program was implemented.
33. "Household Salvage Fat Program" meeting notes.
34. "Household Salvage Fat Program" meeting notes.
35. Russell, "The Fat Salvage Campaign," 254.
36. Russell, "The Fat Salvage Campaign," 254.
37. "Bad News!," *Chicago Daily Tribune*, November 10, 1944.
38. Advertisement, *Chicago Daily Tribune*, May 30, 1945.
39. "Fats Salvage Drops Sharply After V-E Day," *Chicago Daily Tribune*, June 18, 1945.
40. Russell, "The Fat Salvage Campaign," 254–5.
41. Alexander Williams, Information Director, AFSC, to Publishers and Editors Cooperating with Department of Agriculture, OPA, WPB, Fat Salvage Campaign, September 26, 1945, File General- Fat Salvage, Box 808, Entry R 112 Rationing Department, National Office, Food Rationing Division, Meat Rationing Branch, Papers Concerning Fat Salvage, 1943–45, Record Group 188, Records of the Office of Price, National Archives at College Park, College Park, MD.
42. Letter from Chester Bowles to Lever Brothers, quoted in Russell, 252; "Third Annual Report of Fat Salvage Campaign, September 1945," File Fats- Salvage Campaign- Report, Box 1749, Entry 1, RG 179 WPB Policy Documentation File, National Archives at College Park, College Park, MD.
43. Russell, 235; Rockoff, "Keep on Scrapping."
44. Thomas C. Blaisdell Jr. to Thurman Arnold, May 30, 1942, File Soap & Glycerine- Supply, Box 1753, Entry 1, RG 179 War Production Board Policy Documentation File, National Archives at College Park, College Park, MD.
45. Russell, "The Fat Salvage Campaign," 254.
46. Russell, "The Fat Salvage Campaign," 250–51.
47. Letter from Chester Bowles to Lever Brothers, quoted in Russell, 252; "Third Annual Report of Fat Salvage Campaign, September 1945," File Fats- Salvage Campaign- Report, Box 1749, Entry 1, RG 179 WPB Policy Documentation File, National Archives at College Park, College Park, MD.
48. Russell, "The Fat Salvage Campaign," 249.
49. Rockoff, "Keep on Scrapping," 15.

50. C. W. Lenth, "The Fat Salvage Campaign and the Situation on Hard Fats," February 21, 1944, quoted in Russell, "The Fat Salvage Campaign," 250.
51. Russell, "The Fat Salvage Campaign," 235.
52. "Third Annual Report of Fat Salvage Campaign, September 1945," File Fats-Salvage Campaign-Report, Box 1749, Entry 1, Record Group 179, Records of the War Production Board, National Archives at College Park, College Park, MD.
53. Strasser, *Waste and Want*, 262; Rockoff, "Keep on Scrapping," 15.
54. See Mark H. Leff, "The Politics of Sacrifice on the American Home Front in World War II," *Journal of American History* 77, no. 4 (1991).
55. "Copy Policy Information Campaign on Soap Rationing," Box 682, Entry R 67 Food Rationing Program, Office of the Director, Soap Rationing Program RG 188, Records of the Office of Price, National Archives at College Park, College Park, MD.
56. Russell, "The Fat Salvage Campaign," 248.
57. Russell, "The Fat Salvage Campaign," 248.
58. "Drive to Save Fats Starts in Chicago," *New York Times*, June 25, 1942.
59. Orson D. Munn, "Science in the News," *Heppner Gazette Times* (Heppner, OR), July 22, 1943.
60. Munn, "Science in the News."
61. Munn, "Science in the News," File American Fat Salvage Committee, Box 583A, Entry 88, Records of the Deputy Director for Production and Manpower Correspondence Relating to Manpower and Salvage Programs, May 1944–May 1945, Record Group 208, Records of the Office of War Information, National Archives at College Park, College Park, MD.
62. "Copy Policy Information Campaign on Soap Rationing," 35, Box 682, Entry R 67 Food Rationing Program, Office of the Director, Soap Rationing Program RG 188, Records of the Office of Price, National Archives at College Park, College Park, MD.
63. "Copy Policy Information Campaign on Soap Rationing," 18–19, Box 682, Entry R 67 Food Rationing Program, Office of the Director, Soap Rationing Program RG 188, Records of the Office of Price, National Archives at College Park, College Park, MD.
64. "Fats Salvage Drops Sharply After V-E Day," *Chicago Daily Tribune*, June 19, 1945.
65. Advertisement, *Chicago Daily Tribune*, January 3, 1946.
66. Mary Meade, "Recipe Book Collected in Prison Camp," *Chicago Daily Tribune*, November 9, 1946.
67. "Women in War Work," *Chicago Daily Tribune*, August 21, 1942.

SECTION III

Urban Nature

CHAPTER 6

Guerrilla Gardening?

Urban Agriculture and the Environment

Rauno Lahtinen

During World War II, almost all of the European and Asian countries at war were threatened by a severe decline in agricultural production. Securing the procurement of foodstuffs for the civilian populace in particular was critical in order to maintain public health as well as overall morale and military production. Many countries saw the establishment of various programs intended to increase agricultural production. In England, the catch phrase "Dig for Victory" became a call to urban residents to import agriculture into the cities. The moat of the royal prison, the Tower of London, was adopted for cultivation, a step that symbolized the importance of agriculture for the entire commonwealth. In the United States, the "Victory Garden" planted by Eleanor Roosevelt in the White House gardens inspired the cultivation of small garden plots to support households and the wartime economy.[1] Slogans such as "Eat what you can, and can what you can't eat" steered citizens toward self-sufficiency. In countries destroyed by the war, urban agriculture continued to prove important even after hostilities came to an end. In post-war Germany, the residents of Berlin claimed the land fronting the destroyed Reichstag to

R. Lahtinen (✉)
University of Turku, Turku, Finland

© The Author(s) 2019
S. Laakkonen et al. (eds.), *The Resilient City in World War II*,
Palgrave Studies in World Environmental History,
https://doi.org/10.1007/978-3-030-17439-2_6

grow potatoes.[2] Such cultivation continued in the war-ravaged country, and it was not until the resurgence of the economy that citizens who had survived the war were faced with a completely new phenomenon: the overproduction of foodstuffs, obesity, and the "throwaway" society.[3]

Swedish historian of urban agriculture, Annika Björklund has emphasized that urban agriculture can be characterized as an anomalous condition, even a contradiction in relation to the urban-rural dichotomy. This is so because the location of agriculture in towns and cities is in opposition to the general function of urban economies' focus on non-agricultural provision. Urban agriculture is also in opposition to the characteristic high settlement density in towns and cities because large-scale agriculture demands access to substantial areas of undeveloped land. However, according to Björklund, there are a number of other factors that affect how widely urban agriculture is practiced. On the one hand, social safety nets, fast and cheap transport, settlement expansion, and market economy discourage urban agriculture, while insecure or irregular income sources, transport difficulties, subsistence economy, and environmental concerns encourage it.[4] In conclusion urban agriculture is highly dependent on the overall development of societies, their booms and busts, and here warfare steps in (Fig. 6.1).

Research on wartime agricultural history has focused on larger nations and has, understandably, particularly stressed the perspectives of troop munitions and the military economy.[5] But what sort of role did urban agriculture play in smaller countries during the war, and what sort of effect did agriculture have on the environment and its conservation? This article delves into these issues in Turku, a coastal town of southwestern Finland and the nation's third-largest city during the world wars. Although this article focuses on the period of World War II, it also briefly touches on the World War I era, because the experiences of this previous war were still fresh in people's minds during the later war and played a role in how the populace adapted to wartime conditions.[6]

Agriculture has been practiced in Finnish and other Nordic cities for centuries.[7] As late as the end of the nineteenth century, herds of cows being moved to their summer pastures were a common sight on city streets, while horses were needed to transport people and goods. Cows, pigs, chickens, sheep, and later also rabbits were raised in the courtyards of urban buildings. In addition, herds of cattle on their way to the slaughterhouse as well as plenty of stray dogs and cats wandered the streets. City-dwellers cultivated fields that were located right at the edges of the gridiron

Fig. 6.1 Public institutions supported urban agriculture in all belligerent countries. Victory gardens located across from San Francisco City Hall during World War II. (Source: San Francisco History Center, San Francisco Public Library)

zone. In addition, plants for household use were grown in the yards of urban buildings. Manure men transported waste from the city to nearby fields, from where it was hauled off to farmers' fields for fertilizer.[8] Prior to the rapid industrialization of the late nineteenth century, then, cities were in many ways simply large villages. In the early 1900s, Finnish cities were still very small in terms of surface area, encompassing little more than the zoned areas.

Domestic animals were commonly kept in cities at the turn of the twentieth century, but tightening hygiene requirements made animal husbandry less desirable in the eyes of the authorities. In 1890, Turku residents owned 655 horses, 344 cows, and 308 pigs within the gridiron zone. By the end of the decade, the number of domestic animals in urban areas was clearly on the decline, a development that accelerated in the early years of the 1900s.[9] Horses, on the other hand, were needed as traffic increased. In Turku, the number of horses was greatest in 1909, when streetcars had just been introduced. At that time, there were 758 horses in the gridiron zone.[10] As automobile and streetcar traffic spread, horses became less common. Agriculture within the limits of Finland's largest cities was almost entirely abandoned during the 1930s, primarily for reasons of hygiene. But when the acquisition of foodstuffs grew more difficult during the world wars, agriculture flooded back in urban areas. Agriculture gradually disappeared from the big cities following World War II, but in smaller towns it continued to exist; for instance, cows were still kept in the center of one small town in the 1970s.[11]

Urban agriculture and its effects have not been properly studied in Finland—or anywhere else, for that matter—from the perspective of environmental history, as periods of war have often been dismissed as brief and, in terms of nature, insignificant. With this article, I attempt to demonstrate that war and urban agriculture have had significant, long-term effects on urban streetscapes, land use, and resident lifestyles.[12]

As my primary source of data, I have relied on Turku-area newspapers. I went through Turku's largest Finnish-language newspapers (in terms of chronology, *Aura*, *Uusi Aura*, and *Turun Sanomat*) from 1890 to 1950 and gathered articles and editorials related to the urban environment. During both world wars, the newspapers were shaped by military censorship, which limited writing on topics that were considered negative. Although censorship more clearly affected reporting on military developments, it also affected discussion of environmental issues, because the censors did not want to bog readers down with problems in their local environment.[13] In reviewing archival sources, it is remarkable how little information has been preserved on many wartime events. For instance, clearly reliable sources on the numbers of city residents' domestic animals do not exist.

WORLD WAR I AND THE POST-WAR PERIOD

Inspired by central European models, Finnish cities saw the promotion of gardening as a new ideal in the early twentieth century. In line with foreign antecedents, allotment and school gardens had already been established in Finland prior to World War I. Small-plot gardening was considered a necessity for the nature-estranged residents of stone cities and an instructional pastime for children, teaching them useful skills and love of nature while keeping them off the streets and out of trouble. Another aspect that was emphasized was the increase of self-sufficiency that ensued when a portion of a family's vegetables came from its own plot.[14] The first garden plots for children and schoolchildren were founded in Turku in 1914.[15] The necessity of allotment gardens had been discussed by the Helsinki City Council at as far back as the turn of the century, and during the period 1915–1918, they had already been established in several other cities as well.[16] In addition, so-called garden districts were also planned, offering residents the opportunity to live more expansively and to garden at home. Such neighborhoods had been planned elsewhere in Finland; for instance, in Turku, a Kupittaa "park village" was long planned right at the edge of the gridiron area, which according to the conceptions of the time would have meant its being located in a completely natural area, far from the dust and corrupting temptations of the city.[17]

During World War I, the city awakened to the need for urban cultivation due to a food shortage in the spring of 1916, at which time Turku residents were encouraged to cultivate edible foods on even small patches of land.[18] Many rented small plots from landowners. At this point, the issue did yet not inspire broader public discussion. It was a different matter the following year, however, when city officials finally grasped the gravity of the situation. At that time, the Food Administration encouraged municipal food committees to take measures that would allow as many residents as possible to cultivate potatoes and other root vegetables themselves.[19] In May 1917, 1200 people applied for plots of land being leased by the city of Turku. Dozens of acres of leased plots were established, in practice everywhere in the center of town and at its fringes, among other places Kupittaa field, which had traditionally been parkland.[20] In addition, agricultural plots were reserved within city limits for the children of indigent families and for a variety of clubs and associations (Fig. 6.2).

Initially the newspapers were enthusiastic about this agricultural activity. It was wonderful to find a practical use for areas that were previously considered almost useless. In addition, household waste was being recovered more effectively, as it was now being carefully gathered for practical use as fertilizer and food for stock.[21] At the same time, residents were reminded of the fact that potatoes do not grow just anywhere, that it was better to eat seed potatoes than to plant them in places where growing them was hopeless. The following spring, things had changed, and people who ate the seed potatoes they had been given were prosecuted.[22] There was a general lack of potato and rutabaga seeds, but there continued to be enough carrots, beets, and sugar beets. In the spring of 1918, after an exceptionally severe winter,[23] even the old lawns of Kupittaa Park were

Fig. 6.2 Elderly people, women, and children took care of wartime urban agriculture. Mrs. Rodney George irrigates her victory garden in Golden Gate Park, with her daughter Eleanor looking on. Scores of urban youngsters learned a lot about gardening during war years. (Source: San Francisco History Center, San Francisco Public Library)

tilled for cultivation. Prisoners from the nearby prisoner of war camp were ordered to carry out the tilling work; it was said that their effectiveness as laborers was questionable, as it was the prison camps in particular that suffered most greatly from lack of food. According to the newspaper, plenty of middle-class and even upper-class women, for whom planting potatoes was not a familiar task, were witnessed joining the working class in cultivation.[24]

Although animal husbandry and urban agriculture decreased after World War I, the ideology of gardening, in contrast, strengthened. The Kupittaa Park potato fields were restored to lawns in the 1920s, but new lands were prepared for cultivation elsewhere. Enthusiasm for school gardens grew significantly during the 1930s.[25] At that time, as a matter of fact, it was near-mandatory for pupils to participate in growing foodstuffs, even though many found the work unpleasant, and at times caring for the plot ended up in the parents' hands. Still, there was a firm belief in the educational influence of gardening. Many associations arranged patches of field for their members, especially young ones, to cultivate. An allotment garden that had been planned for decades was finally realized in Turku in the summer of 1934.[26] At the same time, living in houses rather than apartments was growing more common. Gardening became a popular hobby, and gardening competitions were organized frequently in such residential areas.[27]

World War II

World War II interrupted the modernization that was well underway in Finland, censorship once again prevented the free communication of information, and the rationing of foodstuffs began immediately in the autumn of 1939. All of this was familiar from the previous world war, but from the start, the nature of the warfare was completely different from what it had been two decades previously. In the following urban agriculture will be discussed in terms of the three separate but related wars that Finland faced during World War II, that is, Winter War (1939–1940) against the attack of the Soviet Union, Continuation War (1941–1944) as an ally with Nazi Germany to regain lost territories during Winter War, and Lapland War (1944–1945) against German troops in Northern Finland.

In Turku as elsewhere, the war began with a new innovation brought by industrial warfare, nighttime air strikes, which continued throughout the war. In addition to the bombing, the nightly blackouts and covering of shop windows changed the general appearance of the city. As in many cities, bombings severely damaged the building stock during World War II; over 800 buildings were destroyed or damaged, or approximately 3.3% of the entire city's building stock. Amazingly enough, only 79 people died in bombings.[28] Abundant building waste resulted from the destroyed buildings and was put to use as firewood and landfill. Empty lots scarred the streetscape for years, acting as prime habitats for weeds, stray cats and dogs, and rats. In addition, the craters left by the bombs were commonly used as dumps.

In terms of foodstuffs, Finland was much better prepared for war in the late 1930s than it had been a couple of decades earlier.[29] Coffee and sugar rationing began as early as the fall of 1939, and many foodstuffs lasted through the Winter War. In the spring, once the war ended, it was possible to start sowing crops again.[30] The shortage of foodstuffs did not grow acute until the years of the Continuation War; the winter of 1942 was particularly bad. The shortage was most acute in the biggest cities. Fortunately, the population of Turku declined to some extent, as the majority of able-bodied men left for the front and a part of the civilian population moved to the countryside. In addition, hundreds of children were sent to Sweden and Denmark during the war due to the lack of food and other resources.

In order to procure foodstuffs, the city residents often resorted to urban agriculture. With regard to agriculture, the situation in cities was much better during the 1940s than it had been during World War I. In Turku, plots at the edges of town had been cultivated during the entire interwar period, and after the Winter War, even more land was set aside for cultivation. In spring of 1940, the Turku social welfare office acquired plots for 446 families as well as 450 compulsory school pupils, and many more residents would have been eager to lease land. Various associations also distributed agricultural land in the spring of 1940; for instance, 700 individuals received a plot through the Turku–Maaria agricultural club.[31]

The surface area of cultivated land increased from year to year during World War II, and in the spring of 1945, 5500 people received a plot from the city of Turku. At that point, the city distributed a total of approximately 370 acres of plots in the immediate vicinity of the city.[32] In addition, many industrial plants, associations, parishes, and schools turned over plots to city residents to cultivate. It is now no longer possible to identify all of areas that were cultivated at the time, but agriculture was doubtlessly practiced wherever it was possible within city limits. Only parks were not, at least officially, divided up into potato plots (Fig. 6.3).

The enthusiasm of residents for cultivation was likely at its greatest in 1946, yet it died out rapidly thereafter. By the spring of 1947, it was complained that, due to the improved food situation, agriculture no longer interested Turku residents. The rationing of potatoes ended in the autumn of 1947, by which time the most severe shortage had ended.[33] In the fall of 1948, the potato harvest was so great in Finland that the price of potatoes collapsed, and part of the harvest remained in the ground.[34] There were so many other things to take care of in the war-torn city that

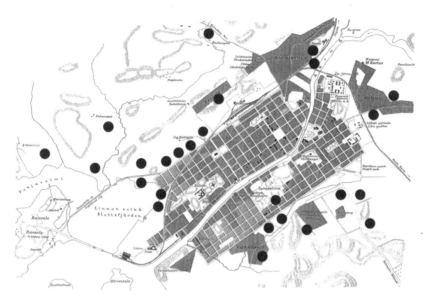

Fig. 6.3 A potato war? A map based on newspapers and archival sources shows the extension of potato fields in Turku town during World War II. (Source: Rauno Lahtinen)

most people no longer felt like wasting time on agriculture; after all, potatoes were available in the shops. It is not likely that city residents who had been forced to turn to cultivation recalled their experiences in the potato patches with much warmth.[35]

In addition to urban agriculture, animal husbandry also increased during World War II. During the war, animals were raised right in the center of town, in both back yards as well as cellars. Story has it that in some Helsinki residences, pigs were even kept in bathrooms or living room corners.[36] Chickens, rabbits, pigs, and sheep were raised in almost every backyard. Cows, on the other hand, were not often kept in cities. The number of horses in the cities presumably declined, as they were taken to the front in great numbers.

The easy-to-raise and rapidly growing pig was the salvation of many a family. Because of the war, in 1940 the authorities in Turku authorized the establishment of pigsties even within the gridiron zone. There was no great wave of enthusiasm to apply for permits, however, as officially there were only 4 pigsties in the city that year, and in 1945 they numbered 47.[37] In practice, most of the animals were raised without permits, and attempts were made to hide them from the authorities as well as possible, as otherwise a portion of the production should have been turned over for sale to the public. According to the newspapers, violating the regulations was widespread. This is no wonder, as pork became the clearly most expensive good in the black market.[38] The authorities did not generally interfere with the raising of individual illegal animals, as long as the activity did not aim at large-scale black-market trade. Now and again, large amounts of animals were discovered in the cities; for instance, in the spring of 1942, the papers reported an illegal herd of 24 pigs had been found in Helsinki practically next door to the Ministry of Supply.[39] In 1943, city veterinarian Leo J. Fabritius estimated that in the center of Turku there were at least 1400 pigs, thousands of rabbits, and dozens of sheep.[40]

Following the war, up until 1949, national counts of cattle, pigs, sheep, and horses were carried out in early spring to prevent black-market trade.[41] According to the newspapers, it was nevertheless clear to all that the true number of animals was not revealed in these counts of domestic animals.[42] In addition, the majority of animals were slaughtered in the fall, meaning their numbers varied significantly according to the seasons. It is evident from the lists of the authorities that animals were legally kept in the very centers of cities, in Turku in the courtyards of the central downtown blocks. The newspapers openly promoted animal husbandry; for instance, *Turun Sanomat* encouraged the raising of pigs in apartment building boiler rooms.[43]

Rabbits became especially popular during wartime, and it was not required to report them during the counts. The rabbit adapted superbly to urban conditions, and rabbit-care was a suitable task for children as well as adults. Rabbits were raised in almost every yard and courtyard, in some by the hundreds. Patches of wasteland that had previously been considered worthless, ditch banks, and stands of forest experienced a rise in value when both animal feed and wild plants not normally gathered for human consumption were harvested from them. The boldest rabbit keepers took their sickles and invaded the grain fields of nearby farmers.[44] The hardest-working cared for herds of dozens or even hundreds of rabbits and also performed the slaughtering and skinning themselves. There was plenty of demand for the meat and the skins, and many did well for themselves with the rabbit trade.

The end of the Continuation War in fall 1944 did not immediately bring about a change in the food situation, and shortages remained acute for years. The peak years in terms of urban agriculture did not arrive until the end of the Lapland War in spring 1945, when a bigger workforce was available, but at the same time, there were more mouths returning from the fronts to feed.

In 1947, the Turku city veterinarian Fabritius found 121 pigsties in downtown Turku, but luckily the situation changed rapidly: in the 1950 inspections, only 2 urban pigsties were recorded. The situation returned to the level of the early 1930s after an approximately decade-long urban agriculture rush inspired by the war.

THE RISE AND DECLINE OF ANIMAL PROTECTION SOCIETIES

The animal protection movement arrived in Finland from England and Sweden, becoming active in the latter half of the 1800s. The conditions of the animals of Turku had been monitored for a long time by the *Turun Eläinsuojeluyhdistys* (Turku Animal Protection Association), founded in 1871, and since 1911 by the *Turun Eläinten Ystävät* (Turku Friends of Animals). With funds it collected, the Animal Protection Association hired extra police during the horse markets to reduce abuse of the animals. The leadership of the association considered police action insufficient, however, and in 1908 the chief of police assigned one officer exclusively to animal protection tasks. During the upheavals of World War I, this officer did not, however, have much time for animals. For this reason, the Turku Friends of Animals hired its own animal protection officer during the war years.[45]

To those involved in animal protection, it was clear that the increased animal husbandry and hardened attitudes of wartime led to animal abuse. Old or temporary animal housing was often in bad condition, there was a lack of feed, and animals were treated brutally as tools of production, nothing more. Especially the poor condition and treatment of horses at the market grabbed the attention of the animal rights activists.[46] Another indication of the decrease in hygiene and the increase in the number of animals during the war was that fact that only then, in 1916, was the post of city veterinarian established in Turku. Duties included, among others, monitoring the sale of meat, but also "ensuring that the barns, cow yards, pigsties, and other shelters intended for housing animals are built and equipped in an appropriate manner and are kept in such condition that they meet demands for hygiene and sanitation."[47]

The period of World War II led to an increase in animal abuse in the cities, as had already happened during World War I. Part of the reason was simply the rapid growth in number of animals after the onset of war. "Cases of animal abuse have increased," announced the *Turun Sanomat* in a spring 1941 interview with the police office's animal protection inspector. Constable Herman Lagerström, who then worked as the inspector, reported that horses in particular were often treated cruelly. According to him, the horses that had remained in town were malnourished and in bad shape, as the best animals had been taken to the front and the city was plagued by a shortage of feed. He also indicated that the treatment of slaughter animals had grown more brutal. Lagerström described the situation in 1941 thus:

> In the morning around 7 o'clock I start my day off by going over to the customs house to inspect the arrival of the horse traders in the city. (...) I look over the condition of the horses, even though in this issue we have to slide pretty far from earlier requirements due to the current shortage of food. As an overall observation, the horses are much thinner than before the war and the stock of horses has grown worse. (...) In general I'd have to say, when it comes to treatment and handling of animals, that the situation has clearly declined since the [Winter] War. And so the number of abuse cases has increased quite significantly recently.[48]

As the war dragged on, the press preferred to remain silent about such issues. There was a desire to not needlessly depress the mood of the populace with unpleasant news from the home front.

When we take into consideration the increase in the number of animals being raised in urban settings, the squalid conditions, and the animal kill patrols, it is remarkable that wartime led to a near-complete collapse of animal protection activities in Finland. Still, during the interwar period, the two animal protection organizations in the city had been active, and their stances publicized prominently in the city papers. During World War II, their activity decreased and focused primarily in the distribution of honorary certificates. The majority of active male members were at the front, which makes this development partially understandable. The situation did not improve after the war, however: at the end of the 1940s, the animal protection associations disappeared almost totally from the public eye, and in the 1950s, animal protection in Turku was in a state of utter paralysis. This activity did not revive until the late 1960s. This same pattern was repeated, for all practical purposes, in all Finnish cities.[49]

Hygiene Issues

A rage for hygiene made powerful headway in the early 1920s. Cleanliness, health, and hygiene were fashionable terms in Finland during the interwar period, spouted continuously to women in advertisements as well as newspaper advice columns. If you opened a newspaper or magazine, it was impossible to avoid advertisements for soap, detergent, and rat poison. As plumbing and sewage systems were in place in Finland's biggest cities by this time, standards regarding cleanliness in the home grew even stricter than before. After the war, standards that had previously been applied to primarily the homes of the upper class spread to apply to the homes of middle and working-class families as well.[50] The city veterinarian, whose post began during World War I, and inspectors from the board of health made rounds inspecting the cleanliness of outhouses and yards.[51]

The tightening demands for hygiene made animal husbandry less desirable in the eyes of the authorities. The raising of domestic animals came to an almost complete end during the interwar period, as the authorities no longer allowed animals to be kept in the cities for reasons of hygiene. As vehicular traffic increased, the number of horses declined rapidly. By the end of the 1920s, no more than about 100 horses were housed in downtown Turku, and 20 in Helsinki.[52] The decline in the number of horses and the end of other domestic animal husbandry also meant the tidying of the streetscape, as animal waste was no longer tolerated in the streets. In photographs taken in the 1930s, the streets generally appear fabulously neat.

There had been plans to renew the waste management system in Turku at the beginning of the 1900s, but World War I put a stop to them. In the early 1920s, the city's waste management was considered completely outdated and new solutions were sought, primarily from Sweden. New sanitation regulations came into force in 1926, but in practice the city's manure men were still allowed to drive their loads to the dumps and fields on the edges of town as they had before.[53] And yet the atmosphere was changing. Unofficial dumping grounds and reeking open sewers were no longer looked on kindly in the vicinity of town, and complaints about them were continuously published in the papers. Particularly in the 1930s, water pollution was covered in the press, as by this time the increasing stench of the Aurajoki river that ran through town was considered utterly intolerable—especially since the river water had also been used as the city's piped water since 1923. A plan commissioned by the city council for a wastewater purification plant was ready in 1934.[54]

The enthusiasm for cleaning up the city and all-around improvements to its appearance came to a peak in 1939. Helsinki had unexpectedly been selected to host the 1940 Summer Olympics, which were originally scheduled to be held in Tokyo, Japan. Consequently at least the residents of Turku assumed that large numbers of Olympic tourists arriving from nearby Stockholm, Sweden by boat would be entering Finland in Turku and the harbor there. The city, thus, had to be spruced up to demonstrate to foreign tourists that Finland was still as clean and hygienic as other Western countries. The spring and summer of 1939 saw prominent nationalistic campaigning on the issue: the dumps in the city surroundings, the disorderly landscapes along the sides of railways and highways, and the poor condition of the streets inspired outrage, and improvements were demanded.[55] The year of 1940 may well have become a real year of environmental activity, but things ended up going in the exact opposite direction. The breakout of war in Poland in the autumn of 1939 put an end to all such campaigning.

In 1946, the Turku city veterinarian made public his desire to rid the city of all domestic animals as rapidly as possible. One of the primary drawbacks of urban agriculture, the rat problem, had grown quite significant:

> The war on rats carried out in 1944–45 was for the most part successful, even though the results did not entirely meet expectations, due to certain building owners. Despite the guidelines, in some yards – especially those where rabbits or pigs are kept – all manner of food waste exists for rats. Among the rat population there are also cunning old individuals who know how to avoid poison. And because the number of domestic animals has increased rapidly in recent years, meaning that now household pigs, sheep, or rabbits are kept in the backyard of almost every building, usually without the requisite permits and in utterly temporary outside shelters that do not meet even the most primitive health requirements, it is no wonder that rats have greatly increased.[56]

The rat problems did not disappear when World War II ended, because many yards were still full of food scraps perfect for rats but actually intended for domestic animals.[57] Not all building owners bothered to set out the poison bait distributed by the city health authorities due to its cost and the significant labor required: the uneaten poison doses had to be gathered up in the spring so that they would not end up in the wrong mouths or kill migratory birds. Many doubtless found setting poison out in the yards, within the reach of children and dogs, questionable. According to Fabritius, the indifference of city residents to the rat wars simply grew at the end of the 1940s.[58]

For the city's feral pigeons as well, the onset of World War II meant harder times. Part of the population had been killed every year prior to the war: for instance, in 1938, over 900 pigeons had been destroyed in Turku. In the winter of 1940, the cities of both Turku and Helsinki began to pay a kill fee of one Finnish Mark per bird. They wanted to get completely rid of pigeons, because it was feared that they spread diseases. As far as is known, feral pigeons were used during wartime for human food in the cities as well.[59] In addition to pigeons, the authorities also reserved discretionary funds for the killing of stray cats.[60] Stray dogs were already a problem during wartime, and after the war it was said there were more than ever of them dodging through the traffic. In the spring of 1948, over 300 dogs were shot in Turku and the problem was gradually eradicated.[61]

The condition of the surroundings of Turku had, during the war and afterwards, grown extraordinarily bad in many ways. In order to reduce problems, the main dump near the center of town was moved further away in 1943, as its stench and packs of rats disturbed not only nearby residents but also the operation of the hospital that was located next to it. A dump at the north end of town was closed, however, when a village of portable buildings built by German soldiers in Finland during wartime, "Little Berlin," was erected on the site. The new official dump was founded about five kilometers south of the city center.[62] However, because of the shortages of workforce, horses, and fuel, it was not possible to transport all waste so far, and so illegal dumps were born all around the edges of town, in abandoned potato fields and elsewhere. Poisons, paints, and solvents from industrial plants that today would be considered hazardous waste were also poured into these illegal dumps without much consideration. After the war, industrial and economic revival was a goal that was not criticized, and industrial plants were given free reign to dump their waste almost wherever they saw fit. Nor was it possible to criticize industrial polluters, as industry produced goods that were used to pay heavy war reparations to the Soviet Union.[63]

The collapse in the level of hygiene in the cities did not remain unnoticed by their residents. Problems were caused specifically by the domestic animals kept in the yards and the rats, mice, flies, and vermin they brought with them. Also the level of hygiene in grocery shops and restaurants is said to have hit rock bottom, as food was bought during the shortages regardless. For years there was a shortage of soap and other goods required for cleaning.[64] Nor was it easy to change attitudes after the war, as people had grown accustomed to a lower standard of cleanliness. Complaints about the matter were published in the newspapers once censorship ended in the late 1940s, but the state of affairs did not improve for a long time. The secretary of the Finnish provincial association for public health, Leo Kaprio remarked in an interview with a local newspaper in the early spring of 1950 on the cleanliness of Finland's cities:

> The superficial cleanliness of cities is satisfactory, and a quick visit the visitor may consider our circumstances tidy and clean. This is, however, simply surface gloss, and in the summertime snow will not cover heaps of garbage, untended compost piles, stinking open sewers and uncovered dumps. The hygiene of enterprises involved in the selling of food is often completely unsatisfactory, and vermin control non-existent. (…) The reason for these failings is not economic, but a lack of will.[65]

Many of the problems of the immediate environment, such as the illegal dumps that appeared during the war and the years that followed, were not written about in the Turku newspapers until the 1950s. In 1953, it was admitted in a local paper that conditions in the city were squalid. In addition to the fact that dirty water and smells spread to the surroundings from the city dump, numerous and completely unmonitored dumps continued to exist in the city's residential areas.[66] And so the rat problem continued to remain topical for a long period following the war. Despite a few complaints, several of these dumps remained functioning in the 1960s, and a few into the 1970s. Some are empty lots nowadays, while parks and even residential areas have been built over others.

Conclusions

Urban agriculture decreased in Finland significantly after the world wars. From a global perspective, the situation in Finland and the other welfare states is, however, exceptional. Animal husbandry and agriculture are everyday events in all villages, towns, and cities in developing countries where residents deal with a shortage of foodstuffs. Domestic animals are kept wherever a bit of space can be carved out; vegetables and other edible plants are grown in backyards and town squares. In the megacities of the third world, up to two-thirds of urban families practice agriculture.

Currently about 800 million people practice urban agriculture, and one-fifth of all food is produced in towns.[67]

In wartime Finland, the situation was precisely the same as it is now in developing countries. Agriculture born of shortages was a fundamental wartime phenomenon that shaped the urban environment, along with bombings, population movements, energy economics, and changes in transportation. Small-plot gardening had begun in many cities prior to World War I. In wartime conditions, the hobby was soon taken much more seriously than before, and large areas of the city's parks and meadows outside the downtown area were cleared for potato cultivation. Although the greater part of these plots was abandoned soon after the war, small-plot gardening was still in favor during the interwar period. The plots were farmed by pupils and adults, for whom allotment gardens were established. Residential gardening also became more popular during the interwar period. In the conditions of World War II, cultivation spread rapidly throughout the urban environment, and in terms of food acquisition, rose to a position of significance for many city residents. Temporary potato plots were abandoned rapidly after the war. Gardening was never completely abandoned, however, as detached houses and having one's own garden outside of the downtown area started being considered the ideal form of living, and more people were capable of realizing this dream.

The keeping of domestic animals came to an almost complete end in the largest cities during the interwar period, but during World War II and especially the Continuation War the number of animals became sizable, and the problems they caused grew notably. The war hardened attitudes, and the conditions faced by animals continued to decline due to lack of workforce and feed, as well as professional skill. The world wars affected animal protection in different ways. During World War I, animal protection was still on the rise, and the suffering of animals received much publicity. During World War II, the most active members of these associations went to the front, and the movement experienced a considerable setback. In the difficult conditions of the post-war years, the animal protection issue was no longer considered as significant as before, and the activity of these associations faded for decades.

Also, newspaper coverage of environmental topics, which had grown more prevalent during the 1930s, declined distinctly and immediately upon the onset of the Winter War and ended almost entirely during the Continuation War. Negative news about wartime events and the polluting of hometowns was avoided. The self-censorship practiced by journalists was effective. It is more than likely that environmental topics were considered insignificant at a time when the independence of the entire nation hung in the balance.

Despite the material shortages cities experienced during the war—or actually, precisely because of this—could this, then, be considered an ecological society from a contemporary perspective? The actions of city residents lend strong support for this notion. City residents raised their food themselves by means of using local resources. People switched to eating more vegetables and cut back on meat and fat. All possible waste was gathered and recycled, leading to a dramatic drop in the amount of household waste. Even small scraps of food were fed to domestic animals or used for fertilizer. Agricultural poisons were scarce, and artificial fertilizers were for almost all intents and purposes impossible to find. Consequently, urban wartime agriculture was in practice organic farming.

Despite all of the positive aspects, the ecological nature of wartime cities was only one side of the coin, as shortages led to negative phenomena, such as undernourishment, animal abuse, declines in hygiene, the poaching of fish and animals, the stripping of parks and forests, and crime. The urban agriculture of wartime Finland was literally a "loot-based economy." During World War II, urban agriculture was not guerrilla gardening, that is, optional and occasional small-scale farming practiced by small groups of people, but rather a military type large-scale operation that was to a high extent realized as a compulsory part the total war.

Notes

1. For North America, see Cecilia Gowdy-Wygant, *Cultivating Victory: The Women's Land Army and the Victory Garden Movement* (Pittsburgh: University of Pittsburgh Press, 2013), 3–6, 107–115; Ian Mosby, *Food Will Win the War: The Politics, Culture, and Science of Food on Canada's Home Front* (Seattle: University of Washington Press, 2014); Anastasia Day, "Productive Plots: Nature, Nation, and Industry in the Victory Gardens of the U.S. World War II Home Front," Ph.D. dissertation, University of Delaware, 2018.
2. Lea Baumach, *Urban food production: A contribution to urban resilience in Berlin?* (Hamburg: Diplom.de, 2014), 24–31.
3. Jacob Darwin Hamblin, "The Vulnerability of Nations: Food Security in the Aftermath of World War II," in Simo Laakkonen, Richard Tucker, eds., A Special Issue on "World War II, the Cold War, and Natural Resources,"

Global Environment. A Journal of History and Natural and Social Sciences no. 10 (2012): 42–65.
4. Annika Björklund, *Historical Urban Agriculture. Food Production and Access to Land in Swedish Towns before 1900* (Stockholm: Stockholm University, 2010), 214–215, Figure 35.
5. Wendy Goldman & Donald Filtzer, eds., *Hunger and War: Food Provisioning in the Soviet Union during World War II* (Bloomington: Indiana University Press, 2015); Brian Short, *The Battle of the Fields: Rural Community and Authority in Britain during the Second World War* (Woodbridge: Boydell Press, 2014).
6. Lizzie Collingham: *The Taste of War: World War Two and the Battle for Food* (London: Allen Lane, 2011), 4–5, 16, 22–27, 50–51, 76–77, 83, 181, 220–23, 325, 380, 501.
7. Björklund, *Historical Urban Agriculture*, Chapter I.
8. Eino Jutikkala, *Turun kaupungin historia 1856–1917* ("History of the City of Turku, 1856–1917") (Turku: Turun kaupunki, 1957), 45.
9. Jutikkala, *Turun kaupungin historia 1856–1917*, 572. In 1890, the population of Turku was approximately 30,000.
10. Marita Söderström, *Ratatieosakeyhtiöstä keltaiseen vaaraan. Sata vuotta Turun joukkoliikennettä* ("From Tramway Shares to the Yellow Danger: One Hundred Years of Public Transportation in Turku") (Turku: Turun maakuntamuseo, 1990), 14–15; Jutikkala, *Turun kaupungin historia 1856–1917*, 43.
11. Veikko Anttila and Matti Räsänen, "Kansankulttuurin murros" ("The Turning Point in National Culture"), in Paula Avikainen et al., eds., *Suomen Historia 7* ("History of Finland 7") (Espoo: Weilin+Göös, 1987), 44.
12. Rauno Lahtinen and Timo Vuorisalo, "'It's war and everyone can do as they please!' An environmental history of a Finnish city in wartime," *Environmental History* 9, no. 4 (October 2004): 675–96.
13. Reino Lento, *Sellaista oli elämä vuosisadan vaihteen Turussa* ("That's What Life was Like in Turn-of-the-Century Finland") (Juva: WSOY, 1979), 190–195; Jussi Kuusanmäki, "Ensimmäisen maailmansodan lehtisensuuri" ("Newspaper Censorship in World War I"), in Pirkko Leino-Kaukiainen, ed., *Sensuuri ja sananvapaus Suomessa* ("Censorship and Freedom of Speech in Finland") (Helsinki: Suomen sanomalehdistön historia –projekti, 1980), 95–101; Alpo Rusi, *Lehdistösensuuri jatkosodassa. Sanan valvonta sodankäynnin välineenä 1941–1944* ("Press Censorship in the Continuation War: Monitoring the Printed Word as a Tool in Warfare 1941–1944") (Helsinki: SHS, 1982), 17–22, 359–369.
14. Sami Tantarimäki, "Puutarhan antimilla yli pulavuosien" ("Surviving the Years of Shortage with Gardens"), in Sakari Tuhkanen et al., eds., *Tutkimusretkistä paikkatietojärjestelmiin – matkalla kulttuurimaantieteen maailmoissa* ("From Explorations to GPS – Travel in the Worlds of Cultural

Geography") (Turku: Turun yliopiston maantieteen laitoksen julkaisuja 164/2001), 148–155.
15. Mirja-Riitta Siivonen, Pirkko Salonen and Tuija Kuchka, *Siirtolapuutarha – kaupunkilaisten paratiisi* ("Allotment Gardens – Paradise for City Dwellers") (Helsinki: Tammi, 1999), 31–45; Tantarimäki, "Puutarhan antimilla yli pulavuosien," 194–197; *Uusi Aura* (*UA*) July 31, 1914 and August 20, 1916.
16. Siivonen, Salonen, and Kuchka, *Siirtolapuutarha – kaupunkilaisten paratiisi*, 31–45.
17. Harri Andersson, "Huvilakaupunki" ("Villa City"), in Susanna Hujala and Päivi Kiiski-Finel, eds., *Näkymätön kaupunki. Toteutumattomia suunnitelmia 1900-luvun Turussa* (The Invisible City: Unrealised Plans in Twentieth-Century Turku) (Turku: Wäinö Aaltosen museo, 2002), 80–87; *UA* February 17 and April 24, 1921.
18. *UA* May 14, 1916.
19. Lento, *Sellaista oli elämä vuosisadan vaihteen Turussa*, 13.
20. *UA* May 9, 13, and 15, 1917.
21. For UK, see Peter Thorsheim, *Waste into Weapons: Recycling in Britain during the Second World War* (Cambridge: Cambridge University Press, 2015).
22. *UA* May 9 and 27, 1917; *Turun Sanomat* (*TS*) June 5, 1918.
23. Various authors have noted the harshness of that winter. See, for example, Alice Weinreb, *Modern Hungers: Food and Power in Twentieth Century Germany* (New York: Oxford University Press, 2017).
24. *TS* June 13, 1918.
25. Tantarimäki, "Puutarhan antimilla yli pulavuosien," 150–151.
26. Siivonen, Salonen, Kuchka, *Siirtolapuutarha – kaupunkilaisten paratiisi*, 42–43; *TS* May 10, July 15, and August 30, 1934.
27. For example, *TS* August 9, 1933 and June 23, 1938.
28. Sinikka Uusitalo, *Turun kaupungin historia 1918–1970* ("History of the City of Turku 1918–1970") (Turku: Turun kaupunki, 1982), 113–115.
29. Ilkka Seppinen, "Talvisodan talous" ("The Winter War Economy"), in Lauri Haataja et al., eds., *Suomi 85. Itsenäisyyden puolustajat. Osa 2: Kotirintamalla* ("Finland 85: Defenders of Independence. Part 2: the Home Front") (Porvoo: Weilin+Göös, 2002), 30–31.
30. Erkki Pihkala, "Kansanhuollon aikaan," in Haataja, *Suomi 85*, 90–91.
31. *TS* May 19 and 22, 1940.
32. *TS* March 30, 1946.
33. Pihkala, "Kansanhuollon aikaan," 93.
34. *TS* March 12, 1947, October 22, 1948 and May 5, 1949.

35. See for instance Anni Polva's memoirs of life in wartime Turku. Anni Polva, *Elettiin kotirintamalla... Lehtiä päiväkirjastani* (Hämeenlinna: Karisto, 1995), 28–136.
36. Jouni Kallioniemi, *Kotirintama 1939–1945* ("Home-front, 1939–1945") (Turku: Vähä-Heikkilän kustannus, 2001), 82. Turku-area newspapers published several reports of illegal domestic animals found in Helsinki apartment buildings, but not a single corresponding case was reported in Turku.
37. TKK 1940–1945; activities of the city veterinarian.
38. Pihkala, "Kansanhuollon aikaan," 88.
39. *TS* August 29, 1942; Liisa Nummelin, Maatalous kaupungissa. Maanviljelys ja karjanhoito Porin kaupungissa 1880-luvulta 1940-luvun lopulle ("Urban agriculture in Pori, 1880s–1940s"). MA Thesis, Turun yliopisto, 1988, 131.
40. *TS* October 30, 1943.
41. Pihkala, "Kansanhuollon aikaan," 96.
42. *TS* February 29, 1944.
43. *TS* January 8, 1942.
44. *TS* June 3, 1943.
45. *UA* November 1, 1917.
46. For example, *UA* November 29, 1917 and *TS* December 12, 1918.
47. The municipal report of the City of Turku (TKK) 1916, section 1, 122.
48. *TS* May 13, 1941.
49. Matti Rasila, *Turun eläinsuojeluyhdistys 1871–1971* ("The Turku Animal Protection Association 1871–1971") (Turku, 1971), 55–57; Ulla Eronen, *Piiskat piiloon, Tunkki tulee! Joensuun eläinsuojeluyhdistys 1899–1999* ("Hide The Whips! The Joensuu Animal Protection Association 1899–1999") (Joensuu: Joensuun eläinsuojeluyhdistys, 1999), 137–139.
50. Kirsi Saarikangas, Asunnon muodonmuutoksia. Puhtauden estetiikka ja sukupuoli modernissa arkkitehtuurissa ("Changes in Housing Forms: The Aesthetics of Cleanliness and Gender in Modern Architecture") (Helsinki: SKS, 2002), 127–133; Riitta Oittinen, Enemmän puhtautta – enemmän terveyttä? ("Cleaner – Healthier?"), in Simo Laakkonen et al. (ed.), *Nokea ja pilvenhattaroita. Helsinkiläisten ympäristö 1900-luvun vaihteessa* ("Soot and Clouds: The Helsinki Environment at the Turn of the 20th Century") (Helsinki: Helsingin kaupunginmuseo, 1999), 143–151.
51. TKK 1917, section 3, 25–27.
52. Nummelin, *Maatalous kaupungissa*, 103; TKK 1933, section 6, 7.
53. Henry Nygård, *Bara ett ringa obehag? Avfall och renhållning i de finska städernas profylaktiska strategier 1830–1930* (Åbo: Åbo Akademis förlag, 2004), 296–300.
54. Jussi Vallin, "Aurajoki avoviemärinä. Ketjuuntuminen ja jätevesiongelman ratkaisut" ("The Aurajoki as Sewer: Concatenation and Wastewater

Solutions"), in Simo Laakkonen, Sari Laurila and Marjatta Rahikainen, eds., *Harmaat aallot. Ympäristönsuojelun tulo Suomeen* ("Gray Waves: The Arrival of Environmental Protection in Finland") (Helsinki: SHS, 1999), 162.
55. *TS* February 17, May 7, May 10, May 14, May 21, June 3, June 10, June 14, July 7, and July 27, 1939.
56. *TS* February 24, 1946.
57. For Helsinki, see Simo Laakkonen, "Asphalt kids and the matrix city: Reminiscences of children's urban environmental history," *Urban History* 38, no. 2 (August 2011): 301–323.
58. *TS* April 26, 1948.
59. Hannu Eskonen, "Kalliokyyhky on pulun esi-isä" ("The Rock Dove is the Ancestor of the Pigeon"), *Kaleva*, February 7, 1981.
60. *TS* January 21, 1940.
61. *TS* May 23, 1948, May 5 and June 22, 1949. After US invasion there were an estimated 1.25–1.5 million stray dogs in Baghdad, a city of some 7 million people. "Baghdad to cull a million stray dogs as rogue canine population soars," *Daily Mail*, 11 June 2010.
62. *TS* September 30, 1942 and January 14, 1943.
63. Rauno Lahtinen, *Ympäristökeskustelua kaupungissa. Kaupunkiympäristö ja ympäristöasenteet Turussa 1890–1950* ("Urban Environmental Discussions: The Urban Environment and Environmental Attitudes in Turku 1890–1950") (Turku: Turun yliopisto, 2005), 58–65.
64. Uusitalo, *Turun kaupungin historia 1918–1970*, 220.
65. *TS* April 19, 1950. Leo Kaprio received his PhD from Harvard University in 1956 and afterwards worked in WHO assignments in Asia and elsewhere.
66. *TS* January 13, 1953.
67. Scott G. Chaplowe, "Havana's popular gardens: sustainable prospects for urban agriculture," *The Environmentalist* 18, no. 1 (March 1998): 47; Nancy Karanja and Mary Njenga, "Feeding the Cities," *2011 State of the World. Innovations That Nourish the Planet* (Washington DC: Worldwatch Institute, 2011), 151–152.

CHAPTER 7

Gaining Strength from Nature

Surviving War in Munich

Thomas J. (Tom) Arnold

Munich, the third largest city in Germany and capital of the state of Bavaria (*Bayern*), sits at the foot of the Alps in Southern Germany along the Isar River. When people think of the city, the most recognizable images are the sunny glades of the English Garden (*Englischer Garten*), the colorful buildings of the Old City (*Altstadt*), and of course the annual *Oktoberfest*, with *dirndl*-clad *fräuleins* holding huge mugs of beer that clank together accompanied by the sweet tootling of oompah bands. It is a picture of prosperity, of abundance and comfort, or as the Germans say it, *Gemütlichkeit*.[1] At first glance, the inhabitants of this pleasant city would not seem to have the strength to endure anything worse than a summer thunderstorm. And yet, there is strength and resilience in Munich's people, as demonstrated by their experiences during World War II and in its immediate aftermath.

From 1940 to 1945, bombs tore up Munich's green spaces and destroyed many landmark buildings along with housing, railroads, and infrastructure. Food and fuel supplies were unreliable, and residents were forced to fend for themselves, seeking food in the surrounding countryside or growing it at home. Munich became distinctively *un*pleasant for its

---

T. J. (Tom) Arnold (✉)
San Diego Mesa College, San Diego, CA, USA

© The Author(s) 2019
S. Laakkonen et al. (eds.), *The Resilient City in World War II*,
Palgrave Studies in World Environmental History,
https://doi.org/10.1007/978-3-030-17439-2_7

residents, seriously altering not only their surroundings but also their way of life. By the time US Army forces took over Munich on April 30, 1945, the city had been bombed 76 times since the first raid in March 1940. The population had dwindled by almost half to 480,000. Roughly 6400 people had been killed and 16,000 injured. Over 300,000 people were homeless.[2] Returning refugees barely recognized the smoking ruin of Munich. Those who had remained in the city told horrible tales of fear, fire, and death.

The transition from the war-ravaged ruin of 1945 to today's prosperous cultural capital took many years, and demanded that Munich and its residents endure many hardships, adapt, and find the strength to rebuild. Many cities, in Asia as well as Europe, faced similar challenges, and many recovered.[3] The question is where did this resilience come from? Potential answers include nonhuman factors based on technology and ecological conditions, or human factors, based on behavior, such as outside help or a willingness to adapt and survive. This essay considers the sources of Munich's resilience from the perspective of environmental history, and argues that they are found in human behavior, specifically what has been called "cultural resilience." In psychology, this term refers to how groups of people deal with catastrophic change (natural or man-made), often holding together despite harsh conditions, but also disintegrating. The concept of cultural resilience is used in many disciplines, including ecology, anthropology, and sociology. This essay focuses on what C.S. Clauss-Ehlers calls "culturally-focused resilient adaptation,"[4] which according to Olena Smyntyna "describes how culture and the sociocultural context have an effect on resilient outcomes."[5] In this essay, "culture" is rooted in the details of everyday life, not in works of culture like art or music.

Unfortunately Munich has yet to be the subject of an environmental history study in any language, but several works in German tell the city's story during World War II.[6] Urban environmental history provides useful models of connections between cities and natural resources.[7] Finally, there are studies of German cities during wartime,[8] and books on bombing that consider the environmental impacts of airborne attacks on cities.[9]

For Munich, the experiences of the war years, specifically 1942–1945, brought enormous changes when the city was subject to increasing damage and disruption from bombing raids. The narrative focuses on how the relationship between humans (Munich's people, aka *Müncheners*) and nature changed during this time in two important ways. First, bombing drastically altered the urban environment and damaged vital connections to natural resources. The attacks made nature an enemy by exposing city dwellers to forces (cold, fire, darkness, hunger, wind, rain, humidity) they had previously managed to keep at bay.

According to interwar bombing theory, these attacks were designed to create misery that should have led to surrender.[10] Instead of giving up, *Müncheners* proved that they could also make nature their ally. To adapt to and survive these forces, they turned back the clock to a preindustrial age, showing that years of city life had not weakened their relationship to nature. Munich's citizens suffered and died, but most adapted and survived. This quality would serve them well in the difficult years of occupation and recovery to come. Those planning the bombing raids however did not foresee such resilience.

Bombing Theory and Resilience

Like many military thinkers in the years between the world wars, Basil Liddell Hart, the (later famous) British military strategist, was looking for ways for his country to avoid repeating the unprecedented carnage of the Great War. In his 1925 book, *Paris, or the Future of War*, he argued that "[t]he enemy nation's will to resist is subdued by the fact *or threat* of making life so unpleasant and difficult for the people that they will comply with your terms rather than endure this misery."[11] Rather than targeting armies in the field, air forces would attack civilians directly to break their morale, achieving victory inside the cities rather than outside them. Bombs aimed at infrastructure targets (railroads, power plants, water and sewer systems, factories) would cut off supplies of food and fuel, making life more difficult. Bombs that killed their friends, families, and neighbors and burned or blew up their houses would demoralize civilians and expose them to the weather. Subjected to all of this "misery," city dwellers would, in theory, convince their leaders to surrender rather than continue to suffer. In the face of repeated bombing raids, their resilience would wear away and eventually disappear.

To Liddell Hart and other military men like the Italian General Giulio Douhet,[12] the American General William "Billy" Mitchell, and British thinkers such as Lord Hugh "Boom" Trenchard, using air power to target cities and civilians and force surrender represented the best way to avoid the seemingly endless slaughter of trench warfare. They realized that in an era of "total war," nations could not fight wars without the material, political, and moral support of civilians. City dwellers, the theory went, were less hardy than country folk, and thus more emotionally fragile. Make their lives difficult enough, and they would stop going to work, preferring to cower in fear. As Lord Hugh Trenchard argued in a 1917 memo, "experience goes to show that the moral effect of bombing industrial towns may be great, even though the material effect is, in fact, small."[13]

These new interwar military theories, viewed from the perspective of environmental history, targeted infrastructure and civilian morale to fundamentally disrupt the relationship between humans and nature. A modern city expressed human control over nature. The transport system of roads, rails, power lines, and water pipes brought food, electricity, and fuel to the city dweller, who could forget his or her dependence on natural resources. Maintaining this control increased dependence upon connections to natural resources outside the city that kept nature at bay and made city life comfortable, let alone possible. As cities grew, these connections became longer, more extensive, and thus more vulnerable to attack.

Driven by the ideas of the interwar theorists, the RAF and USAAF targeted Germany's people with "precision" and "area" bombing to starve the German war machine and demoralize its people. "Precision"[14] bombing targeted infrastructure like railroads, factories, and shipping to disrupt supplies to armed forces and civilians. "Area" bombing sought to undermine morale by targeting cities and their inhabitants. Both strategies aimed to damage or destroy the connections between cities, people, and natural resources.

Allied bombing planners intended on doing this to many German cities, Munich included. As the rail center for southern Germany and the place where the Nazi movement began, Munich was both an infrastructure and morale target. Due to its geographic location, Munich was spared any major raids until 1942, when this story begins.

1942: THE WAR "HITS HOME" IN MUNICH

When war erupted in September 1939, few people foresaw the damage and chaos to come, though some were apprehensive. Munich resident Margarete Konetzky, a 17-year-old schoolgirl, was on vacation in the nearby Alpine resort of Lenggries when she heard the announcement of war on the radio. She related in her diary that she heard "[c]ontinuous rousing music and marches, and again and again the sentimental song 'Rosemarie, Rosemarie', continually interrupted by political reports," but that she also felt "great fear and worry." When she arrived in Munich the next day, she found that the mood in the city mirrored her own experience, describing it as "[a] mix of shock, distraction, and depression, at least among the older people and those who had participated in the last war, and of euphoric hubbub among the younger people and especially among Nazis."

At first it seemed that such worries were unfounded. From 1939 to fall 1942, the RAF attacked Munich seven times, but the city had suffered minimal damage. While high on the list of bombing targets, Munich was not easy to reach early in the war, and commanders deemed the cost of raids to be too high. Bombers would have to fly 570 miles by air, mostly without fighter escort, through the heart of Germany's air and ground-based defenses.[15] This changed the night of September 19/20, 1942, when 79 RAF bombers carried out their first large raid on Munich, dropping 169 tons of bombs on the city center, causing 143 deaths and 413 injuries.[16] The raid provoked a mixed reaction of outrage and defiance, at least in the official propaganda. In a speech given hours after the raid on September 20, *Gauleiter* (Regional Leader) Paul Giessler said:

> We can say that here in Munich, we are just like our soldiers. We are hard and brave and true like them. That is our pride. Munich has been thrown into the front of the war, and she will think and act just like her sons out there [on the battlefield].[17]

While seemingly full of false bravado, Giessler's words expressed the new role of the urban dweller in warfare. Air power forced Munich's citizens to become combatants, and subject to the same risks (death, injury, deprivation) faced by the frontline soldiers. As the war progressed, the city became like the trenches of World War I: a place of darkness and danger short on modern comforts and exposed to natural forces. Margarete Konetzky described the city in her diary as she returned to Munich on September 22:

> From the train, I could already see many destroyed houses and factories in Munich. My heart was very heavy. My aunt picked me up and told me of terrible things. Our neighborhood looked ghastly. The street is covered with rubble and glass. The area between Prinzregentplatz and Prinzregentenstrasse is a wasteland.[18]

While the physical damage Konetzky saw was just a taste of what was coming, the less visible damage to the city's connections to natural resources had already begun. As the war progressed, citizens would have to adapt to both kinds of damage to survive. It began with coal.

Coal in 1942: Act Locally

Coal played a key role in daily life in Munich, and the city depended on sources outside of Bavaria, mainly the Ruhr and Germany's Eastern provinces. As a source of fuel for locomotives, power plants, household heating and cooking, gas production, and in the city's factories and food production facilities, coal's indispensability made the city particularly vulnerable to shortages. Coal was also vital to the war effort, which ramped up significantly with the invasion of the Soviet Union in the summer of 1941. Armies demanded more coal from the same sources (mainly the Ruhr) Munich relied on, so local sources (mines in Bavaria) and local actions (conservation) had to make up the shortfall. Appeals for more production reflected the situation. During a speech at the Penzberg mine in Upper Bavaria on February 5, 1942 Adolf Wagner, *Gauleiter Staatsminister* (Regional Leader State Minister), called on local miners to produce more coal:

> Each ton more that you produce is of immeasurable value for victory… Produce more coal so that our women and children do not freeze and that our work comrades in the armament industries can make the weapons and munitions that the Führer needs for his soldiers…to achieve victory.[19]

Try as they might, local mines were not large enough to fill the demand. This meant less reliable coal deliveries for the residents of Munich, changes in everyday life, and challenges to local leaders. For example, in the fall and winter of 1942, residents saw Munich buses starting to use wood-fired engines. Schoolchildren had time off school for "coal holidays" (*Kohlenferien*) to reduce coal consumption.[20] At their meeting on January 18, 1943, the Munich City Council meeting discussed these problems. Lord Mayor Karl Fiehler complained, "[N]ot only in Munich, but other cities, estimations of consumption are too low." School superintendent Bauer reported that despite four weeks of *Kohlenferien* some schools had to be closed. Fiehler reported that the military was getting all the high-quality Ruhr coal. Rationing ensued, with hospitals, childcare facilities (nurseries and hospitals), and homes for the aged getting first priority for coal deliveries.[21]

Because coal was used to generate electricity, the city encouraged its citizens to conserve electricity.[22] On December 20, 1942, a caricature of a rodent-like robber began appearing in the Nazi Party newspaper *Volkischer Beobachter* (VB)[23] and other newspapers to represent the concept of "Coal thievery" (*Kohlenklau*). The figure, keeping one eye out for the police as he snuck away with a sack of coal, also worked his way into ads for products such as razors, bandages, and carbon paper. Animated cartoons, posters, and board games also appeared.

The *Kohlenklau* campaign inaugurated part of a larger nationwide energy conservation effort that began with a September 8, 1942, speech from Hermann Göring. He urged:

> German housewives! Each of you can contribute to armament and thus for our victory by voluntarily reducing electricity and gas consumption. Remember that gas and electricity are almost exclusively produced from coal. The fruits of the coal miner's hard labor cannot be wasted by you thoughtlessly burning lights.[24]

In the months to follow, the effort expanded with a series of voluntary conservation measures. With the German armies on the defensive after their defeat at Stalingrad, mandatory measures would soon outnumber voluntary ones. "Total war" had arrived.

1943: Total War

In his famous speech on February 18, 1943, Joseph Goebbels asked a cheering crowd "do you want total war?" and would they "give your all for victory?."[25] He received a resounding "Ja!". Along with this shift in priorities, other military developments made 1943 a turning point for Munich's citizens. Coal and fuel became even scarcer, and the city was subject to more frequent and damaging bombing attacks. The Allied invasion of Italy brought the impending threat of bombers based in that country, overcoming Munich's geographic advantage and allowing easier access to Munich as a target.

The USAAF would not carry out its first major daylight "precision" raid until March 1944, but the RAF continued to deliver major night "area" raids in 1943. These raids escalated the damage done to the city by now cutting off public utilities such as electricity, gas, and water and forcing citizens to take cover to survive. Heading to a shelter, either outside in a trench or underground in basements or neighborhood shelters, became a frequent occurrence. Franz Weber, who spent his adolescence in Munich during the war, recalled how he spent only one night in neighborhood shelter, as it was too dark, stuffy, and packed with people. He spent other raids in a basement shelter, but these were less safe. One neighbor family was killed in their basement during a raid.[26] Basically it was a choice between comfort and security. Franz's choice could have killed him. In another part of Munich, six-year-old Theo Rosendorfer was actually enjoying the air raids. He recalled that the war was exciting for a child. Air raid trenches in green spaces near Isar were a "playground...the raids were like an adventure: you sat in the cellar."[27] Adults soon became inured to the danger and inconvenience, at least for a while.

The editors of the *VB* remained adamant that the enemy's bombing goals were less material than political: "With these methods of murder and burning [they] would turn Europe into a land of rubble and make it ripe for Bolshevism."[28] While designed as propaganda the editorial summed up how people in resilient cities do not give up: "There are many ways to live. With and without comfort...we will not be defeated by this terror."[29] The damage to the gasworks was significant because most Müncheners used gas stoves for cooking (electric stoves being relatively rare), so the nearby Dachau concentration camp[30] began cooking not only for the inmates, but for city dwellers as well. Such interruptions became more and more prevalent, especially as coal supplies grew harder to obtain.[31] Altogether, the gasworks suffered ten major attacks during the war, with whole neighborhoods cut off for days, sometimes weeks.[32]

Fire was becoming a more effective way of destroying a city, and also altered the urban environment by igniting fires that lit streets in an eerie glow or even created artificial daylight. While no reliable statistics on the total damage from fire over the course of the war exist, the statistics from one major raid (March 9/10, 1943) tell a grim tale. That night 218 RAF bombers dropped 599 tons of bombs, killing 208, injuring 438, destroying 408 homes, and damaging 1684 buildings and 20,000 homes.[33] The fires also strained local resources, as firefighters from Munich and 15 surrounding municipalities fought 320 fires until the afternoon of March 10, and interrupted essential services such as gas and electricity.[34] And yet, there was also an odd beauty to the fires, as Theo Rosendorfer remembered a specific raid in 1943:

The Paulaner brewery on the Knocherberg was bombed heavily, there was an alarm, I remember to this day, it was simply sensational, how the hill overlooking the Isar (river) where the brewery was located the way it burned; this bright glowing fire; for me as a child this made an unforgettable impression; I can still see it today-the fire, the smoke clouds, how the flames were whipped up into the sky.[35]

Humans could also create their own disasters with firestorms, as demonstrated by the Hamburg raid that same year, which combined high-explosive bombs with incendiaries, large numbers of bombers, warm weather, and dry conditions. Repeating this phenomenon would be difficult, and accuracy was a serious challenge. To generate a firestorm, high-explosive bombs would create corridors and provide fuel for the flames by destroying walls, windows, and doors.[36] To effectively channel the flames specific buildings, especially in the older parts of a city, needed to be hit by for both types of bombs. In blacked-out cities, targets were hard to identify. To remedy this problem, the RAF tested the use of markers in raids on Munich. During the September 6/7 raid on Munich bombers dropped small magnesium bombs (attached to parachutes) that would ignite on impact and illuminate the target area. Because of cloudy weather, the markers were of little use, but that did not stop fires from breaking out in several parts of the city. While this raid did not ignite a firestorm as in Hamburg, the bombs nevertheless made an impression on the citizens. One Munich resident described the effect as "a circle of light as bright as daylight. It was so bright that you could have read a newspaper using its light alone."[37]

While condemning the raid as a "night of horror," the *Münchner Neueste Nachrichten* (*MNN*) put a positive spin on the damage and hinted at the benefits of returning to a preindustrial past. Three days after the raid Eugen Roth, the editor of the *MNN*, wrote, "The war has made our cities quieter; yes, even the cosmopolitan city of Berlin has become provincial. Less traffic, early curfews, no lit-up advertisements...more than in the early war years, people retreat to their homes and families."[38] Munich, it seemed, was returning to an earlier, "simpler" time.

The September raid, along with a second big raid on the night of October 2/3, showed that bombing could make the distinction between night and day almost meaningless for urban dwellers. This is another example of how warfare changed the relationship of humans to nature in Munich by altering conditions (darkness and light) and expectations (city lights keep the darkness away). Writer and commentator Wilhelm Hausenstein, who lived in the nearby (about 40 kilometers) village of Tutzing but visited Munich often, experienced these changes firsthand. He wrote in his diary on October 3: "First some greenish-white stars over Munich, enlarged; there were greenish-white and red bodies of light the whole way from the foothills of the Alps over the reflecting lakes to Munich and beyond; it was as bright as on a night with a full moon."[39]

Helmut Dotterweich spent much of the war at school in nearby Augsburg, and during a different raid, the teenager had a more intense experience from firebombing:

> We literally were walking through flames...We walked 20 to 25 kilometers west to escape the fire. I remember seeing the moon, but it was actually the sun. It was so dark during the day-I thought it was still nighttime, but it was midday.[40]

Hausenstein was seeing the light from anti-aircraft fire and exploding bombs, while Dotterweich was seeing the sun shrouded in smoke from the burning city. Both phenomena, created by bombing raids, changed the urban environment by producing light when it wasn't expected (Hausenstein's experience), or shrouding light and making it difficult to tell night from day (Dotterweich's). People saw the city differently. It became more dangerous and less familiar. Even when no raids occurred, blackouts and other defensive methods ensured that nothing looked the same. The physician and poet Hans Carossa wrote about how blackout measures affected him emotionally in his diary in 1943:

> Since the war began the streetcar windows had been made of bluish blackout glass, in which part was left transparent. This blue half-light seemed to dampen any feeling of happiness. It robbed the parks of their green freshness, and made it seem as if even when smiling people stood together that it was only a dream.[41]

Darkness was a natural phenomenon, like being exposed to weather, that the city dweller could normally control. Bombing made the city more frightening, depriving citizens of the technological advances (electricity, indoor plumbing, lighting) that made the city seem more "civilized." In Munich during the war, light came not from "modern" electric lights, but from a much older source, fire.[42] Just as coal was replaced by wood, electricity, at least at night, was replaced by fire.

Electricity remained important, however. While people began to leave Munich in large numbers in 1943,[43] demands for electricity conservation begun in 1942 became more intense. At the time, the most disruptive and damaging bombing raids were still to come, so most of the city still had power. On April 6, 1943, the hospitality industry was ordered to reduce electrical consumption from lights by 30 percent.[44] For household users, energy-saving measures began in May 1943. Announcements began appearing in the *VB* and *MNN*. For example, the *MNN* May 14 issue listed the following compulsory and voluntary measures:

> In the future, all households with more than ten rooms can consume only 80 percent of the electricity use during the same period (May–October) of the previous year… a 10 percent reduction is expected in the remaining households, as well as an overall reduction in gas consumption of 10 percent.[45]

Citizens could also help on other ways. In July, Albert Speer announced that citizens could earn 50–500 Reichsmarks for energy-saving suggestions.[46] The December 3 issue of the *VB* urged households to go beyond voluntary measures and save more.[47] This was the last of the voluntary measures, however. On January 21, 1944, *Gauleiter* Giessler ordered mandatory energy-saving measures, which forbade the "overheating of rooms" and the use of gas and electricity to supplement coal heaters. The order also restricted electrical use between the hours of 7–9 am and 4–7 pm. Violators would be prosecuted. This was done because "the riches of the Greater German Empire in coal and electrical energy must first serve war production. The sacrificial struggle of our soldiers compels the homeland to sacrifice on its own side, and to conserve in all areas of daily life."[48]

Conservation images were now linked to water consumption and the lives of soldiers on the front, not just the consumption of coal. An article entitled simply "Save Water!" in the November 9 issue of the *MNN* stated: "Each liter of water that flows in our pipes is pumped using a measurable amount of energy from gas, electricity or steam. For this we need coal, and therefore poorly used water is an unnecessary demand for coal, which is to be absolutely stopped."[49]

In 1943, the increased frequency and ferocity of air raids and rising demand for resources for the war effort intensified the transformation begun in 1942 after the first big raids. Bombers brought the concept of "total war," where an entire society and its resources are combatants, home to the citizens of Munich. Like the soldiers on the battlefields, Munich's citizens suffered food and fuel shortages, and threats of destruction and fire from above were rendering life in the city more challenging and dangerous. The process accelerated in 1944, as the USAAF began bombing Munich and the *Wehrmacht* retreated on all fronts, cutting access to resources from conquered territories. The city and people of Munich were to experience the greatest shocks during the war to date.

1944: The Year of Fire

As 1944 began, Munich's citizens lived in a city resembling the site of a continuing natural disaster, as more intense bombing raids destroyed buildings and set fires throughout the city. Journalist Ruth Andreas-Friedrich wrote in her diary on January 4, 1944: "We sweep up rubble... We sit without water, without transport, without electricity. The telephone is also dead, and one only finds out through the grapevine if friends living far away are alive. An auspicious beginning to the year."[50] The damage Andreas-Friedrich referred to was only a taste of what was to come. Inspired by the success of the Hamburg raid, British bombers began to drop large amounts of incendiaries during night raids on Munich in 1944. The experiments and successes of fire raids in 1943 shifted bombing policy and fire became the primary weapon of the area bombing strategy of raining terror from above.

1944 also brought more bombs of all kinds, as Munich's role as a transportation hub now became the focus of American bombers. The USAAF's Fifteenth Air Force began flying missions from Italy to carry out "precision" daylight raids. Like the fires RAF bombs created, American bombs altered the urban environment's dynamics and the meaning of daylight for its citizens. Müncheners had grown used to raids at night, usually between the hours of 10 pm and 2 am. They knew that, while they would likely spend the night in an air raid shelter, during the day they could go about their business in relative safety. The appearance overhead of American bombers during the day disrupted this routine. Journalist Kurt Preis remembered a saying repeated in Munich at the time: "Wenn's hell wird, müssens ja wieder dahoam sei, sonst kommt koana mehr hoam" ("When it gets light, one must be back home, otherwise no one ever comes home").[51] Just as bombing at night made it hard to distinguish night from day, American daylight raids made *real* daylight no longer safe.

Raids during summer and fall 1944 demonstrated the double danger now posed by RAF night attacks and USAAF daylight attacks. In July, the USAAF's Eighth Air Force carried out seven big raids on Munich. Targets included factories, the inner city, airfields, and railroads. These included the heaviest raid of the war on July 12, when 1124 USAAF B-17's dropped 2451 tons of bombs, killing 631 and injuring 1711.[52] The raids interrupted gas and electricity service, damaged a sewage facility, and destroyed water pipes, hampering water distribution, so that trucks had to distribute water to residents.[53]

As the damage increased, so too did Müncheners' exposure to natural forces (fire, darkness-often during the day due to smoke-weather) and retreat from "civilization" that began with the first big raid in 1942 continued. Wilhelm Hausenstein wrote in his diary on July 16: "[T]he fourth raid on Munich. One gets the impression that Munich will be bombed to the ground." Later, on July 30, he wrote of the decline in living standards outside the city in Tutzing:

> The 'technical achievements' begin to decline again – it is like we are returning to an earlier time…every moment the water is cut off, because there is no electricity to power the pump. Every couple of days there is no light; the telephone only works half the time; the trains run very late.[54]

Despite all the damage and misery caused by exposure to nature, reactions to these raids reminded readers that nature could heal as well as harm. An *MNN* editorial from July 22/23 portrayed the natural world as a balm for damaged psyches:

> And then the warning and the all-clear came; we climbed out again into the light of day, into the summer, and saw that the flowers bloomed, full and luxuriant, in the parks we heard the birds singing and trilling and Nature carried on as if nothing had happened. Nature is kind.[55]

Hausenstein also took comfort from the natural world. On September 12, 1944, he wrote, "Overall there is rubble composed of trash, lead, glass, broken bricks, burned rafters; the impression is not only sad, but also disgusting. The most comforting thing is that green things—grass, nettles and others—grow on the heaps."[56]

Bombing raids also led to people seeing their city as a body or organism. In the July 22–23 *MNN* editorial, Munich became something that had a "countenance" that "gets even more distressed" and "once again bleeds from new wounds."[57] Munich had essentially begun bleeding internally. The assault on Munich's "body" continued in fall 1944, when the USAAF launched a series of attacks on Munich's railroads, the "veins and arteries" connecting it to natural resources and other cities.

Severing the Arteries: USAAF Railroad Attacks in Fall 1944

In keeping with their strategy of attacking infrastructure during daylight "precision" raids, the USAAF attacked Munich's rail stations, marshaling yards, and repair shops. Munich's four railway stations (Munich Main Station, Munich East, Munich-Laim, Munich-Ludwigsfeld) received both passengers and freight. This made Munich a key center for moving supplies to and from the Italian and Balkan Fronts. In addition to the rails' military function, these facilities also allowed the citizens of Munich to import coal, food, and other necessities. The reports of the Regional Railroad Authority (GLB), which included most of Bavaria as well as the Munich Railroad District (RBD), captured and translated by the US Army in 1945, give a good insight into the impact of the raids on railroads. Using this data to assess the railroad raids after the war, the US Strategic Bombing Survey concluded that the October 4 raid on Munich's marshaling yards, "put operations on a permanently lower base," that is, they would never reach the same level of activity for the rest of the war. The bombs did substantial damage including 3800 meters of track destroyed, 4950 meters displaced, 26 switches destroyed or damaged, plus damage to signal lines and equipment. Despite the damage three tracks were operational two days later, and by October 25 all of tracks in the main marshaling yard were up and running. As the rail center for Bavaria, Munich was more vulnerable because it forged connections to other cities and regions. Raids on railroad facilities outside the Munich area, such as in Regensburg, Rosenheim, and Innsbruck, created congestion, reduced the ability to send out trains, and reduced overall efficiency.[58] To an environmental or urban historian applying the "organism" metaphor, the Americans had demonstrated that the "organism" of Munich's transportation system could be wounded not only by attacking the main facilities in the city (the heart), but also facilities in the periphery (the arteries).

For Munich's citizens and leaders, railway destruction meant more adaptations and headaches. Recalling his visit to the city on December 21, 1944, Wilhelm Hausenstein highlighted the damage to transportation: "Hard trip: back and forth took eight hours, including the stops at railway stations and the lengthy transfers on the way…The city is now absolutely destroyed; it exists practically only as rubble."[59] As food became more scarce, urbanites headed to surrounding countryside to forage for mushrooms and roots, or trade with local farmers, known as "Hamstering." Franz Weber's family tore out the rosebushes and planted strawberries, and kept chickens for eggs and meat.[60] Track damage forced local officials to find alternative ways of moving people and goods. A steam-powered miniature railway (*Kleinbahn*) originally constructed in August to haul rubble, now carried streetcar passengers, as many of the electrical wires serving the streetcars had been destroyed. The age of steam had returned to Munich's streets.

To further add to present difficulties, the Allies continued to find new and more efficient ways of damaging buildings and making citizens miserable. The RAF now unveiled so-called blockbusters, high-explosive ordnance such as the 12,000-pound "Tallboy" bomb designed specifically to destroy buildings. Alfred Haussner wrote in an article on the November raids entitled "The Rubble and the Remains" that "with every bomb dropped a small piece of the area that we understood as belonging to our homeland, crumbles into rubble and dust."[61] Like earlier editorials, the one entitled "Munich today" in the November 24 edition of the *MNN* revealed the degraded living conditions:

> The autumn wind howls, rain and snow whip through the streets and one's feet are suddenly in water, in one of the thousands of small holes that lurk treacherously after fire and bombs in asphalt and cobblestones next to patched streetcar tracks…When firestorms rage through the streets, when bodies lie buried under ruins, when electricity, gas, water, streetcars (and) railroads fail.[62]

The "firestorms" referred to in the editorial did not approach Hamburg's ferocity, but nevertheless badly damaged life and property. The December 18 RAF raid on Munich caused the war's highest casualty toll on the city.[63] Hausenstein wrote afterwards, "The noose grows tighter and tighter around us here…at 11 o'clock at night the sky in the northeast was red like a bleeding wound. The alarms never stop."[64]

Nature also joined in the attacks on Munich in December. Cold weather and frozen water pipes hampered efforts to put out fires. Ironically, staying warm now bedeviled people living in a city that was often on fire, as the coal and fuel situation deteriorated in late 1944. Increasing coal shortages deepened the suffering of Munich's residents, and forced them to rely more on wood as a fuel resource. In the *MNN* for October 14, citizens were advised that for the time being, to save coal for the winter. Heating government offices, Nazi Party offices, law offices, banks, insurance offices, cinemas, inns, and department stores was forbidden. In the countryside, people were to "collect harvested wood and pinecones to substitute for coal."[65]

As 1944 ended and the final winter of the war began, Munich was in a very bad state, and about to get worse. Its connection to coal was now nearly severed, as the Red Army advanced westward and the transport system shuddered under constant interruptions. As the coal supply continued to shrink, officials now actively encouraged citizens to substitute wood for coal, and allowed them to cut their own wood.[66] Coal dealers in the city were now authorized and encouraged to distribute charcoal (*Holzkohle*) and wood to power wood-burning generators and cars.[67] By January 1945, even the landscaping of the inner city was at risk: a new rule allowed older trees in parks, rivers, and lining city streets to be cut.[68] This acceleration in the process of returning to wood as a fuel source dramatized Müncheners' disconnection from their outside sources of fuel. Now consuming wood from within its urban innards, the starving organism of Munich was beginning to feed on itself.

1945: Collapse

As New Year's Day 1945 arrived, the city had reached a new low in its modern history. An editorial in the *MNN* from December 30, simply entitled "1945," summed up the situation:

> We have all become poorer, our city has become poorer, more distressed; now we see buildings torn apart that once created a sensation in Europe. A toast-wherever it is possible with the "nervous schnapps," the special distribution of liquor can and will not belie the seriousness with which we greet the year 1945.[69]

Giulio Douhet (see above) described just such a scenario in "Probable Aspects of Future War" in 1928, offering a disturbingly prophetic vision of Germany in early 1945. Germany, a belligerent that had lost the "command of the air," would have to fight an unequal fight and resign itself to endure implacable offensives. Its army and navy would have to function with bases and communication lines insecure, exposed to constant threat. All the most vital and vulnerable points in its territory would be subject to cruelly terrifying offensives.[70]

For the people still living and working in Munich, "cruelly terrifying offensives" of 1945 brought more terror and disruptions, and meant that the city bore more resemblance to a corner of Hell than the beautiful city and cultural center it had once been. The January 7/8 raid for example, was a "double attack," as the second wave of bombers struck an hour or so after the first, when citizens had been given the "all-clear" and left the shelters.[71] Karl Ude, serving as a soldier in Munich, remembered that night, when he was on fire watch:

> I had never experienced such a threatening and unnerving night during the air war...The power was out – and as far as we could see the area was in flames. Our boots crunched on broken glass, phosphorous licked at the supports and made the softened asphalt sticky...sparks flying and poisoned air.[72]

The coal and fuel situation in 1945 continued to reflect how the retreat of the *Wehrmacht* and Allied bombing had changed the lives of Munich's citizens. Coal scarcity worsened in February and March, as trains had difficulty getting through. Wood continued to substitute for coal. On March 12, the City Council reported that

> [t]he coal situation is as serious as ever. We should be getting 140,000 tonnes ... It is not all certain that (it) will come at all. We will try to get coal from Upper Bavaria for household heating. 100 wagons full of wood have arrived, and been partially distributed.[73]

Above all other factors, the railroad situation had become critical. New raids destroyed more tracks and locomotives, cutting the connections in the rail arteries serving the organism of Munich. The GLB reported on February 17, 1945: "Despite all efforts, empty open freight- and coal-trains are stopped again and again because of air raid damages." Munich became isolated as "systematic enemy attacks broke up temporarily the railroad system into individual islands, causing complicated detours for the remaining freight and empty trains." Munich's four railway stations, plus Nuremberg, Augsburg, and other junctions, ceased to operate. Damage to telephone lines meant Munich could not be contacted for three days following the March 3 raid. As soon as these connections were reestablished, advancing Allied armies and continual bombing raids would soon cut them again. By April 14 there was "no through passage from north to south and east to west." Rail managers found "the southern districts of the Karlsruhe and Stuttgart, RBD Augsburg and the west part of RBD Munich constituted a closed island."[74] Isolated, cut off from its supplies of food and fuel, Munich was ready to expire as a fighting city.

Conclusions

Sixteen days later, on April 30, American forces arrived. Munich surrendered, but not because of the misery brought by bombing. As in the past, the city only gave up with an army at its gates. Munich at the end of April 1945 was a vastly different place than it had been at the outbreak of war in September 1939. High-explosive and incendiary bombs had radically altered the urban environment. Clouds of dust from rubble and mixed with the lingering odor of burned out buildings and broken sewer lines. The streets were choked with rubble and full of holes. The railroads did not run, and streetcars barely functioned. Electricity, water, and gas services were haphazard at best, and coal could not reach the city. People did find stocks of food, but once they had been looted the possibility of feeding the people without the aid of the Americans was distant at best. Red Cross nurse Maria Schneider's words accurately reflected the situation faced by Munich's remaining residents, "I did not reckon with this level of warfare. I was doubtful. I said to myself: 'this will never end, we have no cities, we do not have any more left. Everything is bombed out. We have no housing. Many of our children are crippled.'"[75]

But other sights and smells served as reminders that Munich had endured, surviving six years of war and four years of bombing. As citizens climbed out of underground bunkers, the smell of wood-burning stoves and vehicles meant that the city had successfully returned to preindustrial fuels. The sight of gardens and henhouses in suburban backyards and city dwellers returning from the surrounding countryside with food from farms demonstrated that urbanites could find food on their own. Despite the damage, the Isar still flowed, the nearby Alps were still snow-capped, and plants sprouted in the rubble.

Despite the interwar theorists' promises, changes in the urban environment and disrupted connections to natural resources had not forced Munich's people to demand their leaders surrender. Munich and its people endured because they were able to adapt to the changes in the relationship between humans and nature that the war brought. By returning to preindustrial practices such as wood-gathering, farming, and adapting to darkness, Munich's urban dwellers became closer to nature (i.e. the natural world), but the main source of their resilience was in their *ability* to do these things, an ability dwelling in their own human nature, which was not as far removed from the natural world as the bombing theorists had thought. The strength that formed the basis of its resilience was found mostly within Munich's *people*, who were themselves important elements of the city and its natural environment.

Müncheners would need this resilience in the first years of occupation. Key connections to natural resources serviced by the transport system present in 1939 had been severely damaged, even amputated. A vengeful enemy, the Soviet Union, now occupied the Eastern provinces, sources of coal and food. The Allies controlled the Ruhr coalfields and the Bavarian farms. As if that was not enough, the winter of 1946–1947, one of the harshest in recent memory, would be followed by a severe drought the following summer. The Zero Hour (*Stunde Null*) had arrived.

Notes

1. The city's relatively late industrialization diverted the traditional "dirty" industries such as steel plants, to other areas. Lacking a traditional industrial base, Munich welcomed specialized, high-tech companies such as BMW in 1917.
2. Irmtraud Permooser, *Der Luftkrieg über München 1942–45: Bomben auf die Hauptstadt der Bewegung* (Oberhaching: Aviatic Verlag, 1997), 367. Statistics on deaths from air raids vary from 6200 to 6600; I have split the difference. It is unclear from the statistics as to how many of those made homeless subsequently left the city.
3. Most famously, Londoners during the "Blitz." For a nuanced look at the reality of living in London during that time, see Amy Helen Bell, *London Was Ours: Diaries and Memoirs of the London Blitz* (London: I.B. Tauris, 2011). Berlin, despite heavy bombing, only fell when Soviet troops took the city at great cost. German language readers may enjoy *Backfisch im*

Bombenkrieg (Berlin: Matthes & Seitz Verlag, 2013) a surprisingly breezy diary of a teenage girl who spent the war in Berlin.

4. C. S. Clauss-Ehlers, "Re-inventing resilience: A model of "culturally-focused resilient adaptation," in C.S. Clauss-Ehlers and M.D. Weist, eds., *Community Planning to Foster Resilience in Children*, (New York: Kluwer Academic Publishers, 2004), 27.

5. Olena Smyntyna, "Cultural Resilience Theory as an instrument of modeling Human response to Global Climate Change. A case study in the North- Western Black Sea region on the Pleistocene-Holocene boundary," *RIPARIA* 2 (January 2016): 1–20. Smyntyna's study should interest environmental historians. See also A. Himes-Cornell, and K. Hoelting, "Resilience strategies in the face of short- and long-term change: outmigration and fisheries regulation in Alaskan fishing communities," *Ecology and Society* 20, no. 2 (2015): 9.

6. Eva Berthold and Norbert Matern, eds., *München im Bombenkrieg* (Düsseldorf: Droste, 1983). Two other books (Permooser and Richardi) are discussed below.

7. The classic work on cities' connections to natural resources is William Cronon, *Nature's Metropolis: Chicago and the Great West* (New York: W.W. Norton & Company, 1991).

8. See Roger Chickering, *The Great War and Urban Life in Germany: Freiburg 1914–1918* (Cambridge: Cambridge University Press, 2007) for a comprehensive study of a city at war before widespread use of strategic bombing.

9. For a theoretical assessment see W.G. Sebald, *On the Natural History of Destruction*, translated by Althea Bell (New York: Random House, 2003). For good overview of German cities in World War II, see Hermann Knell, *To Destroy a City: Strategic Bombing and its Human Consequences in World War II*. (Cambridge, MA: Da Capo, 2003). For the German view of bombing, see Jörg Friedrich, *The Fire: The Bombing of Germany 1940–1945*. Translated by Allison Brown. (New York: Columbia University Press, 2008).

10. The idea was that urban residents would rise up against their leaders. This fails to take into account the fact that some regimes (especially the Nazis) were very effective in creating fear and repression, punishing anyone who protested. It would have been extremely difficult for any city, especially Munich, which had the title "Capital of the Movement" to demand surrender from their leaders.

11. Basil H. Liddell Hart, *Paris, or the Future of War* (New York: E. P. Dutton & Co., 1925), 37.

12. Douhet's 1921 book, *Command of the Air*, is considered one of the classics of interwar bombing theory. Douhet himself started as an artilleryman, but

as early as 1909 was speculating on the impact of the airplane in warfare. He later commanded the first Italian air battalion during World War I. See Giulio Douhet, *The Command of the Air* (Washington, D.C. Air Force History and Museums Program, 1998).
13. Quoted in Tami Davis Biddle, *Rhetoric and Reality in Air Warfare: The Evolution of British and American Ideas About Strategic Bombing, 1941–1945* (Princeton: Princeton University Press, 2002), 37.
14. Given the accuracy (or lack thereof) of bombing during World War II, the term is misleading. "Precision" bombing is better understood as having specific targets in and around a city, whereas "area" bombing simply targets the entire city and its surrounding area. For a complete guide to bombing in World War II, see Charles Webster, and Noble Frankland, *History of the Second World War: The Strategic Air Offensive against Germany 1939–1945* (London: Her Majesty's Stationery Office, 1961.)
15. *The Bomber's Baedeker: Guide to the Economic Importance of German Towns and Cities: Second Edition* (Enemy Branch, Foreign Office and Ministry of Economic Warfare, 1944), 487.
16. Permooser, *Der Luftkrieg über München*, 379.
17. Hans-Günter Richardi, *Bomber über München: Der Luftkrieg von 1939–1945 dargestellt am Beispiel der "Hauptstadt der Bewegung"* (München: W. Ludwig Buchverlag, 1992), 5.
18. Richardi, *Bomber über München*, 102.
19. "Mehr Kohle für Heimat und Rüstung!" *Volkischer Beobachter*, February 6, 1942.
20. "Holzgasgenerator-Anhänger bei den Münchener Omnibussen," *Volkischer Beobachter.*, November 11, 1942. By the end of the war, Germany had over 500,000 such vehicles. See http://www.lowtechmagazine.com/2010/01/wood-gas-cars.html, accessed April 30, 2017.
21. Stadtarchiv München, *Stadtratbesprechungen*, January 18, 1943.
22. Cities in Allied states had similar concerns. See Matthew Evenden, "Lights Out! Conserving Electricity for War in the Canadian City, 1939–1945." *Urban History Review* 34, no. 1 (Fall 2005): 3–124.
23. Since the *VB* was official Nazi Party propaganda, it can be problematic as a source. However, the flowery and patriotic language often only thinly veils the truth. It is useful in seeing how the Nazi Party *wanted* the people of Munich to react to the privations of war. Often people had no choice but to tow the party line. Near the end of the war, those considered "unpatriotic" were executed.
24. "Spart Strom und Gas: Ein Aufruf des Reichsmarschalls-Alle Energie für den Endsieg." *Völkischer Beobachter*, September 8, 1942.

25. "Total War": Excerpt from Goebbels's Speech at the *Sportpalast* in Berlin February 18, 1943 in Joachim Remak, ed., *The Nazi Years: A Documentary History* (Englewood Cliffs, NJ: Prentice-Hall, 1969), 91–92.
26. Franz Weber, interview by author, Munich, Germany, February 6, 2010.
27. Theo Rosendorfer, interview by author, Munich, Germany, February 10, 2010.
28. Interestingly, the American occupiers would worry after the war that poor conditions would lead to Germans rejecting democracy in favor of communism.
29. "Eine Stadt hält Stand: nach dem Terrorangriff auf München," *Völkischer Beobachter*, March 31, 1943.
30. For more information on Dachau, see International Dachau Committee. *The Dachau Concentration Camp, 1933 to 1945: Text and Photo Documents from the Exhibition.* (Dachau: Comité International de Dachau, 2005). For a more recent work on concentration camps in general, see Nikolaus Wachsmann, *KL: A History of the Nazi Concentration Camps* (New York: Farrar, Straus and Giroux, 2015).
31. Richardi, *Bomber über München*, 150.
32. *100 Jahre Gas in München*, 100–102.
33. Permooser, *Der Luftkrieg über München*, 378–379.
34. Richardi, *Bomber über München*, 147. Shortages of equipment (ladders, hoses, gas masks, protective clothing), plus the refusal of many volunteer fire departments to help made the damage worse.
35. Theo Rosendorfer, interview by author, Munich, Germany, February 10, 2010.
36. Max Hastings, *Bomber Command* (London: Pan Books, 1979), 197.
37. Quoted in Richardi, *Bomber über München*, 10. Later the RAF would experiment with low-level marking using the Mosquito, a two-engine bomber constructed mostly out of wood. While such raids earned pilots medals for bravery, the technique was considered ineffective against heavy defenses. See Hastings, *Bomber Command*, 284.
38. Ibid., 176–177.
39. Wilhelm Hausenstein, *Licht Unter dem Horizont: Tagebücher von 1942 bis 1946* (München: Bruckmann, 1967), 161–162.
40. Dotterweich, interview.
41. Richardi, *Bomber über München*, 46.
42. There was one artificial light source. Theo Rosendorfer remembers that at night "there were markers that glowed in the dark so one could see the other passersby, so that people did not run into each other in the dark; people used to make jokes about this." Rosendorfer, interview.

43. The firebombing of Hamburg, with its unprecedented level of casualties in June 1943 caused many people to flee Germany's larger cities. See Permooser, *Der Luftkrieg über München*, 161.
44. "Einschränkung des Lichtverbrauchs," *Völkischer Beobachter*, April 6, 1943.
45. "Strom- und Gasverbrauch," *Münchner Neueste Nachrichten*, May 14, 1943.
46. "Sparsamer Energieverbrauch," *Völkischer Beobachter*, July 21, 1943.
47. "Es können noch Lichtstunden gespart werden," *Völkischer Beobachter*, December 3, 1943.
48. "Spart mit Kohle, Gas und Strom!" *Münchner Neueste Nachrichten*, January 21, 1944.
49. "Spart Wasser!" *Münchner Neueste Nachrichten*, April 9, 1943.
50. Richardi, *Bomber über München*, 238.
51. Richardi, *Bomber über München*, 217.
52. Permooser, *Der Luftkrieg über München*, 381–382. It is important to point out that Munich suffered fewer casualties than more political targets like Berlin or more industrial targets like cities in the Ruhr.
53. United States Strategic Bombing Survey, Medical Branch, *Report 65: The Effect of Bombing on Health and Medical Care in Germany*, Chapter 10: Environmental Sanitation, US National Archives RGB 260: Records of the United States Strategic Bombing Survey, folder 200a/148, 237–238.
54. Hausenstein, *Licht unter dem Horizont*, 260–262.
55. "Wieder Einmal!" *Münchner Neueste Nachrichten*, July 22/23, 1944.
56. Hausenstein, *Licht unter dem Horizont*, 271.
57. "Wieder Einmal!" *Münchner Neueste Nachrichten*, July 22/23, 1944.
58. US Strategic Bombing Survey, Transportation Division, *Report #202: Effects of Bombing on Railroad Installations in Regensburg, Nuremberg, and Munich Divisions*, Second edition, January 1947, U.S. National Archives RGB 260: Records of the US Strategic Bombing Survey, folder 200a/148, 2. Innsbruck was the capital of Tyrol in Western Austria and an important link to rail traffic coming from Italy.
59. Hausenstein, *Licht unter dem Horizont*, 287.
60. Weber, interview.
61. Alfred Haussner, "Der Schutt und das Bleibende," *Münchner Neueste Nachrichten*, November 17, 1944.
62. "München Heute," *Münchner Neueste Nachrichten*, November 24, 1944.
63. Permooser, *Der Luftkrieg über München*, 384–385.
64. Hausenstein, *Licht unter dem Horizont*, 286.
65. "Brennstoffversorgung in Stadt und Land," *Münchner Neueste Nachrichten*, October 14, 1944.

66. "Mehr Waldholz als Brennmaterial," *Münchner Neueste Nachrichten,* November 20, 1944.
67. "Kohlenhändler als Tankholz-Hersteller," *Münchner Neueste Nachrichten,* December 20, 1944.
68. "Erhöhte Holzabführung," *Münchner Neueste Nachrichten,* January 30, 1945.
69. "1945," *Münchner Neueste Nachrichten,* December 30, 1944.
70. Douhet, *The Command of the Air,* 202–203.
71. Permooser, *Der Luftkrieg über München,* 385.
72. Karl Ude, "Soldat in einer verdunkelten Stadt," in Hermann Proebst and Karl Ude, eds., *Denk ich an München: Ein Buch der Erinnerungen* (München: Gräfe und Unzer Verlag, 1966), 255.
73. Stadtarchiv München, *Stadtrat Dezernatbesprechung,* March 12, 1945.
74. US National Archives, RGB 260: Records of the US Strategic Bombing Survey, folder 200a/148.
75. Quoted in Eva Berthold and Nobert Matern, eds., *München im Bombenkrieg* (Düsseldorf: Droste Verlag, 1983), 78.

CHAPTER 8

Resilience Behind Bars

Animals and the Zoo Experience in Wartime London and Berlin

Mieke Roscher and Anna-Katharina Wöbse

INTRODUCTION

Even before the war had broken out, an animal was evacuated from the British capital and relocated to the countryside to ensure its safety. It was an animal with VIP status, *the* attraction and undisputed diva of the zoo—Ming, a young female panda bear. Just months later, in April 1940, Ming was sent back to the heart of the city. She had changed, as the Evening Standard reported. Sequestered in the countryside, she had become more of a bear: she had gained 50 pounds, left behind some of her playfulness, and grown a proper set of fangs, "as sharp as those of a bear and she does not hesitate to use them."[1] Quite simply, away from the turmoil in London, Ming had become a representative of her species. She appeared to be doing well. If it had only been a matter of her well-being, she would have

M. Roscher (✉)
University of Kassel, Kassel, Germany

A.-K. Wöbse
Justus-Liebig-University of Giessen, Giessen, Germany

© The Author(s) 2019
S. Laakkonen et al. (eds.), *The Resilient City in World War II*,
Palgrave Studies in World Environmental History,
https://doi.org/10.1007/978-3-030-17439-2_8

been kept in her rural exile. However, she was ordered back and her own fate married to that of the Londoners, who faced the beginning of clashes with tense anticipation. With the Blitz, the devastating bombardment by the Luftwaffe that began in 1940, the situation in London became more extreme and there was a greater need for animals who could provide relief. The relation between humans and animals thus became more intense and this brought with it a shift that had a direct effect on the inhabitants of the zoo. The public, crippled by the war, their lives under threat, expected their favorite animals would be kept close.

This chapter will, first of all, analyze the real and symbolic position of zoo animals during the war and the manner in which they were used for propaganda. Through descriptions of individual animals, the more concrete roles these creatures played in times of societal crisis will also be illuminated. The example of wartime human-animal relations, furthermore, makes it clear that the meanings ascribed to "threat," "civilization," and "wilderness" are far from static. This is not a strictly British phenomenon. On the side of the German aggressors, human-animal relations during wartime were equally invested with meaning. In Berlin, this took the form of pre-air raid evacuations to bring the "innocent" animals to safety. At the same time, both the London and Berlin zoos arrived at the decision that, as sites of urban normality, they should stay open and the animals remain visible. Zoos were therefore spaces of urban resilience that adapted to everyday life during the war. In this context, we understand resilience as an interspecies effort that took a specific intercorporal form. Our thesis is that, during this time, the boundaries between humans and animals were at least temporarily drawn anew. Faced with the human foes abroad, the zoo animals of the metropolises were seen as belonging to the national "we."

Both zoo animals and bipedal city dwellers were parts of a collective, a network of human and non-human beings, as Bruno Latour has conceived of it, that became connected at various levels and influenced one another.[2] As will be demonstrated here, animals were actors whose agency influenced human behavior. We maintain that the characteristics and the nature of the nature involved and the affected social actors are the object and consequence of reciprocal relational activity within a network.[3] Animals can be seen to function as historical actors, even without proof that they are subjects with autonomous agency. Furthermore, at various points in our investigation we will examine to what extent it is possible to approximate the experience of the animals, as well as how practicable it is to engage in a broader reconstruction of historical relations, formulating a number of desiderata.

The place and time of human-animal encounters play a decisive role in such an investigation. Already in 1827, when the London Zoo opened in Regent's Park, the display of a "wild" animal in the pseudo-naturalistic environment of the zoo, above all, aimed to demonstrate the crown's civilizing influence, satisfy the will of the people, and confer international prestige. Other imperial centers followed suit.

Taking the London zoo as a model, the zoo in Berlin was established in 1844. At this time in both cities, pets, livestock, and wild animals were omnipresent. Nevertheless, as central, public places in the middle of allegedly human cities, both zoos were invested with multilayered meanings bound up with strong emotions and bonds. As a symbolic place where state sovereignty, colonial ideologies, and national competition were played out, as well as the middle-class demand for voyeurism *and* scientific explanation satisfied, the zoo is an ideal object for investigating the many meanings of the animal that can be analyzed historically.[4]

Taking up this prestigious site in two metropolises in wartime, we would like to show how the zoo functioned as a showplace of the real collective suffering, crisis, and danger that were approaching and how these roles were ascribed discursively to the zoo.[5] In this regard, we are particularly interested in the zoo as a tool for propaganda, the manner in which animals provide relief in times of war, and the way in which the shifting of boundaries alluded to above impacted the everyday life of the animals.

The Zoo as a Place of Encounter

As war became ever more probable at the end of the 1930s, the London Zoo, founded more than a century prior, was still regarded as the most scientifically advanced zoological institution in the world. In 1936, a well-known, reform-minded biologist by the name of Julian Huxley became the director of the zoo. Dismantling fences and expanding enclosures, Huxley favored a less rigid separation between zoo animal and observer, which would also serve to improve the living conditions of the animals. After observing the monotonous wandering of the great apes in their cages, he combined a type of occupational therapy with the ideal of English conviviality and organized tea parties for chimpanzees.[6] With these, he not only put the animals' intelligence on display, but also staged a scene highlighting the proximity of the animals to their human counterparts. Huxley would keep up these social gatherings in wartime. They emphasized, first of all, the suffering shared by both human and animal and the connection between them. They were also a way to convey that animals, too, would refuse to allow the war to get them down.

While the political pages reported the growing diplomatic tension in the summer of 1939, Ming, just a cub at this time, dominated the entertainment section. She had been captured with a number of other full-grown pandas in 1938 by an American animal catcher in the mountains of Sichuan and brought to Europe. Almost as soon as they arrived in London, the animals were renamed, discursively rendering them individuals. From the "working titles" Grumpy, Dopey, and Baby, they officially became Sung, Tang, and Ming. Relabeling the animals to reflect their origins, a common practice in zoos, aimed to convey their exoticism and supposed authenticity.[7] Soon Ming made her first public experiences and became a media sensation. Celebrated as the "Shirley Temple of the animal world" and the "it girl" of the hour, the Daily Mail could proclaim: "Ming…is London Zoo's 'Glamour Girl No. 1' these days. And she deserves the title, for there is not a star in Hollywood who has hit the headlines so forcibly, or so often, in six short weeks."[8] Ming appeared to be eager for the spotlight. Tellingly, the photos appearing in the press show her outdoors, as if the trust between human and animal rendered the bars of a cage unnecessary. So great was Ming's attraction that the zoo calculated the number of daily visitors increased by 7500. The market value of the animal was quickly recognized. Soon, Ming could be found as a toy animal, as well as on postcards and brooches. There were panda hats, panda bathing suits, and panda costumes. Sales of black and white accessories boomed. According to Julian Huxley, Winston Churchill, who harbored an extraordinary interest in the animal, himself possessed a hot water bottle cover made of faux panda fur.[9] Not even the royal family was immune to the panda mania. The royal princesses were photographed by the media on several occasions as they played with Ming.

Omnipresent and accessible to all, animals served here as a way of connecting social strata. Before the animals, all Britons were equal; especially during wartime, this was emphasized again and again. Even before Ming, there had been animals at the zoo who had enjoyed a well-documented media presence, for instance, the elephant Jumbo (ca. 1860–1865) and the black bear Winnie (ca. 1913–1934), both of which became the central characters in popular children's books. However, the way in which Ming was marketed, which was highly targeted and had been planned out well in advance, was unprecedented.

During the war, Ming's position in the community of zoo animals would only become more prominent and more singular. The pandas were among the valuable large animals that were quickly whisked away when war broke out to Whipsnade, an extensive animal park 30 km north of London that served as an annex for the zoo. This action made it apparent that the management at the zoo sought to protect these highly symbolically charged animals from the bombs and that the metropolitan animals were in direct danger because of their proximity to people.[10]

After Ming's fellow pandas Sung and Tang died as a consequence of being fed improperly, it was up to Ming to keep up the morale of the Londoners during wartime. In March 1940, she came back to the capital city from her place of exile for the first time.[11] The number of visitors skyrocketed immediately, and the zoo became profitable again despite the war. Ming was now the only living panda in Europe and a symbol for the "Britain can take it" propaganda. In later representations of Ming, she is shown assuming a posture very much like Churchill's, ready to face the enemy head on—"back at the zoo" with a steel helmet in her paw. Ming's return home to her cage at the zoo, in the middle of the devastated capital city, was seen by the inhabitants as a step toward recovering urban normality (Fig. 8.1).

The media coverage of Ming was characterized by a pronounced anthropomorphism. It was suggested that Ming was undecided as to whether to return to war-stricken London to entertain and enliven the populace. The human and animal thus came to be seen as united by a common fate, as evidenced by the behavior of the animal counterpart. This example shows, on the one hand, just how discursively loaded and anthropomorphized representations of Ming were. On the other hand, it illustrates how Ming's deportation affected her everyday experience, the historical conditions of which can be reconstructed: capture, exile, being put on display, sickness, and death thus serve here as the biographical cornerstones in the historiographic presentation of non-human individuals.

Animals as Threat

The London Zoo had thoroughly prepared for the eventuality of war. According to an emergency plan, the big cats would be shot if they escaped from their enclosures during an aerial attack. Not everyone was concerned about this worst-case scenario. Winston Churchill expressed regret over the plan. "What a pity," he said to Julian Huxley, "Imagine a great air-raid

Fig. 8.1 Pandas were naturally on the Allied side. (Source: Poster of the London Zoo 1940. Reprinted from: R. and D. Morris, *Men and Pandas*, London 1966)

over this great city of ours – squadrons of enemy planes dropping their bombs on London, houses smashed into ruins, fires breaking out everywhere – corpses lying in the smoking ashes – and lions and tigers roaming the desolation in search of the corpses – and you're going to shoot them!"[12] It did not, however, come to mass shootings in London, although the poisonous snakes and spiders were killed as a precautionary measure. "Germany invades Poland" appears in the zoo's daily log from September 1, 1939, followed by a list of 35 snakes that were "destroyed owing to war conditions"—among them cobras, vipers, and rattlesnakes.[13]

In Berlin as well as in many other city zoos affected by the war, precisely this threat of distinction being compromised led to disquiet, hearsay, and precautionary measures being taken. People were evidently frightened by the idea of the animals being freed. Through their stable custodianship, zoos had conveyed a sense of the reliability and harmlessness of the animals, but this could no longer be guaranteed. The threat of this boundary breaking down brought with it an immediate end to the anthropomorphizing. In the rumors surrounding possible escape, it was evident that war had once more rendered the wildness of the animals—which had become almost unreal—a threat. At this point it became clear that, in a situation of crisis, the animals would not defer to their human counterparts. When left to themselves, rather, they would intrude into the human sphere and their undomesticated nature would finally be allowed free reign.

Animals as Fellow Sufferers

In order to protect at least some of the animals from the effects of war, an evacuation plan was developed for the pandas, chimpanzees, orangutans, and elephants. The animals departed from Regent's Park on September 1, 1939, amid a feverish bustle.[14] In contrast to the government-run evacuation of the children of London which proceeded chaotically in the first weeks of war,[15] the zoo's War Emergency Plan went smoothly and provided for the safety of the personnel's dependents as well as the animals: wives and children of the zookeepers found their way to Whipsnade along with the animals.[16] Nevertheless, the majority of the zookeepers were obligated to sign on with the armed forces or the home guard. On September 3 at 11 o'clock, "on declaration of war against Germany," the zoo shut its gates.[17] That this was the first time that the zoo had closed since being founded in 1827, and that it stayed closed for two weeks, shows how unsettled public life had become. After this time, however, the zoo would stay open for nearly the entirety of the war, with the exception of isolated days during the air raids.

The zoo thus became also the place to publicly enjoy the presence of animals. As Hilda Kean has recently discovered, hundred thousands of people, especially in the South East, had brought their pets to the animal shelters to be euthanized in the first weeks of September 1939 in fear of them being killed by German bombardment.[18] Although there was no immediate threat to do so and no government order, people saw their dogs and cats rather dead than suffering through the horrors of war. Zoo animals in this sense for many became surrogates for the normalcy and reliability transmitted through people-pet-relationships.

Faced with the prospect of the city being destroyed, the potential suffering of the animals was placed on the same level as that of the human population of London. *The Star* reported, "Many of the creatures in the London Zoo are having to take their chances in possible air raids like the majority of London."[19] Despite the bombardment, the animals, like the population in the southern part of the British Isles, were to "stay put."[20] In these life-threatening circumstances, both human and animal were treated equally, and they were both expected to contribute to the war effort. In the London Zoo's magazine, there were photographs of chimpanzees helping the zookeepers to fill sandbags, camels carrying heavy loads, and other animals endeavoring to do their best to distract and support the people of London. The London Zoological Society expressly fostered this image of loyal animals who were ready to help (Fig. 8.2).

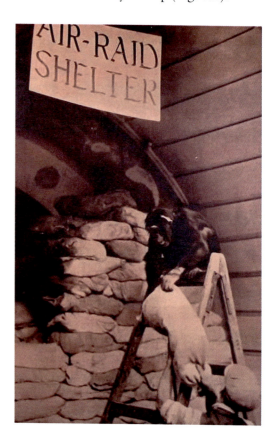

Fig. 8.2 Interspecies resilience in action. A zoo chimpanzee is helping to outfit an air raid shelter in London. (Source: London Zoological Society, Animals and Zoo Magazine, June 1941. ZSL Archive)

In order to guarantee that visitors would continue to come, an air raid shelter for 2000 people was built on the grounds. The opinion was that the zoo "should be kept open and in normal conditions if possible. Apart from its own interest in so doing, the Council felt that the Society would be performing national service by providing healthy recreation and relaxation for the public in wartime."[21] A visit to the zoo was even panned as part of the "Holidays at Home" program, which advertised it as affordable domestic entertainment.[22]

As a place for the welfare of people *and* animals, the zoo acquired a dual function and became a test area for civilizing missions and interspecies resilience. Interspecies resilience can be regarded as a part of an interspecies relationship, which points to the shared vulnerabilities of animals and humans in co-created and co-inhabitated spaces.[23] Jonathan Burt sees the zoo as a microcosmos, a reflection of the state, there to prevent social unrest to protect its "subjects" from illness.[24] Adding to this, Randy Malamud attributes far-reaching meaning to the upkeep of operations at the zoo during the war. He suggests that, in an imperialist framework, a society that manages to provide for and sustain an exotic animal can legitimate its sovereignty over its "imperial subjects."[25] The zoo thus creates a semi-reality of order and reliability, projecting the desired normality through the manner in which the animals are presented.[26] At the same time, it was declared that the animal's sense of normalcy was dependent on the presence of people. They seemed to miss their human counterparts. The chimpanzees in particular were said to be irritated by the sudden absence of visitors.[27]

The zoo here was not merely a one-sided place where humans could gaze at animals. On the contrary, it was assumed that the animals were actively involved in processes of communication. "The animals are bored," the Daily Telegraph claimed, judging the presence of humans to be indispensable.[28] The Evening Standard also chimed in: "Zoo Animals – very bored – wake to welcome every visitor."[29] The regular air raid patrols, on the other hand, were an irritating intrusion into their everyday rhythm, which, while determined by others, was at least familiar.[30] The war thus initially interfered with the everyday life of the animals not in the form of the physical experience of violence, but in the abrupt interruption of their daily routine. There were just short of half a million visitors recorded in 1941, the lowest number since 1864.[31] The animals appeared to be so accustomed to being on display that soldiers were invited to the zoo to make up for the lack of other visitors.[32] The numbers first started rising again in 1943, when the war-fatigued Londoners rediscovered the zoo as a recreational space.

The Zoo Animals' Reality During the War

The animals in the zoo reacted to the bombing in very different ways. They now found themselves in a situation that exacerbated the effects of their confinement. A considerable amount can be deduced from behavioral observations during the war, which shed light on both human and animal behavior.[33] During the air raid alarms, which began in London on June 25, 1940, the animals were driven outside at night, out of their shelters. The wolves howled in unison with the sirens.[34] The apes, on the other hand, were indifferent.[35] The rodents seemed to like the darkness caused by the blackouts.[36]

In general, the animals reacted with either apathy, panic, or aggression, and "among mammals at least, there is considerable evidence to support the claims of those naturalists who regard animals as capable of having definite and distinctive 'personalities'."[37] The first explosions were recorded by the zoo on September 17, 1940. They elicited fear and terror in many animals. A young giraffe named Boxer "died of a dilated heart due to over-exertion caused by fright." Several antelope incurred fatal injuries as they panicked. The zebras, however, used the opportunity presented by a bomb hit to first hide in the basement, then leave their enclosure and bolt. A couple of birds "escaped and have not been recaptured."[38]

Other restrictions affected the way in which the animals were fed. They also had to contend with food rationing, which had been in place since October 1939. Bananas were replaced by carrots. Horse meat coated in cod liver oil served as a substitute for fish. The penguins, for their part, refused to eat until more adequate provisions were provided.[39] With a food shortage looming, members of the Zoological Society declared themselves ready to "adopt" the animals to prevent them from dying from starvation. In this adoption program, donors provided for the upkeep of the animals. In return, these noble gestures were publicly acknowledged with signs attached to the cages and enclosures. By the end of 1939, 180 animals had been provided for in this fashion.[40] The project was a success.

After the war, the Society's Post-war Development Fund was not only put toward securing the food supply for the animals, but also stocking up once again on the animals themselves, whose numbers had shrunk substantially during the war. The animals were not so much immediately endangered by the bombing as by the general circumstances of the war—above all precautionary measures put in place for the security of the populace and the food supply. From 1940 to 1944, the number of animals sunk from 2224 to 1252.[41]

As already mentioned, a number of animals were killed at the beginning of the war for safety reasons.[42] The entry "killed by order" was not uncommon in the Daily Occurrences. The majority of the other deaths, however, were more or less of natural causes. Moreover, the mating of animals was not pursued as aggressively and animals suffering from old age were not replaced. Some animals, among them elephants and lions, were killed toward the end of the war to spare them the agony of death from starvation.[43] Again and again, animals also went missing: "stock missing" was used as the euphemistic label for animals that had found a way to escape their cages. With human and animal sworn to a common fate, it was out of the question to speak of escape, which would have been tantamount to deserting. Some of the less exotic animals fell victim to the food shortage, as well; they themselves became food. The goats and sheep of Regent's Park were sold by the zoo to the Ministry of Food.[44] In Whipsnade, too, zoo animals landed on the menu: first pigs, sheep, and cattle, then deer, and later the occasional buffalo or bear.[45] That said, there were some new arrivals. Private collectors, primarily of exotic animals, wanted their charges to be housed well during the war. Many regiments also brought their mascots to the zoo.[46] There was a long tradition of mascotry in the British army that could be traced to the nineteenth century; mascots ranged from dogs and cats to chickens, pigs, goats, bulls, and even antelopes and lions. The animals were thought to keep up the morale of the troops.[47]

One animal that could no longer sustain the role of keeping up morale was the big star of the zoo: the panda bear Ming. Already in the summer of 1943, her health had begun to deteriorate. She appeared increasingly "dispirited and unwilling to entertain the public." After her death in December 1944, the Londoners nevertheless paid tribute to her contribution to wartime relief efforts. In an obituary in The Times, this meant mourning the death of a living being that had done so much to inspire the imagination of the nation, even if times were desperate: "She could die happy in the knowledge that she had gladdened the universal heart."[48]

THE BERLIN ZOO AND ITS GREATER GERMAN EXPANSION

As has already been demonstrated, zoo animals, far from becoming more marginal during the war, assumed a very important role in metropolitan life. By the same token, the war transformed the life of the animals. How did the human-animal relation and animal-human relation manifest in the German capital? There were unmistakable parallels between London and

Berlin in the way the zoo was stylized as a political place for propaganda. While in London the zoo was presented as an imperial microcosm, Berlin offered a zoological model of the greater German Reich. Over the course of the 1930s, the Berlin zoo had shown itself to be a place that supported and in turn was supported by the Nazi regime and ideology. In 1935, the zoo's director, Lutz Heck—a Nazi-loyalist who was dabbling in a Germanic breeding program that would bring the extinct aurochs back to life— realized a "long-cherished plan" to build a "German zoo" on the premises.[49] Among other creatures, wolves, beavers, brown bears, and eagles were considered "greater German" animals.

Nazi potentate Hermann Göring was enthusiastic about Heck's project, which aimed to reestablish extinct "German" animals[50] and reproduce the wisent (the European bison) and the aurochs, considered "symbols of primal force."[51] In grand style, Heck and Göring planned to create a seemingly untouched landscape with wild megaherbivores roaming.[52] The zoo served not only as a breeding ground for the corresponding animals and theories, but also as a showplace for the related racial science: "Understanding the heterogeneity of animals will allow them [the human visitor] to understand the heterogeneity of human races."[53]

To edify the public, bands from the SA (*Sturmabteilung*), SS, and Hitler's *Leibstandarte* (personal body guards) played regularly amid the enclosures. The happenings of the war initially seemed far away and were of little importance for the everyday functioning of the zoo. In contrast to London, where the evacuations had begun right away, the Berliners took their time. While the zoo animals in London were placed at higher risk because of their proximity to humans and insufficient means of escape, in Berlin the animals were thought to reduce the risk for humans: the zoo was unlikely to be a target for bombers.

In January 1940, a directive was finally issued ensuring that no animals could escape during an air raid alarm. It guaranteed that enough personnel with "low and high-caliber rifles" would be on site to recapture or shoot any dangerous animals that had broken out.[54] Guidelines for the evacuation of the zoo specified that, for the duration of the war, the zoo was "to keep as few large predators and other animals that would unsettle the public in the event of an air raid as possible." As far as the evacuation of the animals was concerned, a hierarchy was established based on the importance and value of the creatures, just as in London. The first of those to be shipped away would be "species that were close to extinction," which meant animals accorded national significance such as the "wisents, wild horses, Alpine ibex," but also the "mountain zebra and other rare species."[55] Official orders, once again, reflected the desire to maintain the zoo as an urban fixture in tumultuous times of war; it could not itself be the source of any threat. Particularly in times of uncertainty, the wild animals with sharp fangs and large paws had to be kept in check, under lock and key.

Like on the other side of the channel the public display of animals served also to supplement relations people would usually have with their pets. With order from the Wehrmacht's high command dated September 19, 1939, all dogs were to be mustered for war purposes.[56] They, just as horses, were routinely declared comrades in the fight for a Great German Reich.[57] Being also declared part of the *Volksgemeinschaft* (folk community) made them a national asset over which its owner could no longer dispose of as he or she liked. Zoo animals seemed therefore to be the more reliable companions.

Animals in Transit

At last, the changes in the Berlin Zoo began to resemble those in London. The younger keepers were drafted by the *Wehrmacht* and food supplies rationed. These new developments were initially easy to conceal. When the artificial cliffs in the goat enclosure were rebuilt, for instance, an air raid shelter was created for 2000 people. Utmost discretion was exercised when the number of animals had to be reduced due to food scarcity. "Older specimens and those that could not be bred" were gotten rid of so that "no changes were evident in the outward appearance of the zoo."[58] Dietary changes and staff turnover had the largest impact on the everyday life of the animals.[59]

The composition of animals was subject to constant change. Newly captured animals were always to be found in the cages of the capital. Göring, for instance, presented the zoo with a "Spanish wolf brought by the troops stationed in Spain."[60] In 1941, the *Wehrmacht* donated "2 Russian wolves from the area around Minsk, a young bear from Lapland, a gazelle from northern Africa, and a miniature donkey."[61] The zoo regularly received living presents from the *Führer*, such as Chinese pheasants, a lion, and a golden eagle. Berlin provided the units of the *Wehrmacht* with tame offspring to serve as mascots.[62] A menagerie, originating in the capital's zoo, advanced through Europe along with the German troops. In 1940 alone, 70 animals were delivered to the army, including monkeys and apes, dingoes, pygmy goats, and parrots. A raccoon and a coati were given to a destroyer and a minesweeper.[63]

The aggressive expansion of Germany and the waging of aerial warfare set in motion a veritable carousel of animals being shipped out. Each voyage entailed a great deal of stress for the animals and the complete disruption of routines and family ties. In 1940, the female elephant Topsi from the Cologne Zoo, for instance, was briefly quartered in Berlin, then in Vienna, then Munich, and was finally evacuated to Wroclaw, where she died at the end of the war.[64] Animals were forced to take to the road not only because of evacuation, however, but also when they were seized from zoos in annexed regions. Eastern European zoos were especially devastated by this type of plundering. After the invasion of Poland, Lutz Heck helped himself to the stock. Already in 1939, a number of animals from the Warsaw Zoo were "entrusted"[65] to Berlin.

By this time, the Warsaw Zoo had already experienced shifts in animal-human relations due to the war. Directly following the invasion, the zoo became a battlefield; it was too late to evacuate the animals. As the American historian Diane Ackermann reports, cages were destroyed by a German aerial attack. The panicked animals that had escaped—polar bears, big cats, and a bull elephant—were shot by Polish soldiers.[66] While Warsaw burned, the surviving animals fled to the old town.[67] Despite its partial destruction, the zoo remained an important place for local identity formation. The people of Warsaw brought food to the animals to ensure that it could continue to operate.[68] On New Year's Eve in 1939, on the other hand, Lutz Heck organized a hunt on the grounds for the amusement of his friends in the SS, at which a number of animals were shot indiscriminately.[69] "Nazi zoophilia"[70] was thus in no way boundless. With the exception of the stock Heck appropriated for his "Germanic breeding program" in Berlin, the animals of the Warsaw Zoo were clearly worth less than those found in German zoos.[71]

Animals as Threat and Fellow Sufferers

Despite isolated bombing, major evacuations did not begin in Berlin until 1943. Before this, some animals had been shipped to the Vienna zoo. The "most Germanic" animals, the wisents and the aurochs, were also brought to Schorfheide, where Göring had his hunting lodge, along with four Przewalski's horses native to the steppes of Central Asia.[72]

After the multiday bombing at the end of November 1943, the Berlin Zoo was strewn with cadavers, as the zoologist Katharina Heinroth describes in her autobiography:

> The view was particularly ghastly: Seven dead elephants were buried by rubble, a rhinoceros next to them appeared to be unscathed, but had succumbed to a pneumothorax as a result of the enormous build-up of air pressure. In the antelope house, eighteen animals, including two marvelous giraffes, had died; in the monkey house, two great apes and fifteen smaller monkeys also lay dead. Of the two thousand animals remaining at the Berlin Zoo after the evacuations, a third were killed on this night; almost seven hundred animals died. The great madness of this war became apparent to us.[73]

Troops of animal catchers were dispatched to bring the escaped inmates back to the devastated zoo.[74] The poisonous snakes had quickly succumbed to the November cold.[75] Again and again, there were rumors that the exhibition pieces had acquired life-threatening agency, as was the case when the crocodile habitat was hit:

> As usual, we heard our great crocodiles calling out in unison during the thunder of the anti-aircraft guns, which they took for roars from their brethren. This was the last time they called out…two or three animals were still alive and were whipping the ground with their powerful tails. One of them had slithered over to the entrance of the aquarium, but none of them had escaped to Budapester Straße, as later rumors would have it.[76]

In any case, the employees of the zoo did not go hungry during this time. A "very tasty crocodile tail soup was served to the entire staff."[77]

The radical disruption of and intrusion into the everyday life of the animals was perhaps never so concretely evident and historically palpable as in the destruction wrought in those days. The surviving animals were profoundly disturbed and shell-shocked, as the eye-witness Bernhard Grzimek, later a popular zoologist in the German Federal Republic, reported: "A giraffe stood stock-still next to its dead companion. Her warm greenhouse had been turned into a smoldering pile of rubble." A tapir sat next to a huge pile of coke behind the cinema near the zoo's grounds and "warmed its black and white backside."[78] With less to come between them, animals and humans now occupied even closer quarters. Animals that had been "bombed out" were housed in shelters; the men's restrooms at the entrance to the train station were occupied, for instance, by the pygmy hippos. Faced with the cold at the end of November, it was obvious that animals from more temperate climes would not survive long. Animals enthusiasts in Berlin signed up to take charge of the ponies, zebras, parrots, and other animals.[79] Human refugees, on the other hand, were temporarily given shelter in animal houses that had withstood the bombing,[80] where they lived shoulder to shoulder with their animal neighbors (Fig. 8.3).

The restoration of the zoo was a political priority. Reichsminister Speer commissioned a work detail of 750 prisoners of war for the clean-up. A group of veterinarians spent a week dismembering the cadavers of elephants and other large animals, which were then taken to be rendered into

Fig. 8.3 At the end of the war, the Berlin Zoo also became a battlefield. A monumental flak tower built in the vicinity of the zoo, was surrounded by burned trees, stables and destroyed tanks in May 1945. (Source: Wikipedia Commons, Berlińska Victoria, E. Kmiecik, "Ruch" 1972)

meat and bone meal and soap. Despite the increasingly dire circumstances, the zoo held such sway that it not only had the means to procure heating and building materials, but it was also able to evacuate the animals.[81] Against the number of animals dwindling, Heck sought to emphasize that the zoo had not lost any of its importance for the regime. Shortly thereafter, in January 1944, bombs rained down on the zoo once again. Two warthogs and a "strong German wild boar" were among the escapees, attacking anyone who attempted to stop them. They were shot down.

During the final days of the war, the zoo itself was turned into a battlefield, "with bomb craters [...] and crisscrossed by trenches," "human and animal corpses lying between branches that had been shot down."[82] In the end, the battles had claimed human and animal casualties alike; it was a hopeless situation for both the inhabitants of the city, surrounded by the Red Army, and the animals in their cages. Under the supervision of Russian soldiers, humans and animals were buried together on the grounds of the zoo. Of the vast stock of animals the Berlin Zoo had once possessed, only a few survived. The bull elephant Siam, often described as "malicious," was born in 1923 in the wild. After being captured, he toured with the Circus Krone. He ended up the only survivor in a cage in Berlin, with seven members of his herd buried under the rubble. His temperament saved him in the end; because he would not let any keeper near him, he was placed by himself in a corner stall without any shackles.[83]

The Zoo as a Discursive Place

In both metropolises, the zoos were a site at which concepts of patriotism were played out. Both the place and the animals were ascribed discursive meaning. Animals were "vehicles for communicating the military aesthetic."[84] In London, the zoo served as a forum in which the British could position themselves against national socialist Germany. For instance, in a column entitled "The Enemy's Animals," the popular magazine of the London Zoological Society reported on how the Nazi-regime confiscated or killed dogs of a certain size and excluded the rest from receiving rations.[85] The magazine was also incensed by the "holocaust among animals, birds, reptiles and fish in captivity."[86] The zoos in Berlin, Hamburg, and Dresden were all forced to kill their big cats. The community of victims was thus expanded to include not only the British but also the German animals: "While Germans starve, Goering stays fat. While German zoo animals are slaughtered, his lion cubs are spared."[87]

Nevertheless, some animals were excluded from this purportedly universal love of animals[88] and were fashioned as evil per se: one piece, titled "Every House Fly is a Nazi," cast the potential carriers of disease as "faithful allies of the enemy." Depending on the level of threat, certain animals were considered perpetrators, while others were seen as victims. Sometimes the distinction was only nominal. In order to avoid any allusion to the axis powers, the axis deer (chital) was swiftly rebranded the spotted deer.[89]

Zoo animals functioned as placeholder for ideological debate on the other side of the Channel, as well. Heck's back-breeding program was peppered with elements of national socialist racial ideology. The "stalwart representatives of the German *Urwild*"[90] such as the aurochs were held up as symbols of "German power and German courage."[91] Through the concepts of primordial nature and the fauna living within it, humans could be recognized as the pinnacle of evolution, "and, in their formation as a people [*Volk*] more specifically, an ancestral entity in which the individual becomes exultant and finds meaning."[92] The way in which certain political discourses operated became obvious in the treatment of prominent animals such as Ming. Again and again, the animals were also instrumentalized visually, cast as allies as well as fearless icons of hope and British opposition in the face of German hostility.

The symbolic treatment of Pongo, Europe's largest gorilla and the darling of the Berlin Zoo, was equally drastic. The accounts of the final days of the war testify to the diversity of ways in which the animals were interpreted and appropriated. Heck maintained that Pongo—who had survived the bombings in the zookeepers' apartment—was killed "through enemy action" during the "occupation of Berlin."[93] Read and Fisher suggest, however, that he was in fact killed by the Germans. A number of SS soldiers were found dead in front of his cage and may have committed the bloody deed rather than let him fall into enemy hands.[94] At Göring's command, the evacuated wisents, which the national socialist leaders had considered so important, were shot.[95] They were perhaps too German, too symbolically charged, to continue to exist as mere cattle.

Conclusions

The extreme circumstances of the war thus led to shifts in the manner in which animals were read by their human counterparts on both sides of the English Channel: animals were now no longer merely the "other," but had become a part of the resilient community, endangered by a common enemy. This shifting of boundaries took place on two levels: on the one hand, the zoo animal was discursively co-opted for certain ends. On the other hand, owing to threats from abroad, humans and animals were forced into a new intimacy. Zoo animals acquired new roles as working animals, carriers of hope, and food, which they also sometimes refused with their obstinance, running away, and biting. This recalls Latour's notion that the meeting of humans and animals, the building of a collective, to a certain extent always involves a negotiation.[96] This collective was exposed to external forces that required human and animal actors engage in multilateral negotiations, "during which the identity of the actors was determined and put to the test."[97] It was precisely this process that influenced the history of the relationship between these actors, which was informed by the dynamics of the war. From the perspective of practice theory, which takes into consideration the non-communicative elements of a relationship,[98] animals can also be recognized as part of non-discursive or pre-discursive processes of negotiation.

In wartime, the way in which this relationship was woven was temporarily redefined, at least from the human perspective. To protect the captured, and therefore dependent, animals was considered a patriotic and civilizing act in barbaric times. Zoos were meant to be places of humanity. This was made evident, for instance, in the Berliners' assumption that the zoo was safe from aerial attack because it was inhabited by "innocents." After central Berlin was entirely destroyed, people who had been bombed out showed up to help care for "their" animals as a way to reassure themselves of their own dignity.[99] The human relationship to the exotic zoo animals became a yardstick for normality in a situation that was very much out of the ordinary.[100] This is to some degree also true for the immediate post-war situation in which both countries and both cities were still struggling to find their way back to normalcy. Indeed, the Berlin zoo reopened its doors as soon as July 1, 1945, to ensure that the strong bind, the interspecies resilience, that had developed between the zoo and the city would not tear.[101]

As the war claimed more and more victims and humans as well as animals suffered under the non-stop air raids and increasingly severe food shortages, new intersections emerged: zoo animals and city dwellers were both prisoners of the circumstances imposed by the war, which put their assigned roles to the test. The new proximity between humans and nonhumans made the limits of tolerance clear; these limits were overstepped when animals did not comply with human rules, when they refused the roles that were assigned to them. The violence of the war allowed the animals a certain amount of leeway to be just that—animals—but this also meant they came to pose a physical threat to the inhabitants of the city. Ultimately, these concrete fears, which fully supplanted the tendency to anthropomorphize the animals, illustrate the impact the creatures could have on the everyday life of people. It is apparent that there is room in the discipline of animal history for new avenues of thought and interdisciplinary collaboration. With the perspective offered by behavioral biology and phenomenology, for instance, changes in the animals' way of being can be approached. However, concrete and fruitful insight into the human-animal relationship, the animal-human relationship, and how these relationships change over time, can only be gained through biographical and historical contextualization.

NOTES

1. *The Evening Standard*, 29 January 1940.
2. See Bruno Latour, *We Have Never Been Modern* (Cambridge: Harvard University Press, 1993); Bruno Latour, *Reassembling the social: An introduction to actor-network-theory* (Oxford: Oxford University Press, 2005).
3. See Ingo Schulz-Schaeffer, Akteur-Netzwerk-Theorie: Zur Koevolution von Gesellschaft, Natur und Technik, in Johannes Weyer, ed., *Soziale Netzwerke. Konzepte und Methoden der Sozialwissenschaftlichen Netzwerkforschung* (München: Oldenbourg, 2000), 188.
4. Garry Marvin and Bob Mullan, *Zoo Culture: The Book about Watching People Watch Animals* (Urbana: Illinois University Press, 1999). See also: Nigel Rothfels, *Savages and Beasts: The Birth of the Modern Zoo* (Baltimore: Johns Hopkins University Press, 2002); Stephen O'Harrow, "Babar and the Mission Civilisatrice: Colonialism and the Biography of a Mythical Elephant," *Biography* 22, no. 1 (Winter 1999): 86–103; Robert W. Jones, "'The Sight of Creatures Strange to Our Clime': London Zoo and the Consumption of the Exotic," *Journal of Victorian Culture* 2, no. 1 (March 1997): 1–26. Some other historical zoo studies include Matthew Chrulew, "From Zoo to Zoöpolis: Effectively Enacting Eden," in Ralph A. Acampora, ed., *Metamorphoses of the Zoo: Animal after Noah* (Lexington: Lexington Books, 2010). Eric Baratay and Elisabeth Hardouin-Fugier, *Zoo: A history of zoological gardens in the West* (London: Reaktion books, 2003). Christina Katharina May, "Geschichte des Zoos," in Roland Borgards, ed., *Tiere. Kulturwissenschaftliches Handbuch* (Stuttgart: J.B. Metzler Verlag, 2016).

5. For other zoos during wartime see Mayumi Itoh, *Japanese Wartime Zoo Policy: The Silent Victims of World War II* (New York: Palgrave, 2010); Frederick S. Litten, "Starving the Elephants: The Slaughter of Animals in Wartime Tokyo's Ueno Zoo," *The Asia-Pacific Journal* 38, no. 3 (September 2009): 1–18; Elia Etkin, "The ingathering of (non-human) exiles: The creation of the Tel Aviv Zoological Garden animal collection, 1938–1948," *Journal of Israeli History* 35, no.1 (February 2016): 57–74. For a recent take on the interwar years at the London zoo see Jonathan Saha, "Murder at London Zoo: Late colonial sympathy in interwar Britain," *The American Historical Review* 121, no. 5 (December 2016): 1468–1491.
6. Julian Huxley, *Memories* (New York: Harper & Row, 1970), 232.
7. Marvin and Mullan, *Zoo Culture*, xix.
8. *Daily Mail*, 22 February 1939.
9. Huxley, *Memories*, 247.
10. Vuorisalo and Kozlov, this volume.
11. Approximately one month after the beginning of the *Blitz*, Ming was returned to Whipsnade for fear of her physical integrity. She was however to return to London on a regular basis. See "Hopeful future for the zoo," *The Times*, 4 May 1940. Daily occurrences at the gardens, 17 October 1940, *Zoological Society of London Archive* (henceforth: ZSLArch), Vol. 1940.
12. Huxley, *Memories*, 248.
13. Daily occurrences at the gardens, 1 September 1939, ZSLArch, Vol. 1939.
14. Daily occurrences at the gardens, 1 September 1939, ZSLArch, Vol. 1939.
15. Angus Calder, *The People's War: Britain 1939–1945* (London: Pantheon Books, 1969), 35.
16. *Zoological Society Minutes of Council*, Vol. XXXI 1939, Meeting 20 September 1939, 352.
17. Daily occurrences at the gardens, 3 September 1939, ZSLArch, Vol. 1939.
18. Hilda Kean, *The Great Cat and Dog Massacre: The Real Story of World War Two's Unknown Tragedy* (Chicago: University of Chicago Press, 2017).
19. *Star*, 5 January 1940.
20. "Stay put," which aimed to discourage citizens from fleeing in a panic, was a slogan appearing on pamphlets and posters distributed in the coastal regions of southern England. Calder, *People's War*, 129.
21. *Reports of the council and auditors of the Zoological Society of London for the Year 1939* (1940), 5.

22. *Reports of the council and auditors of the ZSL for the Year 1943* (1944), 1. Regarding the concept of "Holidays at Home" see Calder, *People's War*, 366.
23. Dominik Ohrem, "An Address from Elsewhere: Vulnerability, Relationality, and Conceptions of Creaturely Embodiment," in Dominik Ohrem and Roland Bartosch, eds., *Beyond the Human-Animal Divide* (Basingstoke: Palgrave, 2017), 44; Joanna Latimer, "Being Alongside: Rethinking Relations amongst Different Kinds," *Theory, Culture & Society* 30, no. 7–8 (December 2013): 1–28.
24. Jonathan Burt, "Violent health and the moving image: The London Zoo and Monkey Hill," in Mary Henninger-Voss, ed., *Animals in Human History: The Mirror of Nature and Culture* (Rochester: University of Rochester Press, 2002), 262.
25. Randy Malamud, *Reading Zoos* (New York: New York University Press, 1998), 67.
26. Utz Anhalt, *Tiere und Menschen als Exoten. Die Exotisierung des „Anderen' in der Gründungs- und Entwicklungsphase der Zoos* (Saarbrücken: VDM Publishing, 2008), 3.
27. John Barrington-Johnson, *The Zoo: The Story of London Zoo* (London: Robert Hale, 2005), 121.
28. *Daily Express*, 10 November 1939.
29. *Evening Standard*, 16 November 1939.
30. "A rival to Ming," *The Times*, 4 November 1939.
31. Barrington-Johnson, *The Zoo*, 127.
32. Ibid., p. 121.
33. See Howard Barraclough Fell, "Animal Behaviour during Air-raids," *Science* 93, no. 2403 (1942): 62; Anon, "Animals and Air Raids," *Science, New Series* 92, no. 2394 (1940): 446–447; R. D. Gillespie, "War Neuroses After Psychological Trauma," *The British Medical Journal* 1, no. 4401 (May 1945): 653–656.
34. Lucy Pendar, *Whipsnade Wild Animal Park: 'My Africa'* (Dunstable, Beds: Book Castle, 1991), 62.
35. Fell, *Animal Behaviour*, 62.
36. "Rodents are happy in the gloom," *The Evening News*, 13 November 1939.
37. Fell, *Animal Behaviour*, 62.
38. *Reports of the council and auditors of the ZSL for the Year 1940* (1941), 8.
39. *Birmingham Post*, 27 October 1939.
40. *Reports of the council and auditors of the ZSL for the Year 1939* (1940), 7.
41. *Reports of the council and auditors of the ZSL for the Year 1940–1944* (1941–1945).
42. Daily occurrences at the gardens, 26 and 28 September, ZSLArch, Vol. 1939.

43. Pendar, *Whipsnade*, 69.
44. Daily occurrences at the gardens, 14 February 1941, *ZSLArch*, Vol. 1941; Daily occurrences at the gardens, 6 November 1944, *ZSLArch*, Vol. 1944.
45. Pendar, Whipsnade, 68.
46. Daily occurrences at the gardens, 28 July1944, *ZSLArch*, Vol. 1944.
47. Jilly Cooper, *Animals in War* (London: Corgi, 2000), 174, 181.
48. "Ming," *The Times*, 30 December 1944.
49. Translated from Kai Artinger, Lutz Heck, in *Mitteilungen des Vereins für Berliner Geschichte* e.V., 1994. https://www.diegeschichteberlins.de/geschichteberlins/persoenlichkeiten/persoenlichkeitenhn/491-heck.html, visited 25 January 2018.
50. Translated from Lutz Heck, *Auf Tiersuche in weiter Welt* (Berlin: Parey, 1941), 201.
51. Ibid., 221. On the history of the European breeding project see Raf De Bont, "Extinct in the Wild: Finding a Place for the European Bison, 1919–1952," in Raf De Bont and Jens Lachmund, eds., *Spatializing the History of Ecology: Sites, Journeys, Mappings* (New York: Routledge, 2017), 165–184.
52. For a critical historical account on travelling Heck Cattles, rewilding programs, and contemporary approaches toward multiple spatialities of wildness see Jamie Lorimer and Clemens Driessen, "From 'Nazi Cows' to Cosmopolitan 'Ecological Engineers': Specifying Rewilding Through a History of Heck Cattle," *Annals of the American Association of Geographers* 10, no. 3 (May 2016): 631–652; Henrique M. Pereira and Laetitia N. Navarro, eds., *Rewilding European Landscapes* (Heidelberg: Springer, 2017).
53. Heck, *Tiersuche*, 289.
54. Luftschutzmaßnahmen in Zoologischen Gärten (Air raid measures in zoos), Runderlass vom 29.01.1940, National Archive Berlin (henceforth *BArch Berlin*), R 2/4749, Sheet 6-7.
55. *Richtlinien für die Durchführung der Räumung zoologischer Gärten* (Guidelines for the evacuations of zoos), 10.04.1940, *BArch Berlin*, R 2/4749, Sheet 9-10.
56. *Runderlass: Erfassung von Hunden für Kriegsverwendung bei Wehrmacht und Polizei*, (circular decree on the registration of dogs for war purposes in the service of police and Wehrmacht) State Archive of Hessen Marburg (HStaM), Frankenberg, No. 2652.
57. Mieke Roscher, "New Political History and the Writing of Animal Lives," in Hilda Kean and Philipp Howell eds., *The Routledge Handbook of Animal-Human History* (London: Routledge, 2018), Chapter 3.

58. *Geschäftsbericht für das Jahr 1939 des Aktienvereins des Zoologischen Gartens Berlin* (Annual report of the stock corporation of the Berlin Zoological Garden (AZGB) for the year 1939), 7.5.1940, State Archive Berlin (henceforth: *LAr Berlin*), C Rep. 105, Nr. 4675, 6.
59. Brief Lutz Hecks an den Oberbürgermeister Berlins (letter from Lutz Heck to the Mayor of Berlin), 07.02.1941, LAr Berlin A Rep. 032-08, Nr. 296.
60. Annual report 1939 of the AZGB, 07.05.1940, *LAr Berlin*, C Rep. 105, Nr. 4675, 6.
61. Annual report 1941 of the AZGB, 15.05.1942, *LAr Berlin*, C Rep. 105, Nr. 4675, 7.
62. Annual report 1939 of the AZGB, 07.05.1940, *LAr Berlin*, C Rep. 105, Nr. 4675, 7.
63. Annual report 1940 of the AZGB, 9.5.1941, *LAr Berlin*, C Rep. 105, Nr. 4675, 7.
64. Bernhard Blaszkiewitz, *Elefanten in Berlin* (Berlin: Lehmanns, 2008), 14.
65. Annual report 1939 of the AZGB, 7.5.1940, *LAr Berlin*, C Rep. 105, Nr. 4675, 7.
66. Diane Ackermann, *The Zookeeper's Wife: A War Story* (New York: W. W. Norton & Company, 2007), 53.
67. Ibid., 62.
68. Ibid., 74.
69. Ibid., 95.
70. Ibid., 86.
71. Annual report 1939 of the AZGB, 07.05.1940, *LAr Berlin*, C Rep. 105, Nr. 4675, 7.
72. There is conflicting information about the exact dates of these shippings. See Heinz-Georg Klös, *Von der Menagerie zum Tierparadies, 125 Jahre Zoo Berlin* (Berlin: Haude und Spener, 1969), 116–117.
73. Translated from Katharina Heinroth, *Mit Faltern begann's, Mein Leben mit Tieren in Breslau, München und Berlin* (München: Kindler, 1979), 131.
74. Heck to the supervisory board of the zoo, cf. Klös, *Tierparadies*, 119.
75. Bernhard Grzimek, *Auf den Mensch gekommen. Erfahrungen mit Leuten* (Stuttgart: Bertelsmann, 1974), 188.
76. Translated from Heinroth, *Mit Faltern begann's*, 131–132.
77. Heck to the supervisory board of the zoo, cf. Klös, *Tierparadies*, 122.
78. Translated from Grizmek, *Auf den Mensch gekommen*, 188.
79. Lutz Heck, *Tiere – mein Abenteuer* (Wien: Ullstein, 1952), 118.
80. Ibid., 113.
81. At least this is how Klös cites Heck, See Klös, *Tierparadies*, 122.
82. Cited from: Katharina Heinroth, in: Klös, *Tierparadies*, 129 f.

83. Blaszkiewitz, *Elefanten*, 14.
84. Ramon Reichert, "Die Medialisierung des Tieres als Protagonist des Krieges," in Rainer Pöppinhege, ed., *Tiere im Krieg von der Antike bis zur Gegenwart* (Paderborn: Schönigh, 2009), 269.
85. Carl Ollson, "How Germany Conscripts Her Dog Population," *Animals and Zoo Magazine*, May 1940., n.p.
86. Carl Ollson, "Fate of Continental Zoos," *Animals and Zoo Magazine*, November 1939, n.p.
87. Ibid.
88. Mieke Roscher, *Ein Königreich für Tiere. Die Geschichte der britischen Tierrechtsbewegung* (Marburg: Tectum, 2009).
89. Pendar, *Whipsnade*, 64.
90. Urwild refers here to the large animals initially populating the so-called Germanic lands.
91. Translated from Heck, *Tiersuche*, 195.
92. Ibid., 285.
93. Heck, *Tiere*, 94.
94. Anthony Read and David Fisher, *Der Fall von Berlin* (Berlin: Aufbau Verlag, 1995), 463.
95. Frank Uekötter, *The Green and The Brown: A History of Conservation in Nazi Germany* (Cambridge: Cambridge University Press, 2006), 107.
96. Latour, *We have never been modern*, 109. See also: Pascal Eitler and Maren Möhring, "Eine Tiergeschichte der Moderne. Theoretische Perspektiven," *Traverse* 15, no. 3 (2008): 95.
97. Schulz-Schaeffer, *Akteur-Netzwerk-Theorie*, 195.
98. On this topic see Clemens Wischermann, "Der Ort des Tieres in einer städtischen Gesellschaft," *Informationen zur modernen Stadtgeschichte* 2 (2009): 11.
99. Heck, *Tiere*, 118.
100. Of the 300 animals of the private zoo in Aleppo, Syria only 13 survived thanks to the efforts of the military of both parties. "See How Syrian Zoo Animals Escaped a War-Ravaged City," *The National Geographic*, 7 October 2017, https://www.youtube.com/watch?v=rgww07Aylw8
101. Mieke Roscher, "Curating the Body Politic: The spatiality of the Zoo and the symbolic construction of German nationhood (Berlin 1933–1961)," in Jacob Bull, Tora Holmberg and Cecilia Åsberg, eds., *Animals and Place: Lively Cartographies of Human-Animal Relations* (London: Routledge, 2018), 120.

CHAPTER 9

Where Have All the Pigeons Gone

Wildlife in Wartime Cities

Timo Vuorisalo and Mikhail V. Kozlov

INTRODUCTION

For a long time, naturalists tended to explore pristine habitats, and only rarely documented the occurrence of wildlife in human settlements. Ironically, World War II itself was one of the factors that facilitated the development of urban ecology. First, bombed sites attracted attention of botanists, because ruins offered warmer and drier conditions than habitats of undamaged cities. As the result, many plants that were previously rare became permanent members of the urban flora in war-damaged European cities.[1] Second, the division of Berlin in the aftermath of World War II and the erection of the Wall in 1961 isolated ecologists living and working in West Berlin from the surrounding countryside. As a result, many of them focused their fieldwork on sites located in West Berlin and, thereby, significantly contributed to the understanding of ecology of urban habitats. In 1970s, West Berlin became one of the centers of the development of urban ecology.[2] World War II and the Cold War thus indirectly contributed to the origin of modern urban ecology.

T. Vuorisalo (✉) • M. V. Kozlov
Department of Biology, University of Turku, Turku, Finland

© The Author(s) 2019
S. Laakkonen et al. (eds.), *The Resilient City in World War II*,
Palgrave Studies in World Environmental History,
https://doi.org/10.1007/978-3-030-17439-2_9

Only in the early 1970s, urban ecology arose as a scientific discipline, and urban ecosystems became legitimate subjects of ecological study.[3] At the time of World War II, there was no scientific tradition to study urban wildlife, and only a handful of professional biologists clearly understood the importance of documenting the impact of war on plants and animals. Therefore the scientific data on wildlife in wartime cities are scarce.

A central concept in our chapter, and in urban ecology in general, is resilience. This concept was originally introduced to ecological literature by the Canadian ecologist Crawford Holling (b. 1930), who was mainly interested in aquatic ecosystems' ability to recover after man-made changes such as eutrophication.[4] Although there is no consensus on the exact definition, resilience broadly refers to the ability of ecosystems to return to a predisturbance state after human-mediated disturbances, either with or without human help.[5] Quite obviously both warfare and peacetime urban constructions constantly disturb natural ecosystems and even threaten their existence. During World War II, a common observation however was that after urban bombings some forms of plant and animal life quickly recovered. It depends on the viewpoint how we interpret this apparent resilience of wildlife during war. One might argue that the undamaged built-up area was by itself the primary disturbance, and bombings only recovered or released the natural processes that had lain hidden underneath urban structures. Alternatively and more conventionally, we could consider the bombings as the main disturbance, and the prewar urban condition as the original state.

Both of these views make sense from the urban ecological perspective. The main difference between them is that the former does not require human intervention while the latter inevitably does: a built-up area will not be restored to its original state without laborious postwar reconstruction. In this chapter, we illustrate the resilience of urban ecosystems by describing impacts of war on urban plants and animals by data from three European cities that differed in both their size and fate during the World War II: Viipuri (now Vyborg), Leningrad (now St. Petersburg) and London. Although the cases differ considerably, certain shared patterns and processes observed in them may indicate the existence of general patterns in changes of wildlife in war-damaged urban areas.

We approach these complex issues and the poorly known history of scientific studies of urban wildlife from the point of view of biographical studies. We will focus on scholarly chronicles, that is, wartime works of three naturalists. These examined scholars are Viljo Erkamo, a Finnish botanist, Alexandr Promptov, a Russian biologist, and Richard Fitter, an English economist. They all were also keen ornithologists.

Urban Vegetation and Birds in Viipuri, Finland/Soviet Union

During World War II, military actions as well as movement of troops and materials were arguably the greatest agents causing changes in plant distributions worldwide. In Finland this was especially true for the city of Viipuri, which was before World War II the second largest city of Finland with 72,680 inhabitants in 1939. As a border town Viipuri had experienced war periods several times over the centuries. Viipuri was again hit by war in November 1939 when the Soviet Union attacked Finland. At the end of the Winter War (November 1939–March 1940) Viipuri was transferred to Soviet control. Viipuri was again recaptured by the Finnish troops on August 29, 1941, during the so-called Continuation War (1941–1944), only to be lost again and for good on June 20, 1944. During the Winter War that preceded these events Viipuri was the most heavily bombed Finnish city. Soviet bombers attacked Viipuri 64 times (with a total of 1400 bombers) and dropped on the city 4700 bombs. As the Winter War ended, bomb damages were visible throughout the city.[6] As many as 3807 of the prewar 6287 buildings had been destroyed in the bombings.

The Finnish botanist Viljo Erkamo (formerly Berkán, 1912–1990; Fig. 9.1) was born in Viipuri which was at the time well known for its high-quality schools and their natural history clubs. He started his career as a field naturalist by making, already as a schoolboy, floristic inventories in his home town.[7] He completed his PhD at the University of Helsinki in 1956 and acted as Associate Professor of Botany in Helsinki in 1963–1975. Erkamo's doctoral thesis dealt with the impacts of climatic changes on the Finnish flora, and his principal professional interests were associated with changes in distributions of plant species.[8] As a broad naturalist, Erkamo also published ornithological field observations throughout his career.

Due to his frail health, Erkamo was not recruited to the Finnish army, but he was instead active in the homefront.[9] His duties brought him back to his hometown in mid-July 1942, and he spent a few days making botanical inventories and ornithological observations in the city, which had been largely ruined by the bombings. His botanical surveys on six urban study areas in the city (at that time again in Finnish ownership) resulted in an article titled "Traces of Bolshevik period in the flora of Viipuri."[10] Due to his lifelong interest in Viipuri's flora Erkamo was undoubtedly the right person to survey the changes caused by two main factors, immigration of new species with Soviet troops during the occupation period, and ruining of urban areas by extensive bombings.

Fig. 9.1 Viljo Erkamo (1912–1990) was a Viipuri-born Finnish botanist who studied plant species and birdlife in the war-damaged Viipuri in the summer and fall 1942. (Source: Pekka Kyytinen, Collection of The Finnish Heritage Agency)

Besides Erkamo, also other Finnish botanists had been (and were later) active in the Finnish eastern front. Niilo Söyrinki (later Professor of Botany at the University of Oulu) published in 1941 a detailed study of flora in the small villages of White Sea Karelia, an area newly occupied by Finnish troops.[11] Later, Saarnijoki published a species list of plants in the Kollaa battlefield,[12] and Mannerkorpi on the flora of Uhtua also located in White Sea Karelia.[13] In fact, there seems to have been no lack of (nearly) professional botanists in the Finnish-Soviet frontline. For instance Lieutenant General Karl Lennart Oesch, the Commander of the Karelian Isthmus forces in the critical summer of 1944, was an able amateur botanist.[14] The main issue in these studies was classification of cultural plant species into those that had arrived in the area from different parts of Soviet Union (usually with horses or fodder transport), and to those arrived with either Finnish or German troops. Erkamo himself published later a study of plants imported to Finland along with German troops.[15]

Certain invasive plants benefit from armed conflicts, and field studies in the Finnish eastern front even indicated association of particular plant species with specific types of war damage.[16] According to Mannerkorpi, the open-air former interiors of burnt houses were almost totally covered by the pineapple weed (*Matricaria discoidea*) and the closely related scentless mayweed (*Matricaria inodora*) and interestingly, only one of these species was usually present in a particular house ruin.[17] Plants typical for the Stuka bombing craters in White Sea Karelia included wood horsetail (*Equisetum sylvaticum*), European water plantain (*Alisma plantago-aquatica*), thread rush (*Juncus filiformis*) and six other plant species.[18] The rosebay willowherb (*Epilobium angustifolium*), on the other hand, was a native plant species that greatly increased in abundance due to forest fires caused by warfare.[19]

During his mid-summer visit to Viipuri in 1942 Erkamo investigated the flora of six urban areas: two mills, three railroad sites and one landfill area. Based on his own earlier studies, Erkamo was able to both observe changes in distributions of native plants, and to identify 13 plant species that with a high likelihood had arrived during the preceding Soviet occupation period.[20] These 13 species had been totally absent from the city before the occupation. But the overall picture was far more complex. There were 29 other plant species of eastern origin that had certainly arrived before World War II[21] and cultural species that had their origins elsewhere. Moreover, bombings and occupation had significantly altered the distributions of some local plant species.[22] For instance, Canadian fleabane (*Erigeron canadensis*), willowweed (*Epilobium adenocaulon*) and sticky groundsel (*Senecio viscosus*) had all had rather limited ranges in the city before World War II, but were in 1942 widespread all over the city. The sticky groundsel was especially common in the ruins.

Erkamo was in particular interested in plants growing in ruins. He compared vegetation between high ruins (uppermost remaining floor structures 2–3 m above ground level) and low ruins (height 1 m or lower), and observed clear differences. The species number was lower in high ruins probably due to limited dispersal capacity of some plant species and the more challenging growing conditions. Also species compositions differed between the ruin types. As genuine ruin indicators (typical for both ruin types) Erkamo listed white goosefoot (*Chenopodium album*), rosebay willowherb, common mugwort (*Artemisia vulgaris*), common groundsel, curly plumeless thistle (*Carduus crispus*) and narrowleaf hawksbeard (*Crepis tectorum*).[23] Erkamo was himself surprised by his observation that in spite of the rather devastating human influence in the bombed areas, the proportion of cultural plant species in ruined areas was only 54% (of the total species number).[24] The increased open space, among other factors, had apparently benefitted also some species of native flora that could occupy new habitats created by war.

During his 1942 visit in Viipuri Erkamo also observed birds.[25] Feral pigeons (*Columba livia domestica*), which before the war had been an essential part of the urban landscape, had nearly vanished. He only saw one pigeon flying high up over the city, but was told by local inhabitants that a few individuals still persisted in the city. When he visited Viipuri again in October 1942, he did not see a single feral pigeon in spite of active surveillance. He also cited a later newspaper article (published in March 1943) claiming that pigeons had entirely disappeared from Viipuri. Erkamo assumed that decreased food availability, wartime fires and explosions, widespread emptying of urban locations of inhabitants, and to some extent active hunting of pigeons had all contributed to their disappearance (Fig. 9.2).

On the other hand, some other bird species seemed to thrive in the ruined urban landscape. These included the northern wheatear (*Oenanthe oenanthe*) and red-backed shrike (*Lanius collurio*).[26] Especially wheatears were relatively common in the ruined areas. These observations were confirmed by Carl Eric Sonck, who spent the entire summer of 1942 in Viipuri. Sonck was a medical doctor but also a keen naturalist. Sonck further

Fig. 9.2 Viipuri Castle is the symbol of Viipuri, a city founded in the thirteenth century as a military outpost against Novgorod. This photo of burning Viipuri was taken during the Winter War, on March 7, 1940. (Source: Photographer probably Otso Pietinen, Finnish Wartime Photographic Archives)

reported that thrush nightingales (*Luscinia luscinia*) seemed to feel at home in the new nesting sites offered by abandoned and ruined gardens and backyards. In the Neitsytniemi area of Viipuri, that was known for its graveyards and garrison, he heard four or five singing male nightingales. He also remarked that "it is difficult to describe in words the emotions caused by simultaneous hearing of anti-aircraft guns and songs of nightingales."[27]

Erkamo was impressed by the adaptability of nature and especially of plants to war conditions. He wrote that "wherever even modest growing conditions were fulfilled, plant life emerged."[28] He was delighted to observe that the spectacular white poplars (*Populus alba*), a deciduous tree growing in Viipuri parks, had survived the war remarkably well.[29] Erkamo did not consider the floristic changes he so carefully reported as permanent. On the contrary, he predicted that "the impact of Bolshevik period on the permanent flora of Viipuri will remain relatively negligible."[30] The main reason was the presumed shortness of the occupation period, or so he thought. Erkamo was, however, mistaken. In June 1944 the winds of war changed, and Viipuri was annexed to Soviet Union on September 19, 1944.

Wildlife in Leningrad During the Siege

The siege of Leningrad, also known as the Leningrad Blockade, represents an unprecedented history of a city's endurance during the war. A voluminous literature describes this siege from different perspectives, from the history of military operations to survival histories of individual people.[31] However, to our knowledge, impacts of the siege on plants and animals in Leningrad have never been analyzed systematically in the scientific literature.

Before the beginning of the war, the population of Leningrad, including suburbs, was about 3.4 million people.[32] The siege started on September 8, 1941, when the last road to the city was severed. Thereafter, the focal problem for the civilians captured in the city was survival—their own, and of their children and relatives (Fig. 9.3). Food, water and firewood were the key resources for survival, and many people—especially in the winter time—were physically and mentally unable to spend their time and efforts for anything else than survival. In 1943, the population of Leningrad declined to 887,000 people.[33] According to different estimates, from 641,000 to 1.5 million people died during the Leningrad siege, primarily (97%) from starvation. The siege lasted for 872 days, and was lifted on January 27, 1944. "More civilians died in the Siege of Leningrad than in the modernist infernos of Hamburg, Dresden, Tokyo, Hiroshima and Nagasaki, taken together."[34]

Fig. 9.3 Vegetables grown in 1942 in the center of Leningrad, next to the Saint Isaac's Cathedral. (Source: M. Trakhman, National Library of Russia)

Before the war Leningrad was famous for its academic institutions. The larger part of the staff of scientific institutions was evacuated during 1941 or in early 1942, but some staff remained in Leningrad to supervise collections left in the city and to participate in work needed for army, hospitals and the city itself.[35] In spite of starvation, the keepers saved even the edible materials, such as fruit collection of the Komarov's Botanical Institute and seed collection of the Vavilov's All-Union Institute of Plant Industry. The staff of the latter institute took all possible measures to prevent the plunder of the grain by a starving population. Nine of the keepers died from hunger during the siege, but refused to eat a single seed from the collection. Moreover, because more advanced long-term storage was not yet available, the material needed periodic replanting and harvesting to ensure it remained fresh and viable. This became very risky to the scientists, and yet they managed to carry it out secretly every year.[36] However, it should not be forgotten that the punishment for any damage or loss of governmental property was rather strict in the Soviet Union at the time; so the keepers, as well as the personnel of the Leningrad Zoo,[37] may have had diverse motivations to save the collections for which they were responsible.

The information on the impact of siege on wildlife in Leningrad is scattered and of mixed nature. We did our best to distinguish between the city folklore and the documented evidence. We also used the stories on everyday live in besieged Leningrad, which were told to one of the authors of this chapter by his mother and his father-in-law, who both survived the siege. But the key source of the scientific information used by us was the paper summarizing the observations on bird abundance and bird behavior in Leningrad during the siege.[38] This study was conducted by a Russian biologist Aleksandr Promptov (1898–1948).

All his life, Promptov was eager to study birds. He was an excellent naturalist, but fate was unkind to him. He was born in a family of a high-ranking Russian revenue official; but after the revolution, the family lost all their property and needed to earn money to survive. Promptov was disabled since birth (spinal curvature[39]), which affected his entire life. Already as a young boy, he took an interest in birds but due to external circumstances he chose to study experimental zoology rather than ornithology, and until his movement to Leningrad in 1940, his paid duties were generally not related to ornithology. From 1935 to 1940, while employed as a lecturer of the Second Medical Institute in Moscow, Promptov however intensively studied bird behavior in his free time, and for instance published a 400-page manual of field ornithology. Promptov became well known as a geneticist who worked under supervision of Nikolay Koltsov and Sergey Chetverikov for his discovery on mutagenic activity of ultraviolet irradiation and for several works on genetics of bird behavior.[40]

Promptov's scientific activities attracted the attention of Leon Orbeli, director of the Institute of Evolutionary Physiology and Pathology of Higher Nervous Activity, which had been established in Koltushi (at the outskirts of Leningrad) by the first Russian Nobel laureate Ivan Pavlov. Orbeli invited Promptov to join the institute and he became the head of the newly established Ornithological Laboratory. For the first time in his life, Promptov was allowed to work on his favorite research topics; but the war distorted all the plans. Promptov's wife was not allowed to evacuate, because she was mobilized to work in an ammunition factory; and Promptov decided to remain in Koltushi. He was officially put in charge of overseeing the institute's properties and equipment,[41] but also had time to conduct some observations on birds.

During the siege, the German troops shot 150,000 heavy artillery shells and dropped over 107,000 incendiary and high-explosive bombs on Leningrad. Along with shell and bomb explosion, the city was rather noisy due to shooting of artillery and antiaircraft guns. However, bird behavior was unaffected by this extreme noise: birds continued feeding and singing, and did not attempt to escape from the sources of noise.[42] In contrast, animals in Leningrad Zoo were much afraid of the sounds of bombing, and some of them died from chronic stress.[43]

Nevertheless, the wildlife of Leningrad suffered from bombs and shells: over 150,000 trees and 800,000 shrubs were killed,[44] and the total area of green plantings decreased by 14%.[45] During the growth season of 1942, the survived people were eating almost everything, including grass, tree leaves and bark.[46] Special leaflets and brochures were therefore issued to advise people on what they could eat without harming their health (Fig. 9.4).[47] Thus, along with direct killing of birds and mammals, their populations were obviously affected by loss of nesting sites and of food supply.

Fig. 9.4 The poster "Early spring wild herbs suitable for salad," published in Leningrad in 1943. (Source: National Library of Russia)

However, at the same time, abundances of some birds, primarily of small-sized insectivores, increased during the siege. Promptov reported that already in 1942, great tit (*Parus major*), blue tit (*Cyanistes caeruleus*), common swift (*Apus apus*), chaffinch (*Fringilla coelebs*), pied flycatcher (*Ficedula hypoleuca*), spotted flycatcher (*Muscicapa striata*), northern house martin (*Delichon urbicum*) and common redstart (*Phoenicurus phoenicurus*) were in Leningrad more common than before the war. In the center of the city, near the Kazan Cathedral, Promptov observed a nesting pair of common whitethroat (*Curruca communis*). The pied flycatcher, spotted flycatcher and common redstart were seen nesting in damaged buildings.[48] In our opinion, the increase in bird population was, at least partly, due to the damage of a substantial number of buildings, which created nesting sites and shelter, as well as to the decrease in human population and extinction of domestic cats, which were eaten by people during the first winter of the siege. A similar process was documented in Berlin, where wastelands created by the bombing of the city during World War II were quickly covered by dense vegetation that attracted birds and other animal species.[49]

Some birds of Leningrad demonstrated plastic behavior in the unusual circumstances. For example, in autumns of 1942 and 1943 great tits in the outskirts of the city actively searched for food in woody parts of old uninhabited buildings, which were disassembled for firewood. Similarly, northern wheatear already in 1942 nested (as it did in Viipuri) in abandoned trenches and in walls of bunkers made of logs. Promptov reported that significant increases in population densities of these birds, explained primarily by availability of nesting sites, were recorded until 1946, when trenches started to be filled with soil, and were overgrown by dense grasses and herbs.[50]

At the same time, house sparrows and feral pigeons almost disappeared from the city. They not only lost their food source (food waste and grain), but also became the objects of hunting. Promptov observed that feral pigeons were hunted by traps and snares, and they all were killed and eaten before they learned to avoid traps or developed wariness against people. In the spring of 1942, rook (*Corvus frugilegus*) colonies were often ravaged by people. These birds, in contrast to feral pigeons, rapidly learned to perceive people as enemies, but still did not change their nesting locations, in spite of their apparent unsafety.[51] We have no documentary evidence, but city folklore reports that house sparrows were also hunted in the city.

Rats were another group of animals whose abundance increased in Leningrad during the siege.[52] Rat populations grew due to presence of numerous unburied corpses in buildings and on streets of the city, disappearance of domestic dogs and cats, their natural enemies, and the impossibility to apply prescribed measures to control rat density. This increase in rat density was seen as extremely dangerous due to the risk of infectious diseases, so the people were strongly advised to exterminate rodents.[53]

Promptov himself survived the siege and even conducted some research on caged birds during the war. At the time when people were eating all green parts of plants, and even the bark of trees in city center was consumed during the winter time, he kept from 10 (in 1942) to 140 birds (in 1944) in his lab in Koltushi and managed to supply them with some food during the siege. This information looks astonishing, as people died from starvation at that time. Likely, the situation in Koltushi may have been slightly better than in the center of Leningrad, because kitchen-gardens provided some food.[54] Also cats and dogs from the vivarium of the Institute were given as a food to staff members and helped them to survive.[55] Unfortunately, soon after the war in 1948 Promprov committed suicide; available data sources contain no information on possible reasons for this unfortunate action.[56]

THE LONDON BOMBED SITES SURVEY

A London-born young and promising naturalist Richard Fitter (1913–2005) wrote in his first book *London's Natural History* in 1945 on the impacts of the war on "the largest aggregation of human beings ever recorded in the history of the world as living in a single community"[57]:

"Not since London ceased to be subject to recurrent fires, due to the carelessness of its citizens combined with the combustibility of their building materials, has the urban area been subjected to such a cataclysm as the 'blitz', which raged from September, 1940, to May, 1941, and was followed three years later by the 'fly-blitz.' The effect of what is now euphemistically known as 'the original blitz' has been to produce extensive areas of open ground throughout the most heavily built-up parts of London, and this has naturally had a profound influence on the fauna, and more especially on the flora of the area. Apart from the aftermaths of previous fires, there is more open ground to-day within a mile of St. Paul's Cathedral than there has been since the early Middle Ages. There is even cultivation, pigs being kept and vegetables grown on an allotment near Cripplegate

Church, which shows that even this land, built over for seven or eight centuries, still retains its fertility. It is sometimes forgotten that much of Greater London stands on some of the most fertile soil of the British Isles; it just happens that under the present order of society it is more profitable to grow factories than fruit or vegetables."[58]

Although Fitter was an economist by education, he became one of the most well-known British nature writers and wildlife conservationists of the twentieth century. During World War II he joined the Royal Air Force Coastal Command, and after the war became secretary of the wildlife conservation committee of the Ministry of Town and Country Planning.[59] It was in this phase of his career when *London's Natural History* was published. The global war had just ended when the book was published in late 1945. The last chapter of the book was entirely devoted to the influence of war on the wildlife of London, the memories of war still being fresh and dramatically visible in the urban landscape.

After the opening paragraph quoted above, Fitter continued to ponder the impact of Luftwaffe bombings on plants and animals. He considered direct effects on plants and animals negligible; but—from the current perspective—this conclusion may be seen as an understatement. He claimed that comparatively few animals and plants in the London area had fallen direct victims of bombings, and "in no case can the status of any plant have been affected."[60] Animals had due to their mobility suffered even less than plants, Fitter argued. One of the few documented direct casualties was an unlucky mistle-thrush (*Turdus viscivorus*) found dead 30 yards away from a bomb crater at Watford.[61] In spite of the fact that London was a favored roosting place of tens of thousands of starlings (*Sturnus vulgaris*), no corpses of dead starlings were, "fortunately for the starlings," ever reported.[62]

Still during the war attempts were made to document the immediate effects of bombings. The flora of bombed areas was first studied by E. J. Salisbury, director of the Royal Botanic Gardens at Kew, and J. E. Lousley, both of whom documented the pioneer plant colonists on the sites.[63] Lousley made his studies during the first two years after the Blitz (in 1941–1942). Salisbury found in the bombed sites 126 vascular plant species, about 30% of which had probably been wind-dispersed to the area. Based on these wartime surveys, the most important pioneer species of bombed sites were the rosebay willowherb, Canadian fleabane (*Erigeron canadensis*), coltsfoot (*Tussilago farfara*), Oxford ragwort (*Senecio squalidus*), common and sticky groundsels (*Senecio vulgaris* and

S. viscosus), sow-thistle (*Sonchus oleraceus*) and common meadow-grass (*Poa annua*).[64] The rosebay willowherb was, according to Fitter, probably the commonest plant in Central London in the late war years. The plant apparently benefitted from the nearly unlimited access to light that was typical for the blitzed sites, and from its exceptionally good tolerance for soil that had been subjected to heat. Salisbury found rosebay willowherb on 90% of his study sites. An interesting ecological by-product of the invasion by rosebay willowherb was the high abundance of its typical insect herbivore, the elephant hawk-moth (*Deilephila elpenor*) in the wartime London.[65]

Partially based on these preliminary studies, the venerable London Natural History Society (LNHS hereafter), founded in 1858, organized a systematic bombed sites survey in the City over several years in the late 1940s.[66] Monitoring of ecological consequences of bombings was launched by the Ecological Section of the LNHS in 1947 within an area of about 55 acres around St Giles, Cripplegate.[67] The study area contained part of the site of Roman Londinium to the south of London Wall. Already in this first year of the City Bombed Sites Survey, as it was called, observational data were collected on the plants (both vascular plants and mosses), birds, spiders, flies and beetles of the severely damaged area.[68] Soil types in the affected area were also studied already in 1947. Urban butterflies (so-called "Macrolepidoptera") of the bombed areas were monitored somewhat later.[69]

Botanist F. E. Wrighton described the Cripplegate area at the onset of study in 1947:

> In the whole area there are but few buildings standing. Roadways still remain and form solid division walls between the basement areas of the missing buildings. These fast crumbling party walls of brick and mortar. Into some of these basements has been tipped rubble, consisting of brick, mortar, plaster, concrete, etc., especially along the sides of some of the roads, where it forms a loose and dusty scree.

The explicitly stated objective of the LNHS was to follow the recovery of wildlife after the Blitz period: "This area in the very heart of London provides a splendid opportunity to members of the Society for recording the colonization and establishment of a new fauna and flora, which, it is to be hoped, will not occur again."[70]

Wrighton divided the survey area, which was to be monitored in the years to come, into five different types of habitat.[71] The first habitat type and largest in coverage were basement floors, which consisted of slabs of concrete or asphalt laid down on the subsoil. The second type, rubble tips, varied in nature from loose and dry screes to smooth consolidated ramps that had evidently been used for carting rubble into the basements. The remaining three habitat types were vertical walls, the remaining small gardens in the area, and small areas of permanent water. Each of these habitat types seemed to have its characteristic flora. For instance, basement floors were mostly dominated by Oxford ragwort and rosebay willowherb, the rubbletips by perennial wall-rocket (*Diplotaxis tenuifolia*) and eagle fern (*Pteridium aquilinum*), and vertical walls, not surprisingly, by the typical pioneer species rosebay willowherb and especially the upright pellitory (*Parietaria officinalis*). Plant life at Cripplegate was monitored since spring 1948 in 16 areas with permanent quadrants and transects.[72]

In terms of ecology, what was studied at Cripplegate by the LNHS was ecological succession, which refers to a gradual and progressive change in the species structure of an ecological community at a given site after major disturbance, from the so-called pioneer phase to the final climax community. The concept of succession had been developed by botanists Eugen Warming and Henry Chandler Cowles at the turn of the nineteenth and twentieth centuries.[73] Although the climax community was never achieved at the London City study sites due to urban reconstruction in the 1950s and 1960s, the study area did show rather exemplary early phases of succession. The sites were first dominated by pioneer plants such as rosebay willowherb and Oxford ragwort, and only some years later by woody plants that in most sites tend to dominate climax communities. The first woody plants colonizing the study area were several willow species, although also silver birch, black elder and black poplar were early established, although in lower numbers.[74] At the local scale and in the course of succession plant species competed with each other and species replacement took place at a regular basis. In his 1950 paper, Wrighton listed from the study area 27 cultivated forms, 121 natural species and 9 species of mosses.[75] He classified nine plant species already as "possibly extinct"—a rather predictable result of proceeding succession.

Wartime Avifauna in London

It was however the changes in the London avifauna that gained most publicity during the war years. The birds of London had been thoroughly documented since Arthur Holte Macpherson's first comprehensive bird list of inner London in 1929.[76] As Fitter rightly remarked, "[b]irds are the largest creatures which have succeeded in fully adapting themselves to outdoor life in the centres of great cities; their superior mobility enables them to escape from the many dangers of town life by flying up to the roof-tops."[77]

In London bird-watching certainly had long traditions.[78] Stanley Cramp and W. G. Teagle even called the 40 square mile area of inner London as ornithologically "perhaps the most watched in the world."[79] London birdlife in the prewar decades was influenced by several rather dramatic changes. Due to rapid urban expansion, inner London became increasingly separated from the rural areas by "an ever-growing zone of built-up land." This probably resulted in the loss of some bird species, such as the rook that was for the last time observed nesting in 1916. Moreover, streets became increasingly crowded with car traffic with a simultaneous decline of horse traffic, leaving less food available for birds such as house sparrows and feral pigeons to pick up in the roadways. Also some minor changes in land use, such as the draining of the lake in St. James's Park during World War I, harmed birdlife locally. In spite of these and minor other changes, both the number of breeding bird species in inner London and the number of regular winter visitors increased in the period 1900–1950.[80]

Black redstart (*Phoenicurus ochruros*) became the symbol of London's changing birdlife during World War II. Like the mythical Phoenix, it appeared to emerge from the ashes of Blitz and rapidly colonize the bombed sites in central London. According to Richard Fitter, "[i]t is not often that London acquires a new breeding bird, and still less often that the newcomer elects to breed in the heart of the built-up area. Yet this is what has happened since 1940, when a pair of black redstarts brought off two broods in the precincts of Westminster Abbey." Fitter and probably many others considered the colonization history of this bird species as the most remarkable event in the first half of the twentieth-century London ornithology.[81] London was "the metropolis not only of man but of the black redstart in the British Isles."[82] Although it was later shown that a tiny black redstart population had in fact resided in the London area at least since 1926, the species clearly benefitted ecologically from the extensive areas of waste ground and ruins created by the Blitz (Fig. 9.5).

The *annus mirabilis* of the black redstarts in London was 1942. In that year, at least three pairs of this rare bird bred successfully, and several additional singing males were observed.[83] According to P. W. E. Currie, the birds mainly fed on insects and other arthropods that they "captured on walls, on bare ground, or in flight from some prominent position, and

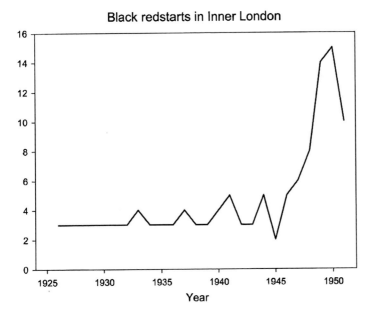

Fig. 9.5 The population growth of the black redstart in the inner London (minimum number of breeding pairs). (Source: Sara Vuorisalo, after The London Natural History Society, *The Birds of the London area*, 252)

they are seldom seen searching on vegetation."[84] The preferred nest sites of black redstarts were upper corners of windows, holes in ceilings or crevices due to fallen bricks, all of them in bombed buildings.[85] Black redstart remained a regular breeding bird in inner London, and by 1950, nestings were recorded not only in the City bombed areas but also in Stepney, Notting Hill, Westminster and at Brompton Cemetery.[86] War conditions made identification of this species easier due to greater audability (less traffic) and improved visibility due to the destruction of buildings.[87] Currie complained that the presence of the black redstarts at the Cripplegate study area was in fact a distracting factor as "the majority of the ornithologists who come on the monthly visits do so principally with the object of seeing this one species."[88] Black redstart was in the war years and especially during the postwar reconstruction period a typical bird species of other bombed European cities as well, including the destroyed capital cities Berlin and Warsaw.[89]

War also changed animal attitudes, at least at the official level. Although Londoners were famous for their fondness for urban birds, feeding of birds with any food fit for human consumption was prohibited due to food shortages. According to Fitter, this regulation was however not very strictly observed, and gulls continued to be fed on the Embankment and ducks in the London parks.[90] Feeding of small birds such as tits anyway became less generous. Some well-established city birds were even hunted for human consumption. The population of feral pigeons in London declined dramatically during World War II due to lack of food and persecution. On the other hand, pigeon population started to increase again in postwar years after the abolition of bread rationing.[91] Even the famous feral pigeons in Trafalgar Square were persecuted.[92] Another urban pigeon species, the woodpigeon (*Columba palumbus*), that had been present in the parks of London since 1834, was hunted at the "massacre" level, as Cramp and Teagle put it.[93] For some reason the fish of park ponds were treated in a more humane way. When the Pen Ponds in Richmond Park were drained for military reasons in 1940, 175 carp, 320 bream, 250 pike, 300 eels, 30 perch and 20 dace, among others, were safely transferred to other ponds in the neighborhood. It was estimated that about half of the fish were rescued in this operation.[94]

Conclusions

All these examples reported above jointly indicate that urban wildlife is very resilient to the direct impacts of the war, and generally colonizes the sites disturbed in the course of military actions, bombing in the first line, if they are not used by people. At the same time, indirect impacts have a potential to cause great changes, such as overcompetition of local flora by invasive species, rapid increases in density of some animals due to the loss of control from predators and/or increase in food supply and in the availability of nesting sites, and rapid extinction following changes in human attitude toward some species caused in particular by starvation or food shortages. For instance, feral pigeons became hunting objects in all three studied cities.

In spite of geographical and historical differences, the three cases showed rather striking similarities in their patterns of resilience. In all three cities some bird species clearly adapted themselves quickly to the damaged environments. The most well-known example was the black redstart population of London. On the other hand, Viipuri and London shared several invasive plant species that thrived in ruined areas; these included at least the Canadian fleabane, rosebay willowherb and common and sticky groundsels. Wind-dispersed plant species were particularly common in the ruins of both cities.[95] Erkamo's data from Viipuri could demonstrate a portion of plant species that almost certainly had arrived in the city during the first Soviet occupation period in 1940–1941. As ecological succession proceeded in the ruined areas, some original colonizer plants were replaced by late successional woody species; this was clearly documented in London. Some years later these plant communities were probably mostly lost in the bombed cities due to reconstruction.

The effects of war on urban wildlife were particularly carefully and widely documented in London, largely due to the activities of the London Natural History Society and its specialist members. Viipuri and Leningrad probably represented a more common situation where documentation of wartime wildlife became the private initiative of individual active researchers, Erkamo in Viipuri and Promptov in Leningrad. Both Erkamo and Promptov were professional biologists who were not recruited to the army due to health problems. Also in the ruined German cities botanical surveys were performed cooperatively by amateurs and professional biologists.[96]

The relationship between humans and urban wildlife suffered from the adversities of war that were by far the greatest in famine-stricken Leningrad, where all accessible and edible plants and animals were out of necessity used for human nutrition—with the notable exception of collections of state research institutes and animals from the Leningrad Zoo. But also in London some city birds such as pigeons were systematically hunted for human consumption. Feeding of animals decreased both in Leningrad and London, which forced some bird species to search for food in the ruins. Fortunately, the human-animal attitudes proved to be resilient, and returned to their prewar positive levels after the war, in the reconstruction period.

NOTES

1. Herbert Sukopp, "On the early history of urban ecology in Europe," *Preslia* 74 (2002): 373–393; Jens Lachmund, "Exploring the City of Rubble: Botanical Fieldwork in Bombed Cities in Germany after World War II," *Osiris* 18, Science and the City (2003): 237.
2. Jens Lachmund, "Ecology in a walled city: researching urban wildlife in post-war Berlin," *Endeavour* 31, no. 2 (2007): 78–82.

3. M. J. McDonnell, "The history of urban ecology: An ecologist's perspective," in J. Niemelä, J. H. Breuste, T. Elmqvist, G. Guntenspergen, P. James & N. E. McIntyre, eds., *Urban Ecology: Patterns, Processes and Applications* (Oxford: Oxford University Press, 2011), 5–13.
4. C. S. Holling, "Resilience and Stability of Ecological Systems," *Annual Review of Ecology and Systematics* 4 (1973): 1–23.
5. Rachel J. Standish, Richard J. Hobbs, Margaret M. Mayfield et al., "Resilience in ecology: Abstraction, distraction, or where the action is?" *Biological Conservation* 177 (2014): 49.
6. Jouni Kallioniemi, *Viipuri suursodassa 1939–1944* (Raisio: Kirjaveteraanit Oy, 1990), 108–109, 121.
7. Ilkka Kukkonen, "Viljo Erkamo 24.1.1912–24.10.1990," *Luonnon Tutkija* 95 (1991): 195–97; Anna Ju. Doronina and Mikko Piirainen, "History of the floristic study of the Karelian Isthmus, Leningrad Region," *Memoranda Societatis pro Fauna Flora Fennica* 85 (2009): 52–53.
8. Kukkonen, "Viljo Erkamo," 195.
9. Kukkonen, "Viljo Erkamo," 197.
10. Viljo Erkamo, "Bolshevikkiajan merkeistä Viipurin kasvistossa," *Annales Botanici Societatis Zoologicae-Botanicae Fennicae "Vanamo"* 18, no. 3 (1943): 1–21.
11. Niilo Söyrinki, "Havaintoja kyläkasvistosta Vienan Karjalassa sotakesänä v. 1941," *Luonnon Ystävä* 45, no. 5–6 (1941): 150–164.
12. Sakari Saarnijoki, "Satunnaiskasveja Kollaan taistelupaikoilta," *Luonnon Ystävä* 46, no. 2 (1942): 69. Saarnijoki listed 49 invasive plant species growing on horse manure heaps of Kollaa.
13. Panu Mannerkorpi, "Uhtuan taistelurintamalle saapuneista tulokaskasveista," *Annales Botanici Societatis Zoologicae-Botanicae Fennicae "Vanamo"* 20, *Notulae botanicae* 15 (1944–45): 39–51.
14. Helge Seppälä, *Karl Lennart Oesch, Suomen pelastaja* (Jyväskylä: Gummerus, 1998), 21–22.
15. Viljo Erkamo, "Beobachtungen über die mit deutschen Truppen im Jahre 1944 nach Helsinki eingeschleppten Pflanzenarten," *Annales Botanici Societatis Zoologicae-Botanicae Fennicae "Vanamo"* 21, *Notulae botanicae* 16 (1946): 7–11.
16. Mannerkorpi, "Uhtuan taistelurintamalle," 39.
17. Mannerkorpi, "Uhtuan taistelurintamalle," 46.
18. Mannerkorpi, "Uhtuan taistelurintamalle," 47–48.
19. Mannerkorpi, "Uhtuan taistelurintamalle," 45–46.
20. Erkamo, "Bolshevikkiajan merkeistä," 8.
21. Erkamo, "Bolshevikkiajan merkeistä," 7–8.
22. Erkamo, "Bolshevikkiajan merkeistä," 19–20.
23. Erkamo, "Bolshevikkiajan merkeistä," 15.

24. Erkamo, "Bolshevikkiajan merkeistä," 18. According to Erkamo's own studies, the total flora of Viipuri before the war consisted of 854 plant taxa, of which 418 were anthropochorous (i.e., cultural plants; Doronina and Piirainen, "History of the floristic study," 53).
25. Viljo Erkamo, "Viipurin rauniokaupungin linnuista," *Luonnon Ystävä* 47 (1943): 110–111.
26. Erkamo, "Viipurin rauniokaupungin linnuista," 111.
27. C. E. Sonck, "Viipurin rauniokaupungin linnuista," *Luonnon Ystävä* 47 (1943), 143.
28. Erkamo, "Bolshevikkiajan merkeistä," 10.
29. Viljo Erkamo, "Viipurin Torkkelinpuiston valtavat hopeahaavat (Populus alba)," *Luonnon Ystävä* 47 (1943): 144–145.
30. Erkamo, "Bolshevikkiajan merkeistä," 9.
31. The catalogue of the National Library of Russia (accessed 19 May 2018) included 808 books, the title of which contained the words "блокада Ленинграда" ("Siege of Leningrad," in Russian).
32. *Soviet Military Encyclopedy*, vol. 1. (Moscow: Voenizdat, 1990) (in Russian).
33. *Soviet Military Encyclopedy*, vol. 1. (Moscow: Voenizdat, 1990) (in Russian).
34. M. Walzer, *Just and Unjust Wars* (New York: Basic Books, 1977), 160.
35. K. B. Yuriev, "Historical review," in O. A. Skarlato, ed., *Zoological Institute of the Academy of Sciences of the USSR: 150 years* (Leningrad: Zoological Institute, 1982), 13–41 (in Russian).
36. S. M. Alexanyan and V. J. Krivchenko, "Vavilov institute scientists heroically preserve world plant genetic resources during World War II siege of Leningrad," *Diversity* 7, no. 4 (1991): 10–13.
37. Marina Boitsova, "How the [Leningrad] Zoo was saved during the Siege," http://www.rosbalt.ru/piter/2014/01/17/1220795.html (Accessed 24 May 2018; in Russian).
38. Aleksandr Nikolaevich Promptov, "Some observations on birds during the war and the siege of Leningrad," *Nature protection* 6 (1948): 85–89 (in Russian).
39. Daniil Granin, *The Bison: A Novel About the Scientist Who Defied Stalin* (New York: Doubleday, 1990).
40. Nikolay D. Ozernyuk, *Scientific school of N. K. Koltsov. Pupils and associates* (Moscow: KMK Press, 2012) (in Russian).
41. A. A. Formozov, "Aleksandr Nikolaevich Promptov (1898–1948)," in V. E. Flint & O. L. Rossolimo, eds., *Ornithologists from Moscow* (Moscow: Moscow State University, 1999), 375–399 (in Russian); N. Krementsov, "A particular synthesis: Aleksandr Promptov and speciation in birds," *Journal of the History of Biology* 40 (2007): 637–682.
42. Promptov, "Some observations on birds," 86. As mentioned earlier, this was also observed in Viipuri.

43. Marina Boitsova, "How the [Leningrad] Zoo was saved during the Siege," http://www.rosbalt.ru/piter/2014/01/17/1220795.html (Accessed 24 May 2018; in Russian).
44. A. M. Naumov, "Planting of greenery in Leningrad," *Architecture and Building* no. 2 (1946): 14–15. (http://www.nipigrad.ru/history_06/; in Russian).
45. O. A. Ivanova, "The system of green plantings in Leningrad," *Architecture and Building in Leningrad* no. 2 (1957): 27–32. (http://www.nipigrad.ru/history_08/; in Russian).
46. Mira A. Kozlova (Maslova), personal communication.
47. *Main edible wild plants of the Leningrad Oblast* (1942). (A leaflet published in Leningrad; https://vivaldi.nlr.ru/ll000075340/view#page=; in Russian). *Early spring wild herbs suitable for salad* (1943). (A leaflet published in Leningrad; https://vivaldi.nlr.ru/ll000075344/view#page=; in Russian).
48. Promptov, "Some observations on birds," 87.
49. Lachmund, "Ecology in a walled city," 79.
50. Promptov, "Some observations on birds," 88.
51. Promptov, "Some observations on birds," 87.
52. N. K. Tokarevich and N. A. Stoyanova, "The work of the department of parasitary infections during the Great Patriotic War," *Russian Journal of Infection and Immunity* 5, no. 2 (2015): 175–182 (in Russian).
53. *Exterminate rodents!* (1943) (A leaflet published in Leningrad; https://vivaldi.nlr.ru/ll000074868/view#page=; in Russian).
54. Formozov, "Aleksandr Nikolaevich Promptov."
55. Lev E. Tsyrlin, personal communication.
56. Formozov, "Aleksandr Nikolaevich Promptov"; Krementsov, "A particular synthesis."
57. R. S. R. Fitter, *London's Natural History* (London: Collins, 1945), 1.
58. Fitter, *London's Natural History*, 228.
59. Rob Hume, "Richard Fitter. Conservation expert who wrote fauna and flora bestsellers," *The Guardian*, 28 September, 2005.
60. Fitter, *London's Natural History*, 228.
61. Fitter, *London's Natural History*, 229.
62. Ibid. On London's well-studied birdlife, see, for example, S. Cramp and W. G. Teagle, "The birds of inner London, 1900–1950," *British Birds* XLV, no. 12 (1952): 433–456.
63. Fitter, *London's Natural History*, 231.
64. Fitter, *London's Natural History*, 233.
65. Fitter, *London's Natural History*, 231.
66. http://www.lnhs.org.uk/index.php/about-us/aims/history; visited 12 May, 2018. Apparently rather similar botanical surveys were performed in

German cities in the postwar years. For instance in Bremen, rubble vegetation studies were done from the end of the war to the end of 1953 (Lachmund, "Exploring the City of Rubble," 240).
67. "Progress Report." *London Naturalist* 27 (1947): 44; P. W. E. Currie, "Some Notes on the Birds of the Bombed Sites," *London Naturalist* 29 (1950): 81.
68. "Progress Report," 1947, 44.
69. "Progress Report," *London Naturalist* 29 (1950): 81.
70. Ibid. Such attitudes seem to be typical for biologists studying catastrophes. The volcanic eruption of Krakatau, Indonesia, in 1883, inspired great interest among biologists who wanted to study the gradual colonization of the desolated island by plants and animals; Charles J. Krebs, *Ecology. The Experimental Analysis of Distribution and Abundance. Second Edition*, 32–33 (New York: Harper & Row, 1978).
71. F. E. Wrighton, "Plant Ecology at Cripplegate, 1947," *London Naturalist* 27 (1947): 45–47.
72. F. E. Wrighton, "Plant Ecology at Cripplegate, 1949," *London Naturalist* 29 (1950): 85.
73. Krebs, *Ecology*, 428–429; Lachmund, "Exploring the City of Rubble," 248–249.
74. Wrighton, "Plant Ecology at Cripplegate, 1949," 85.
75. Wrighton, "Plant Ecology at Cripplegate, 1949," 85–88.
76. A. H. Macpherson, "A List of the Birds of Inner London," *British Birds* 22 (1929): 222–244.
77. Fitter, *London's Natural History*, 116.
78. E. M. Nicholson, *Bird-Watching in London* (London: London Natural History Society, 1995).
79. S. Cramp and W. G. Teagle, "The Birds of Inner London," *British Birds* 45, no. 12 (1952): 434.
80. Cramp and Teagle, "The Birds of Inner London," 434; Fitter, *London's Natural History*, 235.
81. Fitter, *London's Natural History*, 120.
82. The London Natural History Society, *The Birds of the London Area* (London: Rupert Hart-Davis, 1964), 251.
83. Fitter, *London's Natural History*, 122.
84. Currie, "Some Notes on the Birds," 83.
85. LNHS, *Birds of the London Area*, 253.
86. Cramp and Teagle, "The Birds of Inner London," 445.
87. Fitter, *London's Natural History*, 126.
88. Currie, "Some Notes on the Birds," 81.
89. Klaus Witt, "Berlin," in John G. Kelcey and Goetz Rheinwald, eds., *Birds in European Cities* (St. Katharinen: Ginster Verlag, 2005), 20; Maciej Luniak, "Warsaw," in *Birds in European Cities*, 391.

90. Fitter, *London's Natural History*, 238.
91. LNHS, *Birds of the London Area*, 211.
92. Fitter, *London's Natural History*, 238.
93. Cramp and Teagle, "The Birds of Inner London," 433; see also LNHS, *Birds of the London Area*, 211.
94. Fitter, *London's Natural History*, 235.
95. Erkamo, "Bolshevikkiajan merkeistä," 17.
96. Lachmund, "Exploring the City of Rubble," 241.

SECTION IV

Urban Society

CHAPTER 10

Partial Resilience in Nationalist China's Wartime Capital

Surviving in Chongqing

Nicole Elizabeth Barnes

INTRODUCTION

Soon after fighting broke out between the Imperial Japanese Army (IJA) and China's National Revolutionary Army (NRA) on July 7, 1937, China's head of state Chiang Kai-shek decided upon Chongqing as his Nationalist government's final redoubt. A bustling river port in Sichuan—China's southwestern rice-bowl province the size of France—Chongqing granted safe harbor to so many refugees that its population nearly tripled.[1] For its part, Sichuan province shouldered a huge burden, sending 3.4 million soldiers into battle, of whom those wounded or killed constituted a full 20% of all casualties in the NRA.[2] Chongqing and its immediate environs provided an even greater majority of soldiers, filling 14.7% of Sichuan province's hefty quota, even though its fair share would have been 5.6%.[3] Sichuan farmers also provided the majority of the country's rice and other food grains throughout the war.[4] Sichuan carried the nation through an eight-year war with aplomb.

N. E. Barnes (✉)
Department of History, Duke University, Durham, NC, USA

© The Author(s) 2019
S. Laakkonen et al. (eds.), *The Resilient City in World War II*,
Palgrave Studies in World Environmental History,
https://doi.org/10.1007/978-3-030-17439-2_10

Known to locals as the War of Resistance against Japan (*KangRi zhanzheng*), World War II began in China two years earlier than in Europe and played a decisive role in the 1949 victory of the Communist armies. Its total death toll—an estimated 18 million soldiers and civilians—far exceeded that of any other belligerent nation save the Soviet Union.[5] Though scholars are still coming to terms with this colossal event, it would be nearly impossible to overestimate the war's overall impact on Chinese society. The Japanese occupied nearly one-third of Chinese territory primarily on the eastern seaboard, while the Nationalist government retreated inland to the southwest, and the Communist guerrilla forces grew to unstoppable size in the north. This country divided in three nonetheless survived eight years of foreign invasion, followed by a tremendously violent civil war.

The Nationalist government's retreat to Chongqing augmented both threat and succor. It increased the citizens' vulnerability because the retreat signaled a new style of warfare—a multi-front war of attrition—in which the IJA adopted air raids as its primary strategy. Chongqing absorbed more enemy bombs than any other city in China, and also more than any other capital in the world; in other words, Chongqing was the most heavily bombed capital in World War II.[6] On the other hand, Chongqing made the perfect wartime capital. The city's unique geography provided magnificent protection. Surrounded on all sides by water and mountains, Chongqing remained impregnable to Japanese foot soldiers (if not the air force) throughout the duration of the war. The soft limestone mountains also provided useful shelter for the besieged civilians: after much digging by hand, they held a vast network of air raid shelters that protected people from enemy bombs. Central government buildings and official residences—including one for Chiang Kai-shek and his wife Song Meiling—took refuge on the most forested mountains, where the dense vegetation provided constant cover from Japanese bombers.

This chapter defines the resilience of cities as their capacity "to function, so that the people living and working [therein] – particularly the poor and vulnerable – survive and thrive no matter what stresses or shocks they encounter."[7] By this definition, Chongqing was both resilient and not resilient. Its lack of resilience stemmed from the fact that the state policed the poor more than it provided services for them. Many of Chongqing's poorest residents lost their homes or their jobs, and all were more vulnerable to disease than were wealthy residents. On the other hand, given the extreme stress that Chongqing endured due to repeated air raids and its ballooning population, overall survival rates prove the wartime capital to have been remarkably resilient. Chongqing was therefore *partially resilient*. This chapter provides detailed analysis of how the people of meager means in Chongqing survived—but failed to thrive—during China's eight-year war with Japan.

On the Brink of Conquest

The same dynamic of surviving but failing to thrive characterized the entire country during the war. In terms of bare survival, the retreat to Chongqing undoubtedly changed the course of Chinese history. In August 1937 Japanese troops occupied the old imperial capital Beijing (called Beiping or "northern peace" at the time to signal its loss of status as national capital in 1928) and quickly moved south to the city that the Nationalists had made their capital: Nanjing ("southern capital"). After losing Nanjing in December 1937, then their first provisional capital Wuhan in October 1938, the Nationalist government's ability to take refuge in the highly defensible city of Chongqing saved the country from conquest. Yet having survived, the state failed to thrive. After losing so much ground to the IJA, the Nationalist state had to support its military and general civilian functions with no more than 3% of the country's gross national product. Meanwhile, many of the most loyal and best-trained military personnel died in the first several months of fighting, leaving a mere 40% of NRA soldiers who were truly loyal to the state. Defections were common not only among foot soldiers but also among generals.[8] Japanese soldiers quickly earned a reputation for fierceness and cruelty that struck terror in the hearts of men sent to defy them on the battlefield. This, plus the poor conditions of life for NRA troops, produced a healthy fear of military service that set conscription officers against the people in a battle of wits and power. Conscription officers took several measures to discourage desertion; they yoked the men together with chains around their necks, stole their clothing at night, and located recruitment centers far from conscripts' homes. This last factor led to very long forced marches, which, according to one scholar's estimate, killed more than a million men before they reached the battlefield.[9]

Generalissimo Chiang Kai-shek, sitting at the helm of both the Nationalist state and the military, was painfully aware of his capital and country's bare survival. The circumstances of total war forced both Nationalist and Communist leaders to demand a lot from the people in their respective jurisdictions, and therefore the two parties competed with one another in offering services to those people in order to win their loyalty.[10] This rivalry took place under the watchful eyes of Chongqing's foreign residents, many of whom were strident critics of Chiang's regime.[11] As the chosen capital of a nation under siege, replete with foreign reporters, ambassadors, and military advisers—particularly after late 1941 when the Pacific War knit together the world's eastern and western belligerent zones—Chongqing's resilience had profound political significance.

Social inequality was the most fundamental cause of the partiality of Chongqing's resilience. Its wealthiest residents lived in massive villas, where they played tennis on private courts (one rumored to be encased in glass).[12] Exclusive shelters garnered such a cachet among the wealthy that real estate ads always mentioned if the residence in question boasted a private bath and air raid shelter.[13] Meanwhile, its poorest residents lived in shantytowns comprised of makeshift shelters known colloquially as "War of Resistance shacks" (*Kangzhan peng*). These houses constructed of cardboard, plywood, and other stray materials had no chance of surviving the annual floods that followed the city's regular torrential rains, or the fires ignited by incendiary bombs. In the infamous spring of 1938, fires followed a flood, killing over a hundred people and leaving more than 30,000 homeless.[14]

The war accentuated this disparity in wealth. In addition to the thousands of poor refugees, wealthy businessmen, bankers, and high officials counted among the city's new occupants. The same conditions of runaway inflation and material scarcities that made life so difficult for the average worker allowed bankers and businessmen to "grow wealthy off the national tragedy" (*fa guonan cai*) through profiteering, hoarding, currency trading, and speculating. High-placed officials feasted at lavish banquets and cavorted with courtesans while civil servants tightened their belts and made do with one meal a day, and day laborers scrounged for scraps and weeds. Although Chiang Kai-shek and Song Meiling (often referred to as "Madame Chiang Kai-shek" at the time) shunned extravagance, some Party officials reveled in luxury. When plane service between Chongqing and Hong Kong was briefly suspended after the bombing of Pearl Harbor, Song Ailing—First Lady Song Meiling's eldest sister and the wife of Minister of Finance Kong Xiangxi (H. H. Kung, 1881–1967)—chartered several private flights, at 10,000 gold shillings each, to transport personal belongings and the family's ten dogs. News of her actions caused great furor in the city.[15]

This kind of press gave Chongqing a reputation of despicable decadence. As the primary site of retreat from the Japanese, many people also associated Chongqing with a cowardly and selfish lifestyle. Cai Jusheng and Zheng Junli's 1947 film, *Yijiang chunshui xiang dongliu* (*A Spring River Flows East*), portrays it in this manner. In the film, life in Chongqing transforms the hero Zhang Zhongliang (Zhang the Loyal and Kind) from a patriotic Red Cross volunteer to a debauched lothario consumed with dreams of wealth and power. In a hackneyed narrative, this transformation begins with an evil seductress; arriving in Chongqing as a filthy and homeless refugee, Zhongliang beseeches an old friend for help. Now the adopted daughter of a local businessman, she weaves him into her black widow's web, offering him a job and taking carnal pleasures in return. Loyal Zhongliang shows up early for his first day of work, arriving at 7:50 a.m. only to find dust on the desks and not a soul in sight. Finally, after ten o'clock several well-dressed and stylishly coiffed men and women enter in small groups, laughing and logging an arrival time of 8 a.m. sharp. Once in the office they listen to operas, read comic books, and smoke cigarettes all day without doing any work, and then go out drinking. Soon Zhongliang joins them in these profligate behaviors, tainted by Chongqing's shameless self-absorption. Screened in 1947, the film's didactic tirade against Nationalist Party corruption spoke just as much to Civil War concerns as to the War of Resistance reality, yet it nonetheless reflected a portion of that reality.[16] Chen Guojun, daughter of a powerful Red Gang member and wife of a wealthy businessman, recalled that "during the war years the wives of rich and powerful people in Chongqing partied harder than ever.... [The] war did not affect our lives much, and I lived a life of luxury and spending." Even Chongqing's notorious air raids posed no threat to Madame Chen, who merely moved her *majiang* parties to her private shelter when the sirens rang out.[17] By 1941, Chongqing had over 1000 private air raid shelters assigned to 127 private citizens, as well as social organizations such as hospitals and government offices.[18] The Chongqing Municipal Hospital had a large and sturdy private shelter just behind its facility, where Bureau of Public Health Director Mei Yilin and his family comfortably sat out air raids on wicker chairs.[19]

While not all residents enjoyed such luxurious underground quarters, overall Chongqing provided a remarkable degree of protection from enemy bombs. Thanks to the rabbit's warren of air raid shelters constructed throughout the city, Chinese pilots' tight pursuit of enemy airplanes, the effectiveness of the capital's seasonal population dispersal, and an effective air raid warning system, Chongqing suffered relatively few casualties: 1.2 wounded or killed per bomb dropped.[20] Air raids so punctuated the rhythms of life that people soon grew accustomed to the new routine. People quickly learned to finish writing letters and eating meals before heading to the air raid shelter.[21] Japan's terror bombings did not kill civilian resolve; Chongqing always rebounded quickly after a raid (Fig. 10.1).

After incendiary bombs ignited interminable fires in the city of predominantly wooden structures, builders once again repaired the damage. In September 1940, the Church Committee for China Relief's newsletter included this assessment of the situation: Japanese air raiders "by persistently destroying [Chongqing's] shoddy buildings are *hastening the creation of a better city.*"[22] In 1941, foreign correspondent Israel Epstein described his wartime residence thusly:

> Two successive summers—two successive [air raid] seasons—had wiped out the Chungking that previously existed. During two successive winters, Chungking had been twice rebuilt, rising phoenix-like from its ashes with straighter, wider roads, with more rational planning of municipal

Fig. 10.1 A man perched on the roof of his house which he is planning to rebuild after its destruction in an air raid. Chongqing was the most bombed capital city in World War II. (Source: Prints and Photographs Division, U.S. Library of Congress, LOT 11511-7, WAAMD #131)

construction, with civic enterprises such as libraries, social service centers, consumers' cooperatives and dollar restaurants on almost every corner.[23]

Some of this reconstruction—such as the wide boulevards filled with department stores, theaters, and fancy restaurants—was the product of a gleeful government that had already worked hard to rebuild its first capital Nanjing on the Haussmann Plan model, and now found a silver lining on the cloud of dust rising from the bombs dropped on its wartime capital.[24] This type of rebuilding obviously benefited the wealthy, but largely disadvantaged the poor. Nonetheless, some poor refugees also rebounded from disaster. Zheng Hongfu, a "downriver" immigrant from Shanghai, arrived in Chongqing as an orphan and eventually built a restaurant, which was destroyed in the May 1939 air raids. Unfazed, Zheng reopened his business in a makeshift lean-to and continued to serve Shanghai cuisine from his "dollar restaurant" to other immigrants from his hometown.[25]

Assessing Vulnerability in Wartime Chongqing

Chongqing's air raid shelters certainly saved many lives, but on at least three well-known and openly reported occasions, the shelters themselves turned into death traps: June 11, 1939; August 13, 1940; and June 5, 1941. Death tolls ranged from 8 to 400 in the August 1940 incident, and untold hundreds—some say thousands—in the June 1941 "18 Steps" incident (named after the location of the shelter's main entrance).[26] In all three cases the victims died not from enemy bombs, but from lack of air in overcrowded shelters: they asphyxiated to death. Sadder still, the lack of air was a result of human error. Chongqing civil servants and civil engineers held direct responsibility for the untimely deaths of their own civilians. One reporter wrote of the June 1939 event, in which he witnessed many women and children who had died, "the Military Police posted at the entryway did not let the sufferers leave, even opening fire and killing people." Not having learned their lesson yet, in August 1940 when people once more tried to leave the shelter for a bit of air during an hours-long air raid, police officers and medical relief personnel stationed at the site "blocked their passage, creating conflict, so people [panicked and] trampled on one another, killing many." These people ran out of air because, although electric fans had been installed in the shelter for better air circulation—authorities *had* learned this lesson—they were all still. The electricity company had shut off power in that area for maintenance but failed to report the scheduled outage.[27] Moreover, during this unanticipated, nighttime raid, the shelter had accommodated more than twice its maximum capacity of 5000 people.[28] After the disaster, relatives who could locate their loved ones claimed their bodies and gave them proper burials, while hired burial teams lay unclaimed bodies in mass graves.[29]

Though deadly, these events were anomalies. Far more quotidian forces posed a much greater threat to the poor of Chongqing. Stories of Chongqing's poor reveal that poor sanitation, seasonal epidemics, hunger, and lack of medical care all caused a great deal of suffering. Xu Chengzhen's family endured hunger after paying medical costs for her younger brother, who sustained a shrapnel wound in his groin when his elementary school was bombed during an air raid. Despite their high price tag, doctors were unable to cure the little boy, and he ended up both crippled and sterile. Cui Xiangyu lost her three-year-old son after he contracted an illness in a crowded air raid shelter. Her husband, a military doctor stationed elsewhere, had little chance even to know his son before his premature death, much less to treat him. Ye Qingbi, who at the age of 13 migrated from Sichuan's Fuling county to Chongqing to find work after the death of her father, became a slave laborer in the Yuhua Textile Factory, never received a single payment for her work, and was forced to buy her way out of the contract when she fell ill.[30] Although they received good pay, arsenal workers in Chongqing frequently contracted lung disease from breathing the dust and particulates at their workplace, as well as the polluted air in the overcrowded and mountainous city filled with heavy, damp air that settled into the basin. Statistics for Arsenal No. 21 in wartime Chongqing attributed 90% of worker fatalities to sickness and disease.[31]

To a certain extent, this vulnerability to disease and other dangers existed despite the best efforts of local health officials. The Chongqing Bureau of Public Health (*Chongqingshi weishengju*, CBPH), established in November 1938 soon after the Nationalist government moved to the city, worked quickly and efficiently to develop a health infrastructure. By late 1940, the CBPH had opened the following hospitals and medical clinics: the Opium Treatment Hospital, the Trauma Hospital, the Infectious Disease Hospital, the Hongji Hospital (*Zhongguo hongji yiyuan*), and eight Maternal and Children's Health clinics.[32] The Chinese Red Cross had its headquarter hospital and a Chongqing branch hospital in the city.[33] Chongqing also boasted multiple private hospitals and clinics, including three missionary hospitals.[34] Three other hospitals came much later: in August 1943 the CBPH opened a Tuberculosis Sanatorium, and in May

1944 the Maternity Hospital (*Chanfuke Yiyuan*), each with 30 sickbeds.[35] Also in May 1944, the NHA opened the Wartime Capital Chinese Medicine Hospital (*Peidu zhongyiyuan*) with 30 staff, where hundreds of patients sought treatment each month.[36] By May 1944, a total of 21 hospitals and numerous clinics served the wartime capital (Fig. 10.2).

Lack of funding constrained all of the state-run hospitals and the financial situation only worsened as the war dragged on.[37] The Infectious Disease Hospital suffered constant financial shortfalls and equipment shortages from its opening in 1940. It had received only half of its originally planned construction budget and had rented some private residences so as to have enough room to accommodate its patients. CBPH Director Mei Yilin deftly juggled his limited resources to maximize health services. He combined the Infectious Disease Hospital with the second branch location of the Municipal Hospital in the Nan'an district and decreed that all poor patients requiring free treatment be sent to this location, thus

Fig. 10.2 Resilient city in action. Female nurses are carrying an air raid victim through the streets of Chongqing. (Source: Prints and Photographs Division, U.S. Library of Congress, LOT 11511-7, WAAMD #123)

cutting operating costs to a point at which the Bureau could cover them on its own without assistance from the municipal government.[38]

This type of juggling happened frequently, particularly during the air raid season when city officials encouraged people to disperse into the outlying areas to get out of the line of enemy fire. When Chongqing people moved into the suburbs, the Bureau still considered them its charges and sent mobile medical teams, health clinics, and even entire hospitals out to the "dispersal region" to provide for their medical needs. In August 1940, there were no less than six health clinics, six hospitals (three Trauma Hospitals, one Maternity Hospital, one Infectious Disease Hospital, and one missionary Hospital), two mobile medical teams, and one Sanatorium in this region.[39] During the biannual vaccination campaigns conducted every spring and autumn, the Bureau also sent roving teams of doctors and nurses into these outlying areas to administer shots to the dispersed population.

Clearly, health officials did their best to serve the needs of the poor, but they faced multiple obstacles. The sheer enormity of need, coupled with the fiscal challenges of living through a war that sparked runaway inflation, would have been enough to stymie good work. Yet health officials faced another major challenge: the need to adapt their work to the demands of central government officials, who placed prime importance on municipal aesthetics (*shirong*). As previously mentioned, Chongqing hosted foreign journalists, ambassadors, military advisers, missionaries, and others, whose frequent reports to their home countries shaped world opinion about China. Chiang Kai-shek believed that negative reports caused grave damage and did not wish to countenance any chinks in his armor while he simultaneously faced multiple enemies: the IJA, communist guerilla soldiers, and leftists within his own Party (including Wang Jingwei, who established a separate puppet government in Nanjing with Japanese tutelage in 1940).[40] Chiang therefore wished for his wartime capital to shine and sparkle, even though it also provided safe harbor to thousands of poor refugees who had already lost their homes, land, and frequently some family members to the invading army, and most of whom arrived in Chongqing on foot with only the possessions they had managed to carry.

These were the people who lived in the "War of Resistance shacks" along the two rivers that shaped the isthmus of downtown Chongqing. In January 1943, a new municipal regulation outlawed the construction of these shacks, citing their obstruction of traffic, shopkeepers' rights, and "municipal appearance." The law displaced an estimated 10,000 people who lived in such makeshift homes built by war refugees, primarily on the riverbanks.[41] Two such women, Wang Shufen and Li Shuhua, resisted this law as best they could. Li Shuhua and her husband joined with other poor families in a group of "guerrilla residents" who stayed in their makeshift bamboo shelters only until a government official discovered them, then moved along quickly to avoid punishment. Li and her husband moved at least 50 times in an ongoing effort to dodge state officials. During this time, Li Shuhua gave birth to six children, but due to her constant hunger and the anxiety of living in extreme poverty in a frequently bombed city whose police officers treated her existence as illegal, only two of them survived. As she put it, "a woman's body was just a machine for giving birth to babies, even though we had to bury most of them."[42] Wang Shufen had much less luck at avoiding reprimand; the security police regularly caught her and beat her with leather belts.[43] The two women's experiences gave testament to the implicit war on the poor within the government's urban aesthetic. The focus on municipal appearance taught state officials to treat the poor as dangers to social order, not as individuals in need of care. Li Shuhua and Wang Shufen experienced Nationalist government officials as adversaries; not only did the government fail to provide relief, it also refused to let them construct the only homes within their means.

Conclusion

Beset by air raids, floods, disease, inflation, and a punitive state, Chongqing's unique situation as wartime capital made it the first-choice destination for many war refugees, many of whom, already of meager means and now unmoored from their communities and peacetime livelihoods, eked out a tenuous subsistence within the unorthodox economy. Nonetheless, Chongqing health officials and police officers failed to recognize the additional struggles that these people faced during the war and continued their pre-war policies of policing poverty and criminalizing the poor. Even some of the available relief—such as mandatory medical care for mendicants—served more to reinforce this dynamic than to provide actual succor to the destitute.

Chongqing's poor occupied a city that granted them very little breathing room. Nationalist Party officials used public health regulations to police the citizens, rather than to serve them. While health officials kept the people's needs in mind, in order to court central government support, they had to couch everything in the preferred language of maintaining appearances. The Bureau of Public Health received a stream of orders from Chiang Kai-shek demanding that staff use their power to enforce specific behaviors and instill a particular aesthetic, rather than to create useful services that the people most needed. In socioeconomic terms, officials deemed middle- and upper-class residents clean and, by extension, modern, while they deemed poor residents filthy and, by extension, disgusting.

None of this was entirely new. The ideological connection between poverty and social undesirability had long pervaded Chinese culture. The Qing (1644–1911), Nationalist (1927–1949), and Communist (1949–) states all treated poor people like criminals "guilty of indigence."[44] 1920s guidebooks to Tianjin recommended that tourists steer clear of the poor Han and Muslim residential areas, where locals were "unsanitary in their habits."[45] However, the war pushed the state to its limit, and this pressure transferred all the way down to society's most vulnerable class: the destitute. Perhaps herein lies the most salient key to the victory of the People's Liberation Army in 1949. As communist guerrillas recruited people to their cause, they championed the right of the poor to take the reins of power in their own hands and fashion a world in which they could live at the top of the social pyramid. Never mind that this same Party now oppresses the poor in much the same way that its successors have done for centuries; in the late 1940s, no one knew that this would come to happen. They did know that the Nationalist Party had only managed to secure a bare survival for itself and its people during a harrowing conflict, all while only the wealthy thrived.

NOTES

1. Chongqing's official population count increased by nearly one million during the war. Many more residents likely evaded counting. Danke Li, *Echoes of Chongqing: Women in Wartime China* (Urbana and Chicago: University of Illinois Press, 2010), 18; Peng Chengfu, ed., *Chongqing renmin dui kangzhan de gongxian* (Chongqing People's Contributions to the War of Resistance) (Chongqing: Chongqing Publishing House, 1995), 125. Each source gives a different population count, mostly due to differences in determining city boundaries.
2. Wan Fang, *Chuanhun: Sichuan Kangzhan dang'an shiliao xuanbian (The Soul of Sichuan: An Edited Collection of Historical Materials on the War of*

Resistance from the Sichuan Archives) (Chengdu: Sichuan Provincial Archives, 2005), 2–3.
3. Yang Yulin, "*Bingli yu liangshi: Sichuan sheng disan xingzheng duchaqu renmin zai Kangzhan zhong de zhuyao gongxian*" (Military Power and Grain: Principal War of Resistance Contributions from the People of Sichuan province's Number Three Administrative Superintendency), in *Sichuan Kangzhan dang'an yanjiu (Research on the War of Resistance in Sichuan Archives)*, ed. Li Shigen (Chengdu: Southwest Jiaotong University Publishing House, 2005), 119, 121. See also Peng, *Chongqing renmin*, 1–17.
4. Tsung-Han Shen, "Food Production and Distribution for Civilian and Military Needs in Wartime China, 1937–1945," in Paul K.T. Sih, ed., *Nationalist China during the Sino-Japanese War, 1937–1945* (Hicksville, NY: Exposition Press, 1977), 181.
5. Diana Lary, *The Chinese People at War: Human Suffering and Social Transformation, 1937–1945* (Cambridge: Cambridge University Press, 2010), 173.
6. Diana Lary and Stephen MacKinnon, eds., *Scars of War: The Impact of Warfare on Modern China* (Vancouver: University of British Columbia Press, 2001), plate 8; Edna Tow, "The Great Bombing of Chongqing and the Anti-Japanese War, 1937–1945," in Mark Peattie, Edward Drea and Hans van de Ven, eds., *The Battle for China: Essays on the Military History of the Sino-Japanese War of 1937–1945* (Stanford, CA: Stanford University Press, 2011), 256–82.
7. Arup International Development, *City Resilience Framework* (New York: The Rockefeller Foundation, 2014), 3.
8. Lloyd E. Eastman, "Regional Politics and the Central Government: Yunnan and Chungking," in Paul K.T. Sih, ed., *Nationalist China During the Sino-Japanese War, 1937–1945* (Hicksville, New York: Exposition Press, 1977), 353.
9. Lloyd E. Eastman, "Nationalist China during the Sino-Japanese War 1937–1945," in John K. Fairbank and Albert Feuerwerker, eds., *The Cambridge History of China, Vol. 13: Republican China 1912–1949, Part 2* (Cambridge, UK: Cambridge University Press, 1986), 572–73.
10. Rana Mitter, *Forgotten Ally*, 173.
11. *Time* magazine correspondents Theodore White and Annalee Jacoby provided the best example of this in their book *Thunder Out of China* (New York: Da Capo Press, 1946).
12. Graham Peck, *Through China's Wall* (New York: Houghton Mifflin, 1940), 164, 167. By 1941, Chongqing had over 1000 private air raid shelters assigned to 127 private citizens, as well as social organizations. See

Chongqingshi tongji tiyao (A Summary of Chongqing Statistics) (Chongqing: Chongqing Municipal Government, 1942), Table 58.
13. Jim Endicott to Jesse Arnup, September 20, 1939, 1983.047C, 8-167, United Church of Canada Archives, Toronto (hereafter UCC).
14. Luo Zhuanxu, ed., *Chongqing Kangzhan dashiji* (Grand record of Chongqing's War of Resistance) (Chongqing: Chongqing Publishing House, 1995), 21; Lee McIsaac, Limits of Chinese Nationalism: Workers in Wartime Chongqing, 1937–1945 (PhD diss., Yale University, 1994), 40.
15. J.R. Sinton Papers, Chungking Diary, December 8, 1941–April 19, 1946, SOAS ASC, CIM Personal Papers, box 2, folder CIM/PP 20 Sinton Papers, 2.
16. Cai Chusheng and Zheng Junli, *A Spring River Flows East* (Shanghai: Lianhua Film Company, 1947).
17. Li, *Echoes of Chongqing*, 82–83.
18. *Chongqingshi tongji tiyao* (A Summary of Chongqing Statistics) (Chongqing: Chongqing Municipal Government, 1942), Table 58.
19. Huang Yanfu and Wang Xiaoning, eds., *Mei Yiqi riji* (*Diary of Mei Yiqi*), *1941–1946* (Beijing: Qinghua University Press, 2001), 35–36.
20. Freddie Guest, *Escape from the Bloodied Sun* (London: Jarrolds Publishers, Ltd., 1956), 172; Xiao Liju, ed., *The Chiang Kai-shek Collections: The Chronological Events* (vol. 43): January–June 1940 (Taipei: Academia Historica, 2010), 556. On Chongqing's air raid warning system, see Huang, *Mei Yiqi Diary*, 35–36. First, a warning flag hung high in the sky, signaling that enemy airplanes were on their way. Next, one ball was raised on the flagpole, signaling everyone to begin preparations to enter the air raid shelters; at the addition of a second ball, everyone headed directly to the shelters, and when enemy planes dispersed, the two balls were lowered to signal the all-clear.
21. Chen Lansun and Kong Xiangyun, eds., *Xiajiangren gushi* (Stories of Downriver People) (Hong Kong: Tianma Book Publishing Company, 2005), 453.
22. "Bombing of Chungking Fails to Break China's Morale," *Have a Heart for China* A-24 (September 1940), 1. Emphasis added.
23. Israel Epstein, "The May Third Chungking Bombing," in *United China Relief Series, no. 1: Undaunted Chungking* (Chungking: The China Publishing Company, 1941), 1.
24. McIsaac, "The Limits of Chinese Nationalism," 61; William C. Kirby, "Engineering China: The Origins of the Chinese Developmental State," in Wenhsin Yeh, ed., *Becoming Chinese: Passages to Modernity and Beyond* (Berkeley: University of California Press, 2000), 137–60; Zwia Lipkin, *Useless to the State: "Social Problems" and Social Engineering in Nationalist Nanjing, 1927–1937* (Cambridge, MA: Harvard University Asia Center Press, 2006).

25. Stories of Downriver People, 453.
26. The first newspaper report of the August 13 incident, printed in *Dagongbao* on the following day, cited rumors that at least 400 people had perished (*Dagongbao* No. 13179 [August 14, 1940]). On August 15, the Chongqing Municipal Air Raid Relief Team released its official report, citing only 8 dead, 36 seriously wounded, and 5 "severely disoriented" people ("Letter from the Chongqing Air Raid Relief Corps," August 15, 1940, Chongqing Asphyxiation Cases file, AH, Taipei). The newspaper responded, citing the Air Raid Relief Team's numbers along with those from another report, which stated that 9 had died and over 100 had been wounded, while also citing eyewitness accounts claiming that "the number of dead and wounded certainly exceeded 100" (*Dagongbao* No. 13181 [August 16, 1940]). On the death toll of the June 5, 1941 asphyxiation case, see Ding, "A Brief History of Air Raid Defense," 11; and Chen Lifu, "*Suidao zhixi an shen weiyuan baogao fabiao*" ("Tunnel Asphyxiation Case Investigative Committee Report"), 1941, AH, Taipei, 8b-10a.
27. *Dagongbao* (*L'Impartiale*) No. 13181 (August 16, 1940).
28. Ding, "A Brief History of Air Raid Defense," 10–11.
29. In June 1941, 151 severely wounded victims were taken to the Trauma Hospital for immediate treatment, and boatmen transported 920 unclaimed bodies to Black Pebbles (*Hei shizi*) for burial in mass graves. Claimed bodies constituted a minority and could not have totaled more than 72 of the reported 992 dead. See Chen, "Tunnel Asphyxiation Case," 8b-10a. An official report from the Chongqing Air Raid Relief Corps filed for the August 1940 asphyxiation case cited only eight dead bodies, which they "sent away to be claimed by relatives." "Letter from the Chongqing Air Raid Relief Corps." Chongqing civilians also voluntarily participated in memorializing the air raid dead; they reportedly piled the remains of over 7000 compatriots into 12 "White Bone Pagodas" (*Baigu ta*). None are extant today.
30. Li, *Echoes of Chongqing*, 102–06, 61–64, 121–23.
31. Joshua Howard, *Workers at War: Labor in China's Arsenals, 1937–1953* (Stanford: Stanford University Press, 2004), 130–31. Arsenals used coal to fire their ovens, exposing workers to "high levels of sulfur and ash."
32. CBPH Work Report, 1938, CMA, Chongqing, 66-1-2, 182; CBPH Work Report, January–June 1940, CMA, 66-1-3, 171; CBPH Work Report, 1940, CMA, Chongqing, 66-1-3: "Table of Private and Public Hospitals in Chongqing Proper and in the Suburbs," August 1940, CMA, 66-1-3, 182–83; and CBPH Work Report, September 1940–February 1941, CMA, 66-1-3, 198.
33. "Table of Private and Public Hospitals."

34. Ibid.; "Survey of Physicians, Pharmacists, and Midwives in Chongqing City."
35. CBPH Work Report, 1943, CMA, 66-1-2, 203–04; and CBPH Work Report, 1944, CMA, 66-1-2, 17, 66.
36. "Wartime Capital Chinese Medicine Hospital" files, 1944–1945, CMA, 163-2-19; 163-2-24.
37. For example, in 1944, the Trauma Hospitals exceeded their collective budget by over 76,000 yuan. "Trauma Hospital Budget" documents, March–June 1945, CMA, 53-19-1920.
38. *Shili chuanranbing yiyuan xianzai qingxing ji jianglai banli gejie* (Current Conditions and Future Plans of the Municipal Infectious Diseases Hospital) October 1942, CMA, 66-1-6, 115–116.
39. *Chongqingshi jiaowai gong sili yiyuan zhensuo yilanbiao* (A Table of Public and Private Hospitals and Clinics in the Outskirts of Chongqing), August 1940, CMA, 66-1-3, 182–83.
40. Mitter, *Forgotten Ally*, 197–209.
41. Zhou Yong, *Chongqing tongshi, vol. 3* (A Comprehensive History of Chongqing) (Chongqing: Chongqing Press, 2002), 1160.
42. Li, *Echoes of Chongqing*, 86–87. Stories like Li Shuhua's corroborate Gail Hershatter's discovery that many rural women in the collective era—another period in Chinese history when women's domestic and reproductive labors were unrewarded—experienced extreme exhaustion during their reproductive years, and therefore supported the One-Child Policy in the 1970s. Gail Hershatter, *The Gender of Memory: Rural Women and China's Collective Past* (Berkeley: University of California Press, 2011), 182–209.
43. Li, *Echoes of Chongqing*, 86, 90, 93.
44. Janet Y. Chen, *Guilty of Indigence: The Urban Poor in China, 1900–1953* (Princeton: Princeton University Press, 2012).
45. Ruth Rogaski, *Hygienic Modernity: Meanings of Health and Disease in Treaty-Port Tianjin* (Berkeley: University of California Press, 2004), 201.

CHAPTER 11

Japanese-Occupied Hanoi

War, Famine, and Famine Management in the Red River Delta

Geoffrey C. Gunn

INTRODUCTION

Astride the Red River of northern Vietnam in one of the most densely populated rural zones on earth. In 1945–46, the city of Hanoi faced down a crisis practically unprecedented in its near-1000-year history.[1] Notwithstanding an uneasy cohabitation between invading Japanese forces that progressively intervened in Vietnam from 1940 and a compliant Vichy French colonial regime, life began to change in Hanoi with the impact of US bombing raids over the city commencing in 1943–44. Simultaneous with a devastating cycle of typhoons and floods that coincided with the Japanese military *coup de force* of March 9, 1945, against the French, agriculture collapsed with devastating human consequences. Famine was no stranger in Vietnam, just as the ancient kingdoms and, in turn, the French colonial administration had developed elaborate hydraulic systems to harness the waters of the Red River Delta in the production

G. C. Gunn (✉)
Emeritus, Nagasaki University, Nagasaki, Japan

© The Author(s) 2019
S. Laakkonen et al. (eds.), *The Resilient City in World War II*, Palgrave Studies in World Environmental History, https://doi.org/10.1007/978-3-030-17439-2_11

of wet-field rice along with coping mechanisms such as the construction of elaborate systems of dikes to protect against seasonal flooding. But with the American bombing of major infrastructure, including the vital north-south trans-Indochinese railroad connecting the rice surplus of southern Vietnam with the food-deficit central and northern zones, and with the Japanese internment of French officials, food relief was paralyzed and/or mismanaged, leading to the deaths of between one and two million people.[2]

Set against the social ecology of the Red River Delta, as studied by the French geographer Pierre Gourou, the population of northern Vietnam faced down a veritable Malthusian crisis.[3] Onerous Japanese military rice requisitions, colonial mismanagement, and the exogenous impact of war all played their part. Neither can urban Hanoi surrounded by protective dikes and a sea of flooded paddy fields be easily compared to Western cities with their far less populous hinterlands. Still, as the capital of a federal Indochina, wartime Hanoi commanded resources, especially military and civilian stockpiles of food. The keys of these stockpiles and rice warehouses were dramatically transferred from Vichy French hands to Japanese military control on March 9, 1945. While up until this point, Hanoi—as with such other Japanese-occupied or Japanese-controlled cities as Canton, or Macau—witnessed a coexistence of poor and needy alongside a privileged class of officials, merchants, and other well-connected individuals who could cope with rationing and rising prices, increasingly the famine began to level the playing field as all food sources began to dry up.[4] Certainly, as the Indian economist Amartya Sen argued with respect to the near-contemporaneous Bengal famine of 1943—and not without reference to our study of urban Hanoi—absolute scarcity, or insufficient food to feed everyone, is rare even in wartime. Much more important is the question of distribution or, at least, inequalities in distribution and who has the power to manage distribution. Ultimately, in this scenario, food becomes a weapon.

As acknowledged in mid-August 1945 by the first arriving Allied forces in Hanoi following the Japanese defeat, nature was also unremitting, unleashing a great flood—the worst in decades of recorded history—that literally turned the city into an island surrounded by an immense expanse of flooded paddy fields. Nevertheless, different urban groups suffered differentially.

With special attention to the social ecology of the Red River Delta, I also seek in this chapter to examine rural-urban connectivity through various stages of the crisis, with urban exodus in the face of wartime bombing and rural ingress as famine stalked the countryside and, at the height of the famine, with corpses lining the boulevards of the once elegant French Indochinese colonial capital. The famine served as a powerful propaganda call by the triumphant rural-based Viet Minh guerrillas, who seized power in Hanoi in August 1945. Just as the French suffered defeat at the hands of the Japanese, so they lost the "hydraulic pact" established between ruler and ruled in the critical management of the dikes, canals, and embankments that crisscrossed the Red River delta. But even if the French had lost the "mandate of heaven" in this Confucianized world, still the insurgent Viet Minh had to win their legitimacy by carrying on the task. The chapter also discusses the theme of resilience against adversity as survivors sought to cope and the new revolutionary order faced down a range of challenges both human and natural.

Environmental Background to the Famine

For Pierre Brocheux and Daniel Hémery, two close students of Vietnam's social and political landscape, the background to the crisis was essentially demographic.[5] Public health programs and vaccination campaigns did control mortality stemming from terrible cholera epidemics. From 1927 onward, Tonkin (northern Vietnam) was spared of catastrophic ruptures of dikes until the dramatic flooding of August 1945 when 230,000 hectares were submerged, the most serious flooding of the century. But, in the course of a century of French contact, the population of Vietnam had increased by a factor of six and cultivated surface by two. The balance of population and grain production therefore became extremely uncertain and the peasants were periodically wracked by agro-ecological crisis. Starting before 1930, vast areas of rural misery expanded in the regions where the ratio of population to cereal production was most strained, namely the Red River Delta and the north-central provinces of Nghe An, Ha Tinh, and Quang Ngai. In 1937, there were from two to three million agricultural day laborers and more than a million unemployed in the Red River Delta. There was also extreme parcelization of land ownership and a rising class of Chinese-style big landlords. Taken together, Brocheux and Hémery argue, the situation approximated that of "agricultural involution" as described by Clifford Geertz in his classic study of late colonial and early postwar rural Java, characterized by increasing labor intensity in the rice paddies, but without increasing output per head.[6]

The basic point is that, while in normal times the overcrowded northern delta region could feed itself and indeed even export a rice surplus, in times of dearth occasioned by floods or other natural disasters, it was reliant upon food imports from central Vietnam or especially the southern Mekong Delta region with its abundant rice surplus. In normal times in colonial Vietnam, north-south transport was assured by the trans-Indochinese railroad along with coastal steamer and junk trade. As the war progressed, China-based US aircraft mercilessly commenced bombing transport infrastructure, sometimes hitting crucial dikes as well as cities. American submarines also took their toll on both Japanese and Vichy French shipping without distinction.

Political Background to the Famine

The background to the Great Famine in northern Vietnam is the increasing scale and character of Japanese military intervention in Indochina from 1940 down to surrender in August–September 1945. The Vichy French regime in Indochina and Japan existed in a tense albeit unequal cohabitation with Japanese forces until matters changed absolutely on March 9, 1945, when Japan mounted a *coup de force*, militarily interned all French military personal who did not escape to the mountains, and sequestered most French civilians. The Japanese military took over full administrative responsibility alongside local puppet regimes, including the Tran Trong Kim cabinet in Annam (central Vietnam), under a pliant Emperor Bao Dai. Economically, Japan had used Indochina under the Vichy administration as a source of industrial and food procurement, from coal to rubber, to a range of industrial crops and, especially, rice from the surplus-producing Mekong delta region. Though notionally under French administration, Japanese military requisitions profoundly distorted the colonial political economy, shattered the import-export system, eroded many bonds across communities and classes, and sowed the seeds of disasters to come. Even with French administrative services continuing, including dike repair, the monitoring of agricultural activities, and the collection of taxes, the rural population, increasingly bereft of cash as market mechanisms collapsed, was obliged to cope in a situation of virtual economic autarky just as Indochina came to be subordinated within Japan's Greater East Asia Co-Prosperity Sphere.[7]

Hanoi Profile

In a classic colonial set-up, the French constituted the official class, administrators, and their families, along with military and business people. As a privileged category, they occupied a discrete quarter of the city, and even today the grandeur of their architecture is on full display. The category "French" may also have masked natives of such French colonies as Pondicherry in India or Réunion in the Indian Ocean. The Chinese were active in a number of trades but were predominant as rice merchants linking with the point of production through the Red River Delta. As residents of the old quarter of Hanoi, they were also connected by networks of solidary organizations. Many Vietnamese, a heterogeneous group, worked in administration and, thereby, as a relatively privileged class were, like the French, eligible for a limited rice ration. But the majority of Hanoi's Vietnamese population, working class or day laborers with one foot in the countryside and one foot in the city, had no safety net and, with their women and children, were the most vulnerable of the city's population to food security. They rallied en masse to the Viet Minh.

Even allowing for some population increase during the war years, the number of Vietnamese evacuating the city to the countryside or back to their natal villages appeared to amount to more than 50 percent of Hanoi's population. The population of Hanoi in 1940 was estimated at 135,000. This number included 4000 French, 6000 Chinese, and 125,000 Vietnamese. In this connection, the Japanese consul in Chiang Mai (northern Thailand) reported that 60 percent of the population evacuated that city as a result of Allied air raids.[8] Then, as today, Hanoi was divided into urban and rural quarters or districts, just as the rural districts far eclipsed the urban in population according to how the boundary was drawn. As Danielle Labbé determined from her research on Hanoi's broader geographical setting, a strong "interdependence" existed between the city's periurban villages and settlements and the inner city in terms of labor, industrial production, and trade. The "margins" of Hanoi also served as a vital "productive belt" in terms of ensuring food security.[9] We cannot track Hanoi's demography exactly through the 1940–45 period, but the urban-rural exodus of native Vietnamese became more pronounced as the US bombing continued to target urban sites.

Population Movements and Food Shortages

In a series of reports through December 1943, Japanese Ambassador Yoshizawa Kenkichi specifically referred to Allied air raids of Hanoi on December 10 and 12, 1943. In one of these reports he focused upon the psychological effects of the raids (unstated in this report but inferring major consequence). Accordingly, on December 15, he instructed all Japanese women (numbers not specified but likely scores) who had taken refuge in the suburbs of Hanoi to evacuate to Saigon—evacuation commenced the following day. Second, all French officials and merchants were ordered to remain in Hanoi but, by December 17, 1500 French had evacuated. Third, by December 15, 40,000 Vietnamese were believed to have fled Hanoi and, by the 19th, the figure had risen to 70,000. A food shortage was developing in spite of an order that food markets be kept open only from 3 to 8 p.m. Prices had already increased from 100 percent to 150 percent and were rising daily.[10] This was a quite early warning of the risks of future starvation if critical food supply was not met. But at this stage the Japanese and French authorities did not take urgent measures to address the problem.

According to Ken MacLean, who interviewed survivors in Hanoi in 2000–03, food shortages were not limited to the countryside. In Hanoi, monthly food rations declined dramatically between late 1943 and early 1945 from 15 to 7 kilograms of paddy per person.[11] Much of that paddy was either stale or moldy and heavily cut with rice husks and bran. Citing Vietnamese sources, as MacLean reveals, the steady decline in the size and quality of the rations fed fears that existing stockpiles were approaching exhaustion.[12] The fear led the black market price for paddy to rise dramatically from 57 to 700–800 VNĐ/kilogram over this period. The price eventually peaked at 2400 VNĐ in July of 1945.

As French sources confirm, the ability to purchase food on the market became increasingly difficult even for the privileged salariat. Corruption was rampant not only on the part of middlemen engaged in the food business but also on part of the role of officials. For example, in January–February 1945, the concerned French civilian authority in Hanoi, Paul Chauvet, responded to actions of hoarders and speculators by sending defaulters to work camps, including the president of the Hanoi Grain Board. Although too little and too late, he also sought to boost stockpiles of rice in Hanoi at a number of locations (as did the French military and the Japanese). He also set up a system of ration cards especially for officials and urban classes, although widely exploited. In early January 1945 he also sought to control the ostentatious consumption of food in Hanoi's restaurants, as with restrictions upon the number of dishes and opening times. Cakes were eliminated from sale and bread became strictly rationed. By this stage, the presence of abandoned children in the streets of Hanoi came to official attention. Without demonstrable success, he also sought to mobilize a fleet of junks to sail south and return with rice. In any case by early March, with the Japanese military takeover, top-level French management, Chauvet included, were interned by the Japanese.[13]

Impacts of US Bombing on Rice Supply

Toward the end of the war, US bombing raids on Vietnam mounted variously from India, Yunnan in China, and the Philippines, as well as from carrier-based aircraft, began to take a toll on infrastructure in Vietnam. The major target was the Trans-Indochinois rail line that linked north and south Vietnam, including spur lines, bridges, and rail junctions. Bombing was at times wildly inaccurate, as bombs falling on urban areas occasionally hit critical dikes. Harbors were mined as submarine raids took a deadly toll upon both Japanese and local coastal shipping (not excluding Vichy French shipping). With all but a few French administrators behind bars, administrative services deteriorated, whether run from Hanoi, Saigon, or Hue. In this environment, customary rural statistical surveys were rarely conducted. Japanese military authorities, moreover, paid scant attention to local needs across Vietnam, not to mention traditionally rice-deficit Laos, and even rice-surplus Cambodia, also ruthlessly exploited of its rice resources. Japan's priority was to fulfill Imperial imperatives designed to feed Japan's own on the battlefronts and at home.[14]

For example, on February 6, 1944, a Japanese consular official in Hue reported that Indochina's coastal railroad had been bombed at a point just north of Quang Tri and, as a result, rail traffic between Hanoi and Saigon had been interrupted. As he pointed out,

> The coastal line is Indochina's only through North-South railroad. It is single track and narrow gauge and its capacity has been hampered by its depreciation of rolling stock. In recent months it has been subject to increasing strain owing to the closing of Haiphong port to large ships and has made it necessary for the Japanese a/to transport badly needed war materials produced in the north overland to southern Indochina for export to Japan and b/to rely upon the railroad for transporting from the south imported goods destined for the North.

As the American analyst commented, "The Japanese have been hoping to alleviate the situation with coastal shipping but so far have accomplished little in that direction."[15] This was true and proved to be a major cause of supply failure during the height of the Great Famine during the Tet or Vietnamese New Year of February–March 1945.

Separately, a summary of Japanese reports describing damage done to railway bridges in Indochina by the US 14th Air Force during February 1944 offers a short list matching with American reports. For example, six planes attacked bridges on the railway line some distance north of Quang Tri (February 5). French engineers estimated that repairs would take up to eight months, with pier repairs requiring diving operations. In the meantime, sampans and trucks were being used for trans-shipment of goods, taking four hours to cover the 25-kilometer gap (February 14). Another bridge was destroyed at Dong Ha, estimated to take one month to repair (February 6).[16]

According to Martin Mickelsen, the Commander-in-Chief of the French army in Vietnam, General Eugene Mordant, was convinced that the 14th Air Force had deliberately bombed Hanoi in December 1943, and again in April 1944 in retaliation for Vichy French Governor General Jean Decoux's policy of surrendering downed American fliers to the Japanese.[17] Notably, on December 10 and 12, 1943, Hanoi (and not the usual target of the nearby Japanese airfield at Gia Lam) had been attacked for the first time, causing 1232 casualties and 500 deaths. On April 8, 1944, Hanoi was hit again by the 308th Bomb Group (H) when the Yersin hospital complex was targeted, leaving 46 civilians dead, with 141 wounded in Vietnamese and Chinese residential areas. Mordant's fears were reinforced by a warning from the 14th Air Force Commander Claire Chennault that all the major towns in Tonkin would be bombed if the act of transferring American prisoners to the Japanese be repeated.

On August 6, 1945, Joseph C. Crew, Acting Secretary of State, communicated to the Secretary of State that military operations in Tonkin and the prospect of serial bombardment of the dikes in the Red River Delta—as evidently contemplated by military planners—would cause "formidable danger to the population of this area." This was no understatement. At risk were the lives of eight million people in densely populated land crosscut by dikes built up over the centuries. The gravity of the situation had been conveyed initially by the French Military Mission in Kunming to the Commander of the US 14th Air Force.[18]

An Eyewitness to Starvation

Even though the famine was reported in local Vietnamese newspapers of the day, surprisingly few French accounts have emerged. In this sense, Michel L'Herpinière's recollections of his Vietnam years are exceedingly valuable.[19] L'Herpinière attended Lycée Albert Sarraut, where the future General Vo Nguyen Giap was among his teachers. As war clouds developed, he joined the colonial army, moving through university in 1944, also becoming a clandestine resistance operative. He experienced the US bombing raid over Hanoi of October 1943, the biggest air raid to date, involving at least 30 US bombers targeting the Gia Lam railway yard and depot, and with the Gia Lam airport bombed as well, a "terrifying experience for all."[20]

As the war progressed it took its toll on nutrition. In L'Herpinière's account, there was no coffee or wheat flour. Bread was made from a mixture of rice and corn flour. With medical imports frozen, the Pasteur Institute attempted to improvise. Inflation was rampant, with prices up 400 percent. Even the black market run by the Chinese was out of reach for the average Vietnamese. Buses were then running on charcoal and with rice-based alcohol fuel strictly reserved for the military. At the time of the *coup de force*, the Japanese were requisitioning rice from the reserve granaries to feed their armies and, with a bad season looming, northern Vietnam was threatened with major food deficit. Those with money could still purchase on the black market, but those without were already in difficulties. By the time of the Japanese *coup de force*, food supplies in the city were becoming "critically low."[21]

Emerging from a five-month period of incarceration in Number One Police Station in Hanoi on August 16, 1945 (at the moment of the Japanese surrender and himself a medical case owing to prison privations), L'Herpinière beheld a staggering scene of the streets of Hanoi. Turning down the normally busy Boulevard Bourgnis Desbordes (Trang Thi), he observed "dead bodies, perhaps a dozen of them scattered." "Some appeared to have died in their sleep, and others lay twisted in grotesque positions where they had fallen. All were Vietnamese, all were emaciated." As he entered another road, "sprawled on both sides of the street were crumpled human forms, dead and dying of starvation." The only sign of life he beheld was a couple of Vietnamese hauling a two-wheeled ox cart, crisscrossing from one side of the street to the other. Two more men were picking up and unceremoniously dumping the corpses into the cart "in a tangle of wasted arms and legs." Others near dead, as L'Herpinière speculated, would doubtless be picked up dead the following day. Further moving on through Hanoi's streets he encountered the frail and dying gathered around "in emaciated little groups, perhaps finding some comfort in their togetherness."[22] Recuperating in a hospital and musing upon the sight of a bird, he remarked that "even the wildlife had suffered terribly in this famine, as every edible creature was hunted down for food."[23] Venturing out in the morning he noted that the streets were clean of corpses, which had been removed by the ox-cart gang. Still, swarms of beggars remained. On the other hand, wealthy Vietnamese were now out and about, apparently untouched by the famine.[24] That condition held for French citizens under virtual house arrest. Although near-starved, the 5000 or more French military incarcerated in Hanoi's eighteenth-century citadel survived. As L'Herpinière remarked, the Japanese had commandeered every grain of reserve rice from government granaries, including that held in reserve to tide over a bad season. But in another turn of misfortune, with the dikes rundown, the latest rice harvest was devastated by floods. With the famine raging over a period of four months, starving peasants flocked from the provinces to the city hoping to find relief. Even so, the Japanese released no rice to feed them. As alluded, the most they had organized to manage the situation "was the conscription of a few hundred youngsters to haul the oxcarts and pick up the day's corpses, and the digging of huge dumping pits on the outskirts of Hanoi."[25]

The Vo Anh Ninh Photos

Still, we lack a picture of the everyday lives of people on the streets of Hanoi. Beside some literary works in Vietnamese and newspapers of the day, the best evocations of the famine as it affected Hanoi and surroundings are actually those captured by the young photographer Vo Anh Ninh (1907–2009). His photographs were first published in the Viet Minh's *Cuu Quoc* (Save the Nation) magazine (No. 133, January 3, 1946), and were quickly reprinted elsewhere.[26] In a later interview, Ninh said his photo-"reportage" constituted a visual denunciation of the French and Japanese crimes, but added that what struck him most was the inescapable smell of the dead and of those about to die.[27]

Many of his photos were shot on the road to Thái Bình (110 kilometers from Hanoi) on one of his bicycle trips through the country. Of these photos half are identifiable urban scenes, undoubtedly Hanoi. Not accurately dated, they pertain to March–August 1945 and with the Japanese still in charge. The photos offer graphic evidence that, during the famine's peak in March 1945, people from other provinces flocked to Hanoi with the vain hope of finding food. Perhaps the most evocative of the photo scenes is the hand-drawn "ox cart" laden with corpses such as described in L'Herpinière's recollections. The photos' captions are also evocative: "Some hungry people attempt to steal rice from the Japanese army. In the photo they are beaten up by the Japanese forces after they were caught stealing". "People from Hai Phong City, Nam Dinh, Thái Binh, Hai Duong, and Ha Nam provinces flood to Hanoi to beg for food. Many of them died while waiting." One photo reveals corpses being stretchered off the streets, another offers a close-up of limbs protruding from the corpse-laden cart. Yet another shows a lorry laden with corpses lain horizontally. And yet another shows the indigent squatting, huddled together on a street with a colonial edifice in the background, again befitting L'Herpinière's description. Still other photos reveal horrific close-ups of emaciated women and children, and still others, piles of corpses. Other photos coming to light reveal women trying to catch crabs and snails, perhaps a typical urban and rural activity, alongside a street scene with people pathetically seeking to pick up grains of rice from the street. "Thousands of starving people go begging for food near Hang Da Market in Hanoi" (still a small market, 300 meters west of Hoan Kiem Lake). "Horrific photos recall Vietnamese Famine of 1945," "Groups of hungry people in front of Cua Nam Market in Hanoi" (today a flower market inside a modern building, a few blocks north of Hanoi's railway station, itself a great gathering point for the indigent in 1945), "Thousands of starving people go begging for food near Ham Da market in Hanoi," "Hungry people gather at a social welfare center in Hanoi," "The skulls of starved victims piled up at the basement of Hup Thien graveyard in Hanoi" (also memorialized with a collective burial in the graveyard in 1951, dubbed famine of 1944–45).

The famine photos also had "political" importance. A separate set were presented by Ho Chi Minh to Archimedes Patti (1980: 86) of the American Office of Special Services or OSS (precursor to the CIA) in early September 1945 with the request to bring the humanitarian disaster to the attention of the American authorities, although without evident response.[28] Desperately, in late November, Ho Chi Minh addressed a series of cables to US President Harry S. Truman calling upon his humanitarianism. As he pointed out in one of these letters, owing to "the starving policy of the

French," two million of his countrymen had died of starvation during winter 1944 and spring 1945. In summer 1945, three-quarters of cultivated land had been submerged by flood, followed by drought and with five-sixths of the normal harvest lost. Moreover, the presence of the Chinese army imposed a further burden and with the "French-provoked" conflict in the south further delaying relief. Although received by the US State Department, Ho Chi Minh's cables were not even passed on to the president.[29] Here Ho Chi Minh is referring to the entry of 152,000 Nationalist Chinese army personnel in northern Indochina under General Lu Han to accept the Japanese surrender. Not only did the Chinese presence compound the difficulty of food supply, the Chinese also had political ambitions such as in succoring political rivals to the Viet Minh.

After the Revolution

The events surrounding the Japanese surrender in Hanoi, the August Revolution of 1945 whereupon the Viet Minh dislodged the Imperial delegate and seized power, the formation of a Provisional Democratic Republic, and the proclamation by Ho Chi Minh on September 2 of the Democratic Republic of Vietnam (DRV) are fairly well documented. This includes not only official Vietnamese histories but such witness accounts as those of Archimedes Patti, who arrived in Hanoi on August 22 via air from Kunming accompanied by the Gaullist French representative Jean Sainteny, charged with ensuring the relief of French citizens. It is striking, as Sainteny recorded, that Hanoi was surrounded by a "lagoon" of floodwater. The Red River was then at an historic high water level of 12.40 meters and with the Banc de Sable village opposite Hanoi fully submerged.[30]

Given the issues of floods and food security we wonder how the Viet Minh coped with the situation. As the L'Herpinière account explains, the most urgent problem confronted by Ho Chi Minh and the newly installed DRV administration was dealing with the famine. "In the rural areas where there was no military presence, the daily struggle for survival had fallen to banditry and anarchy. Ho had all the rice in Japanese-held granaries immediately seized and milled for distribution. He introduced new laws, increasing the land under rice cultivation for the next harvest of 1946."[31] King C. Chen also confirms that the food situation was on the verge of disaster.[32] According to Viet Minh estimates, the 1945 autumn harvest was poor and hardly sufficed to feed eight million persons for three months. To avoid nationwide starvation, the Viet Minh government launched an All-out Campaign Against Famine.

The problem was only gradually alleviated with the arrival of supplies from Saigon. There were also fewer mouths to feed. French prisoners in Hanoi were also on the brink of starvation and speed was of the essence. There was also contest over who would distribute the relief between the French, the Chinese, and the DRV administration. Although these developments are not well documented, the DRV as the de facto administering authority in Hanoi and the countryside would eventually take charge of food distribution. Undoubtedly, the seizure of Japanese-controlled grain stores along with those formerly controlled by the French military was critical for survival, and Hanoi may well have been better stocked in this regard than rural areas. But Hanoi and its hinterland also depended upon external supply of subsidiary sources of food such as would normally stock marketplaces. It is not well documented, but with the Chinese in control of the highland border regions with rice-surplus Yunnan and western Guangxi, it is likely that petty traders and smuggler bands, some of them dealing in foodstuffs, operated with Chinese protection, especially in favor of their clients among Viet Minh political rivals.

How then did Hanoi cope with the crisis? The first external relief arriving in Hanoi actually comprised milk and medical supplies flown in regularly from Kunming by the OSS, at least until late September when their mission was withdrawn. With some pathos, one of the OSS teams compared the plight of children in Hanoi to that of the Holocaust.[33] However, it was not until the end of November that the first major consignment of food relief arrived in Haiphong port from Saigon. This was aboard the French vessel *Kontum*, bearing a cargo of 9000 tonnes of rice. While considerable in itself, the French were obliged to work with the Chinese army in Haiphong, who retained 1000 tonnes for their own purposes and with further amounts siphoned off as bribes before the balance was delivered up to the French-controlled citadel in Hanoi. With the *Kontum* sunk in Haiphong harbor by a Japanese mine as it embarked for the return trip to Saigon, it would not be until January 1946 that the next French relief convoy arrived in the northern port.[34] It is not clear, but we may assume that at least some of this rice made its way into the Haiphong and Hanoi marketplaces.

According to David Marr, although food riots appeared imminent "at one point" (October) "and starvation was not prevented entirely," there was no repeat of the mass deaths that had taken place one year earlier. By May 1946 most districts could report a bountiful rice crop ready to harvest.[35] The turnaround is confirmed by Nguyen The Anh, who asserts that, by the end of 1945, the DRV claimed that the harvest of the winter of 1945 took place under favorable conditions and that the harvest of the fifth month of 1946 produced a yield superior to the prewar period or the equivalent of 2,592,000 tonnes of basic foodstuffs, including supplementary crops.[36] If so, then this was extraordinary recovery, especially given the human losses, flood damage, loss of farm implements (including livestock), and, especially, the question of replenishing vital rice seed needed for replanting.

Conclusions

Although spared the physical destruction of many other European or Asian cities during World War II, as this chapter has highlighted, Hanoi and its population endured a food crisis of historic proportions. This was also at a critical juncture in Vietnamese political history, with the northern Viet Minh leveraging themselves to power at the moment of the Japanese capitulation in August 1945 and with key French colonial administrators charged with the management of the famine either in prison or out of the scene.

Wartime bombing also took its toll upon the population of Hanoi, inducing some 40–50 percent to vacate the city for the countryside, as revealed by Japanese reports. Such a movement also fitted local culture and tradition, as with the practice of urban dwellers, including recent or even temporary arrivals from the countryside, to return to natal villages during festival periods or to lend a hand during times of intense agricultural labor. But in the Hanoi-Red River Delta, when the famine began to bite, it was the destruction of rail infrastructure linking rice-surplus southern Vietnam with the north that massively impacted food supply, creating devastating consequences for rural communities. With major ingress of desperate peasants into urban Hanoi, the city literally became a graveyard for incalculable numbers. This influx must also have been based upon a prevailing felt peasant logic that the city was rich or somehow better endowed with a food surplus relative to the countryside. That was a correct assumption if stockpiles and privileged food hoardings are taken into consideration, but, as revealed by L'Herpinière and confirmed by MacLean, food prices were far beyond the most privileged classes and food relief and charity were also in short supply in this moment of crisis. As a privileged "island," the city may have been resilient to an extraordinary degree, but the countryside and its dwellers also carried the burden.

In other words, Hanoi and the wider Red River Delta region experienced a series of shocks arising out of the war that went far beyond traditional coping measures. Shortages, inflation, and, for the indigent peasant rice producer, forced crop requisitions plunged millions below the survival threshold. Adding to the precarious situation of millions, US bombing also threatened crucial dikes and flood control mechanisms. Coincident with the power vacuum produced by the Japanese eclipse of French colonial power, the population of the Delta faced down a massive human environmental shock engendered by the flood-famine-drought-flood cycle, in turn impacting upon urban Hanoi and its citizens. While the fledgling DRV administration appeared to have managed relief where all other services had run down, a number of unforeseeable consequences served to prolong the crisis. Here, we might include the lingering Nationalist Chinese military occupation, Viet Minh excesses committed against landlords and/or French collaborators in early attempts with land reform, and with full-on war breaking out between the French and the Viet Minh in northern Vietnam in December 1946. Nevertheless, the famine itself would recede as natural conditions returned to some kind of normalcy.

Whether or not the famine contributed to the Viet Minh victory is beyond the scope of this chapter, but it was clearly momentous in ushering in a new era. Here we are not only reflecting upon the political events surrounding the August Revolution in Hanoi, but also reflecting upon new policies that would seek to transform agriculture collectively and to build food security, no matter how contested. However, to attribute the famine outcome to adverse weather conditions, as some Japanese spokespersons put about at the time, misses the point about crisis famine management, including inequalities in food distribution such as theorized by Sen with respect to the Bengal famine of 1943.

NOTES

1. For a study of the city as an imperial capital reaching back to its origins in 1010 as Thang Long, and with maps supplied, see Nguyen Khac Vien, ed. *Hanoi. Volume 1, From the origins to the 19th century* (Hanoi: Vietnamese Studies, no. 48, 1977). For a popular modern history but with a spare paragraph on the wartime famine, see William Stewart Logan, *Hanoi: Biography of a City* (Sydney: University of New South Wales Press, 2000). For a longitudinal study on Hanoi connected to its immediate rural hinterland written by a professional geographer, see Danielle Labbé, *Land*

Politics and Livelihoods on the Margins of Hanoi, 1920–2010 (Vancouver: UBS Press, 2014). Distinctive in the way it attends to the everyday practices of "periurban residents" (p. 44), Labbé's work briefly addresses food insecurity in Hanoi through the Great Famine into the 1950s and down until relatively recent times.

2. Although the Great Famine has attracted wide scholarly interest, no single work offers a focus upon Hanoi. An exception might be David Marr's *Vietnam: State, War, and Revolution (1945–1946)* (Berkeley: University of California Press, 2013), at least in the sense of intertwining the revolution and famine theme. I also attempted the same in my *Rice Wars in Colonial Vietnam: The Great Famine and the Viet Minh Road to Power* (Lanham, MD: Rowman & Littlefield, 2014), but with greater attention to the agrarian setting, colonial policy with respect to famine management, Japanese rice procurement, and French-Japanese mismanagement of the crisis in the critical March–April 1945 period.

The following studies can also be usefully consulted, although the list is hardly exhaustive, namely: Sugata Bose, "Starvation amidst Plenty: The Making of Famine in Bengal, Honan, and Tonkin, 1942–44," *Modern Asian Studies* 24, no. 4 (October 1990): 699–727; Bui Ming Dung, "Japan's Role in the Vietnamese Starvation of 1944–45," *Modern Asian Studies* 29, no. 3 (July 1995): 575–76; Nguyen The Anh, "Japanese Food Policies and the 1945 Great Famine in Vietnam," in Paul H. Kratoska, ed., *Food Supplies and the Japanese Occupation in South-East Asia* (London: Palgrave Macmillan, 1998), 208–226; and the author's "The Great Vietnamese Famine of 1944–45 Revisited," *Japan Focus* 5, no. 4 (January 24, 2011). http://apjjf.org/2011/9/5/Geoffrey-Gunn/3483/article.html

Of special interest, at least for its attempt to collect and analyze data on famine victims, is the Vietnamese-language work published by the state-sponsored Institute of History, namely Van Tao and Moto Furuta, eds, *Nan doi nam 1945 o Viet Nam: Nhung chung tich lich su* [The Famine of 1945 in Vietnam: Some Historical Evidence] (Hanoi: NhaNam, 1995). This and a companion publication are critiqued by Ken MacLean, "History Reformatted: Vietnam's Great Famine (1944–45) in Archival Form," *Southeast Asian Studies* 5, no. 2 (August 2016): 208. https://english-kyoto-seas.org/2016/08/vol-5-no-2-ken-maclean/. As the title of his work suggests, MacLean views these official studies as examples of "how historical evidence regarding the Great Famine is fashioned rather than found," to fit a patriotic narrative. On his part, MacLean drew upon ethnographic and archival research conducted in the northern Vietnamese countryside during 2000–03.

3. Pierre Gourou, *Les paysans du delta tonkinois* (Paris: EFEO, 1936). See also René Bouvier, *Richesse et Misère du Delta Tonkinois* (Paris: André Touron et Cie, May 1937). And see Charles Robequain, *The Economic Development of French Indochina* (London: Oxford University Press, 1944), for a standard text on the colonial economy of Indochina, drawing upon a range of French literature, including a supplement on Japanese military rice procurement.
4. See Geoffrey C. Gunn "*Hunger amidst Plenty: Rice Supply and Livelihood in Wartime Macau,*" in Geoffrey C. Gunn, ed., *Wartime Macau: Under the Japanese Shadow* (Hong Kong: Hong Kong University Press, 2016), 72–93.
5. Pierre Brocheux and Daniel Hémery, *Indochina: An Ambiguous Colonization, 1858–1954* (Berkeley, CA: University of California Press, 2010), 439, note 33.
6. Clifford Geertz, *Agricultural Involution: The Processes of Ecological Change in Indonesia* (Berkeley, CA: University of California Press, 1963).
7. Gunn, "The Great Vietnamese Famine of 1944–45 Revisited."
8. US War Department, Magic Summary No. 652, January 5, 1944.
9. Labbé, *Land Politics and Livelihoods on the Margins of Hanoi, 1920–2010*, 44–45.
10. US War Department, Magic Summary, No. 652, January 5, 1944.
11. MacLean, "History Reformatted," 203.
12. MacLean, "History Reformatted," 203.
13. Gunn, *Ricewars in Colonial Vietnam*, 240–242; 247–48.
14. Gunn, "The Great Vietnamese Famine of 1944–45 Revisited."
15. US War Department, Magic Summary, February 1945.
16. US War Department Magic Summary, February 1945.
17. Martin L. Mickelsen, "A mission of vengeance: Vichy French in Indochina in World War II," *Air Power History* (December 17, 2010).
18. United States State-War-Navy Coordinating Committee (SWNCC), 35 Crew to Secretary of State, August 6, 1945.
19. Mandaley Perkins, *Hanoi adieu: A bittersweet memoir of French Indochina* (Hanoi: The Gioi Publishers, 2012). This is a posthumous memoir published two years after L'Herpinière's death and edited by Mandaley Perkins.
20. Perkins, *Hanoi adieu*, 91.
21. Perkins, ibid., 100, 123–24.
22. Perkins, ibid., 142.
23. Perkins, ibid., 144.
24. Perkins, ibid., 152.
25. Perkins, ibid., 140.
26. "Horrific photos recall Vietnamese Famine of 1945." http://www.thanhniennews.com/arts-culture/horrific-photos-recall-vietnamese-famine-of-1945-37,591.html. The photos can be found at the Museum of

Vietnamese History in Hanoi. See "Rare photos of Vietnam's famine in 1945." http://english.vietnamnet.vn/fms/society/138833/rare-photos-of-vietnam-s-famine-in-1945.html

27. Ken MacLean, "History Reformatted."
28. Archimedes Patti, *Why Vietnam: Prelude to America's Albatross* (Berkeley: University of California Press, 1982), 86.
29. William Duiker, *Ho Chi Minh: A Life* (New York: Hyperion Books, 2000), 330; Gunn, *Rice Wars*, 263.
30. Jean Sainteny, *Histoire d'une paix manquée: Indochine 1946–1947* (Paris: Amiot-Dumont, 1953), 71, 74.
31. Perkins, *Hanoi adieu*, 165.
32. King C. Chen, *Vietnam and China, 1938–1954* (Princeton, NJ: Princeton University Press, 1969), 114, 133.
33. Dixee Bartholomew-Feis, *The OSS and Ho Chi Minh: Unexpected Allies in the War against Japan*, (Lawrence, KS: University Press of Kansas, 2006), 250–51.
34. Gunn, *Rice Wars*, 264–65.
35. Marr, *Vietnam*, 6.
36. Nguyen The Anh, "Japanese Food Policies and the 1945 Great Famine in Vietnam," 223–234.

CHAPTER 12

The Esteros and Manila's Postwar Remaking

Michael D. Pante

INTRODUCTION

Like other coastal cities, Manila, capital of the Philippines and located in the central part of the country's largest island (Luzon), is a city whose fate is intertwined with that of its waterways.[1] Three important bodies of water define its morphology: Manila Bay, the Pasig River, and the *estero* system, a network of estuarial creeks that branched out into the inland districts. By the early twentieth century, the estero system had a total length of 12,000 meters that could be divided into three main subsystems, which branch out from these major esteros: Estero de la Reina, the Estero de San Miguel, and the Balete–Provisor esteros.[2] These waterways afford Manila with international, regional, and intracity linkages and thereby commercial centrality that is undisputed in the country.

The blessings of water, however, come with a curse. Its hydrographic features and low elevation—with its central districts less than two meters (seven feet) above sea level[3]—turn strong rains and powerful tides into a deluge. Human-induced factors compound the geographical conundrum. Inadequate drainage, unchecked urbanization, environmental abuse, incompetent local governance, and widespread poverty have turned the

M. D. Pante (✉)
Department of History, Ateneo de Manila University, Quezon City, Philippines

© The Author(s) 2019
S. Laakkonen et al. (eds.), *The Resilient City in World War II*,
Palgrave Studies in World Environmental History,
https://doi.org/10.1007/978-3-030-17439-2_12

city's waterways into a complex social problem that includes water pollution, water-borne diseases, disastrous floods, and many more.

Many of the aforementioned problems predate the Japanese occupation of Manila (1942–1945). However, this period serves as a helpful signpost in understanding the environmental decay that has gripped the city's waterways for almost two centuries now. It would be an understatement to say that Manila's devastation at the end of the Second World War forever altered its morphology. Indeed, the environmental repercussions of this episode are felt to this very day. Just a quick look at today's esteros reveals two crucial and intertwined postwar transformations that have caused major ecological problems. One, esteros have become the catch basin for the deluge of low-income households that have been rendered homeless by the city's unequal social structure. Two, these estuarial creeks are silent witnesses to the city's overreliance on overland mobility, which has cultivated a highly unsustainable metropolitan sprawl. The stereotypical image that probably comes to mind when an average Manileño thinks of urban poverty is a cramped corridor of shanties straddling over a stagnant, garbage-filled estero. This chapter thus asserts that Manila's esteros serve as an index of the effects of the Second World War on the city's natural environment.

There is, however, much historiographical difficulty in writing an environmental history of Manila, let alone a period as specific as wartime Manila. The foremost reason is the lack of sources: contemporaneous official documents and personal accounts about the Second World War are numerous, but these sources limit their observations to the devastation in terms of life and property, with little regard for the catastrophe that nature endured. Nevertheless, the need to go beyond an anthropocentric view of the city amid and after a devastating war is imperative now more than ever.

Revisiting the Venice of the East

Manila's waterways enabled its urban upsurge in the nineteenth century when it was still under Spanish colonial rule (1565–1898). While Manila Bay hosted an international port through which cash crop exports and manufactured imports flowed, the Pasig River and its esteros facilitated the movement of these goods between the port and the city's numerous merchant houses, storage facilities, and markets. Moreover, these waterways ferried the workers whose labor kept the city's commercial engine running.

Given these conditions, water vessels (*bancas*) were indispensable to Manila for both freight and passengers. And due to the predominance of water routes for the conveyance of products, areas that were accessible via esteros were the first to benefit from trade. Factories, warehouses, and other business establishments were built near esteros to minimize transport costs. In an age before the predominance of carriages, wealthy families saw the practicality and aesthetic value of building their residences along these waterways. Such was an important factor behind the development of the districts of Quiapo and San Miguel as elite suburbs in the nineteenth century.[4] As esteros became places of affluence, Manila solidified its claim to being the Venice of the East.[5] More than just providing mobility, upstream esteros, especially in areas like Makati and Santa Ana, were also sources of drinking water mainly for the lower classes, although such became an increasingly problematic practice as population densities increased in these areas through the years.[6]

The issue of the decline in the quality of potable water from esteros actually points to a more serious concern: the gradual ecological degradation of Manila's waterways, the symptoms of which appeared as early as the mid-nineteenth century. Because Manila lacked adequate drainage and sewerage systems, runoff and household wastes flowed through the estuaries. Slowly but surely, pollution and siltation made the esteros spaces of disease and disasters, as seen in the city's constant bouts with epidemics and floods.[7]

The onset of American colonial rule in Manila in 1898 only highlighted the city's ecological problems, and to an extent worsened by the Philippine–American War.[8] Notwithstanding the racial-colonial anxieties that informed how Americans viewed Manila's environment, their preoccupation with disease in the tropics highlighted how the city's waterways contributed to floods and epidemics.[9] In 1909 the city government inaugurated a system of separate disposal for rainwater and sewage,[10] marking a technological breakthrough for the city. Success in environmental management gave the Americans an "imperial positive feedback loop,"[11] with science and technology assuring the sustainability of empire. Nevertheless, despite infrastructural improvements, the flood problem never left the city, and esteros became even more polluted, as low-income households that were not connected to the sewerage system had no recourse but to dispose their wastes into these waterways.[12] According to the *New York Times*, Manila was no longer Venice of the Orient because of its increasing level of pollution and deteriorating public health situation.[13] In fact, this bleak situation forced city officials to order on 1 June 1931 the closing of 12 esteros by sealing them out from adjacent estuaries and filling them with earth. Officials and ordinary people regarded esteros "as a liability rather than an asset."[14]

Although inland waterways remained in use for intracity transportation during the American colonial period, no significant changes were seen in intraurban, water-based passenger transport during this time. Despite the intended integration of land-based and estero-based passenger traffic in the 1905 Manila master plan of the famed American architect and planner Daniel Burnham, this proposal never materialized, supposedly due to the lack of funds.[15] In contrast, land transport innovations brought about by American colonialism led to the gradual marginalization of traditional water-based transport.[16] Given that the inflexible routes of inland waterways were mostly confined to areas near the bay and the river, changes in land-based transport modes had to be effected to cope with urban expansion. Moreover, navigable esteros suitable for banca traffic were not prominent in many of these suburban areas, unlike in the central districts of the city. Reynolds and Caballero's analysis of city maps from the late nineteenth century to the end of the American colonial period shows how the previously numerous estero branches in Quiapo had decreased in number in the early twentieth century.[17] From transport arteries, these esteros became mere sewers, but "when they can no longer serve this purpose, they are filled up or covered."[18]

The Filipino-led Commonwealth government replaced direct American colonial rule in 1935. It recognized the poor conditions of the esteros and tried to improve their conditions, yet its discourse appeared no different from that of the previous regime. In his message to the First National Assembly on Public Works Program, 1 February 1938, Commonwealth President Manuel Quezon stated:

> Only about one-third of the City of Manila has been provided with a proper system of drainage and the remaining part is dependent upon sea-level esteros or canals. Most of these esteros are inadequate because [*sic*] of poor alignment and insufficient capacity, thus making it impossible to maintain sanitary conditions in the districts affected. It is proposed that these very unsatisfactory conditions be remedied by straightening, widening and paving those esteros that are necessary for drainage purposes and by filling in those that may be dispensed with.[19]

A "Mini-Venice" in Wartime Manila

The Commonwealth government might have had good plans for Manila's esteros, but all these were for nothing following the outbreak of war in the Philippines. The bombing of Pearl Harbor was immediately followed by Japanese attacks on key military targets in the country. Quezon's government declared Manila an open city by December 1941, and Japan's occupation forces quickly took control of the capital. Rather than defend Manila, the American and Filipino armies of the United States Armed Forces in the Far East (USAFFE) concentrated their efforts in two focal points in Luzon: in Corregidor Island, located at Manila Bay, and in the Bataan Peninsula. The power of the Japanese occupation forces, however, overwhelmed the Allied defense, forcing them to surrender on 9 April 1942. Nonetheless, anti-Japanese guerrilla units consequently emerged and continued to harass the new colonizers and their Filipino collaborators, especially in Manila's immediate hinterlands. As a result, although Japan controlled the major cities, its dominance was seriously contested deep in the countryside, with guerrillas penetrating even the Japanese-sanctioned organs of local administration. By 14 October 1943, the Japanese had announced formal Philippine independence under the Second Republic, headed by President José P. Laurel and whose powers were still heavily constrained by the Japanese Imperial Army, thereby making the republic an independent state in name only.[20]

Wartime conditions took a heavy toll on Manila's physical upkeep. The Japanese occupation forces, as expected, focused on military-related functions of the state. Although the Filipino politicians and bureaucrats of the Second Republic had a modicum of power over everyday administration, it could barely function due to the scarce resources at its disposal; the state's assets were, of course, diverted by the Japanese Imperial Army to their war machine. At the same time, widespread guerrilla activities in the countryside meant that the natural resources in these areas were not maximized by the new colonizers, let alone funneled to the urban areas for nonmilitary consumption. Meanwhile, civilians in the capital could no longer be expected to be mindful of their surroundings. The result was a city in decay, and once again esteros were indicative of the state of degradation. The effects of war on the everyday lives of people in Manila were catastrophic, forcing thousands of city dwellers to relocate to rural areas[21] and thereby debilitating the city's basic functions, including garbage collection and estero maintenance.[22] The effect on the esteros was not surprising: "Behind the Divisoria, on the banks of the *estero* choked with assorted refuse and garbage, men and children were lying in the sun, the innumerable flies feasting on their festering tropical ulcer."[23] Manila resident Joaquín García gave a rather interesting remark in his recollection of wartime Estero Tripa de Gallina, "a long and narrow dirty waterway which, when at low tide, stank to high heavens. As a consequence, it was dubbed Canal No. 5 (apologies to Madame Chanel)."[24]

During the war, esteros resumed their role as conduits of passenger and freight traffic, given the disruptions in land-based transportation and the scarcity of gasoline.[25] Unfortunately, despite the sudden increased importance of water-based transportation, esteros remained unimproved during the Japanese occupation. City Engineer Alejo Aquino lamented that "since the inauguration of the present administration no dredging has been done and apparently the Insular Government is not in a position to continue with the maintenance of navigable esteros at their depth."[26] Moreover, he recognized that the present condition of esteros was "deplorable" because of garbage, sedimentary deposits due to constant floods, and foul odor owing to the lack of a sufficient sewerage system in Manila and the accumulation of dead animals and garbage. In the districts of Sampaloc, Pandacan, Santa Ana, and Tondo, "the poor people living along the banks of the esteros dispose of their human excreta by throwing them directly into the esteros."[27] Perhaps reflective of the wartime atmosphere, Aquino recommended a "more militant supervision on the part of sanitary inspectors and the police in order to stop or at least discourage the practice of dumping waste material and human excreta, dead animals, refuse and garbage on the esteros."[28]

Harping on the same sentiments but on a more positive note, journalist I. V. Mallari noted that Manila residents failed to see the "aesthetic possibilities" of esteros, as employed in other countries like Japan: "And now that the Japanese are here, perhaps they could teach us how to utilize our esteros to advantage .... Following the example of Japan, we could make our esteros the nucleus of a wonderful park system that would be a source of health and joy and pride to all of us."[29] Trying to ride on the discipline-centric atmosphere of the period, Mallari suggested that residents living along the esteros "should be made to feel that the parks and the waterways in front of their houses are their personal responsibility," and that they should be amenable to such "social discipline."[30]

As expected, these grandiose plans for the esteros were never implemented during the occupation period. Instead, a perennial hazard exposed the potential for disaster that unimproved waterways could inflict upon a city. In November 1943 a catastrophic typhoon hit central and southern Luzon, causing widespread flooding in these regions, including Manila. The flood, "attaining flood heights higher than any previously recorded,"[31] submerged the capital: "Overnight, the area had been transformed into lake-like proportions and became a mini-Venice."[32] Manila had no supply of potable water and electricity for five days.[33] The impact of the flood went beyond the crippling of the city: it occurred at a crucial time. The flood occurred during harvesting time, and it exacted a grave effect on the already delicate food situation of the country, especially in the cities.[34] Although food distribution was in the hands of Filipino collaborator politicians, the sheer scarcity of resources rendered such wartime bureaucrats powerless. The flood was such a significant event during the occupation period that it "was a defining date of sorts, marking the end of what might be called a period of sullen submission, and the beginning of sharp distress and deprivation—not because of the flood, but simply because it provided a convenient dividing line."[35]

Perhaps the most detailed recollection of the 1943 flood was that of A. V. H. Hartendorp, a wartime detainee at the University of Santo Tomas (UST) internment camp for foreign civilians in Manila's suburban Sampaloc district. According to him, the rains began on 13 November and quickly inundated the campus, 1–6 feet high. In other places in Manila, electricity was only restored a week after the first day of the storm.[36] The floods demonstrated, in a catastrophic manner, the sheer inadequacy of relying on esteros for city drainage and so affected state officials that in the following month, 3 December 1943, President José P. Laurel created the Flood Control Board.[37] But as fate would have it, the documents and studies compiled for the board's flood-control plan were destroyed in the Battle of Manila in February 1945 in the twilight years of the Japanese occupation.[38]

The destruction caused by the November 1943 floods, however, paled in comparison to the tragedy that befell Manila in February 1945. Allied forces led by the USAFFE were already about to occupy the capital when Japanese forces decided to stage a desperate last-ditch defense. As American and Filipino troops surrounded Manila, Japanese soldiers, seeing an inevitable defeat, began to set fire to civilian establishments, kill innocent residents, and rape women; scenes that made the city the stage for the worst urban fighting in the Pacific front of the war. Japan's prolonged struggle in the capital cost thousands of lives—around 100,000 fatalities, according to official estimates[39]—and millions worth of damaged property. The city's utter destruction led some observers to describe the city as the Warsaw of Asia.[40] Amid the wanton killings that characterized the days of February 1945, the Japanese found esteros an expedient tool in disposing civilian victims of their summary executions, especially those found south of the Pasig River.[41] For instance, people who hid at the Vincentian Central House were murdered and thrown into Estero de Balete.[42]

THE WATERWAYS AND WAR'S AFTERMATH IN ASIA'S WARSAW

Clearing the esteros was just one in a long list of tasks that had to be done in the rehabilitation efforts in Manila right after the war. In the aftermath, these waterways carried not just the rubble of destroyed structures, but also much more sordid remnants of the mayhem. Manila also contended with the problem of pollution[43]; but then again, these environmental problems were probably just minor concerns for a society that had more pressing matters to tackle.

Even before the war's conclusion, the newly installed government under President Sergio Osmeña, who assumed the Commonwealth presidency after Quezon's death in 1944, was already planning the rehabilitation of the country's war-torn cities, especially Manila. He entrusted this task to his special adviser on land planning, Louis Croft, who had also been part of a prewar team of architects and planners who prepared the master plan for Quezon City, a purpose-built satellite city of Manila, and was a civilian prisoner of the Japanese at the UST during the war. In February 1945—just days after the capital endured a devastating battle and with Croft just recently released from imprisonment—the president asked him to prepare plans for the reconstruction of destroyed towns and cities.[44] The following month, Osmeña created the City Planning Office as an emergency body to attend to the rehabilitation efforts in the capital. In response to these efforts, a citizens' planning association, the Metropolitan Manila Planning Association, emerged to work with the government.[45]

Croft seemed prepared even in these most uncertain circumstances, for even while he was imprisoned, he improved his plans for greater Manila using "invaluable suggestions" from "civic-minded Americans and Europeans"[46] who were his fellow internees at UST. In 1946 Osmeña appointed him to the National Urban Planning Commission.[47] When Manuel Roxas succeeded Osmeña in the presidency that same year, Croft maintained his influence as adviser on land planning. As postwar Manila's primary planner in the early years of rehabilitation, Croft was confident and optimistic. His plan was a massive "face-lifting" of Manila to restore its beauty. Specifically, he "declared that priority in the construction projects will be given to thoroughfares."[48]

Croft left his imprint in the postwar rebuilding of Manila most notably in his plan of the overall layout of the major thoroughfares for the capital and its surrounding towns. Building upon the existing road network, he envisioned a much larger metropolis that would be connected to the urban core via new roads, based on automobile-centric models in the US.[49] The blueprint was ready and partially built in the prewar period but was only completed in the postwar years.[50] Unfortunately, Croft's overall design for an expanded Manila marginalized the city's estuarial arteries.[51] Indeed, Croft's 1945 plan for the thoroughfares of greater Manila paid no attention at all to esteros, apart from denigrating them as a sanitary menace, in contrast to how Burnham incorporated them in his 1905 design. Mallari lamented how Croft's plan disregarded these waterways. Reiterating his stance on this issue during the occupation years, Mallari stated that these bodies of water could have been turned into "finger parkways" that could have improved sanitation and convenience.[52]

The same lack of attention was also apparent in Croft's thorough discussion of the progress of city planning in 1947.[53] Canal de la Reina, for instance, saw a drastic decrease in its importance as a transport circuit in the immediate postwar years.[54]

The expected outcome of Croft's plan was Manila's dependency on overland transport for mobility, in particular automobility. The wartime destruction of the city's electric streetcar system and the decision of its owner, the Manila Electric Railroad and Light Company (Meralco), not to rehabilitate it after the war ensured the predominance of motor cars in intraurban conveyance. Instead, Meralco imported more buses from the US to revive its transport services.[55]

Croft was also a key player in the immediate plans to address the acute housing shortage in postwar Manila.[56] In January 1946 a three-man federal housing commission from the US left for the Philippines at the request of President Osmeña, who was also coordinating with Croft. The task of addressing the housing situation was daunting, to say the least, especially since it directly affected tens of thousands of poor households in the capital.

The aftermath of war saw the increase in informal settlements in Manila, compounded by an influx of migrants and the skyrocketing cost of living.[57] The representative image of that period was the *barong-barong*, "which came into being as an inevitable result of the war .... a makeshift affair of broken slats, rusty galvanized iron punched with machinegun bullets, flattened cans which once contained corned beef hash, soot-blackened tarpaulins, and thin, mud-splattered posts picked up on the other side of an army compound."[58] The destruction caused by the war did not only render thousands homeless but also disrupted property relations. As housing experts Abrams and Koenigsberger noted: "The sites marked for bombing and those chosen for squatting were often identical."[59] These areas were located at the city's core, near places of work and livelihood, attracting desperate migrants. Moreover, the postwar period witnessed the escalation of urban and suburban real estate prices and speculative practices of landowners and real estate companies, with little government interference, thereby contributing to further homelessness among ordinary people.[60] Lastly, the early postwar years saw the outbreak of another violent episode. In the Central Luzon provinces just outside Manila the Hukbalahap, an anti-Japanese guerrilla group of mostly poor peasants, waged a rebellion against the Philippine government. This peasant unrest intensified rural-to-urban migration to Manila and contributed to the housing crisis.[61]

The extent of Manila's postwar crisis was perhaps most visible in the esteros. Thousands of homeless refugees and migrants were so desperate for shelter that they decided to settle along Manila's already decaying waterways. A "[f]amiliar squatter scene" consisted of "patchwork 'barong-barongs' edging an estero."[62] Slum conditions worsened "especially in the areas around the upper reaches of Estero de Magdalena and Estero de San Lazaro."[63] In the immediate postwar years, the number of houses with toilets was only a third of the number it had been before the war, while the population had tripled.[64] In such a situation and coupled with state neglect, estero dwellers for sure had no other way to dispose of their garbage and sewage wastes.[65]

The emergence of these informal communities was arguably a manifestation of the resilience of the city's economically vulnerable in navigating the vagaries of a metropolis dealing with postwar dislocations and asymmetrical social relations. Unfortunately, the country's political elite saw this situation not in terms of everyday resilience nor as a socioeconomic crisis but as a security risk, a "social cancer."[66] A bill was proposed in Congress that sought a moratorium of two years against the forced eviction of squatters on public lands/places but explicitly excluded those living along esteros. The bill treated the homeless as threats that had to be eliminated[67] rather than as victims of structural inequality. Consequently, the government was rather heavy-handed in evicting these informal communities.[68] In contrast, it showed leniency in dealing with those who had the right connections:

> On the other hand, consider the big structures that have been illegally built by certain powerful individuals in Manila on so-called "prohibited places." Take, for example, the case of one Candida Tan who constructed a strong, two-story commercial building right over an *estero* in Quiapo district, this city.
> The office of the city fiscal of Manila has already issued a ruling to the effect that the construction is illegal. But has the building been demolished? No. It is still standing there and it will remain standing there, congesting the *estero*, as long as its owner who seems to be "just like that" with certain city authorities can continue to pull the right strings among city officials. The big fish have always a way of nullifying the law.[69]

Estero-dwelling households were far from being abusive and parasitical, as authorities would often depict them. Quite the contrary, they were a vulnerable group in the city, made even more so by the state's neglect of the city estuaries, that struggled to keep itself afloat amid socioeconomic uncertainties. One of the clearest effects of neglected esteros was the increased prevalence of floods, which became more acute due to another ecological issue: deforestation, which includes both the loss of upland forest cover due to the lucrative timber trade centered in Manila's immediate hinterland, a problem that was already apparent since the Spanish colonial times,[70] and the destruction of mangroves in Manila's northern peripheries.[71] The early postwar years saw esteros and their inhabitants endure floods made worse by both natural and man-made causes: "When the Pasig River is in flood and the water level rises to bank elevation, the flows in all esteros are reversed and low city areas are immediately flooded. Improvements of the channels of various esteros, without providing flood gates, have only served to aggravate the flood situation."[72] Such was seen when Typhoon Gertrude hit the country in September 1948. In response to Gertrude, the government implemented flood-control measures, including the widening of esteros and lining them with concrete,[73] and, most importantly, the completion of a drainage master plan in 1952.

The 1952 "Plan for the Drainage of Manila and Suburbs"[74] sought a capital-intensive, infrastructure-driven solution to the flood problem. It aimed at the "control of the flood flow of the Pasig river"[75] by constructing river walls and floodgates along its banks and constructing channels to create a flow diversion from the Marikina River to Laguna de Bai.[76] Expectedly, it called for the cleaning and dredging of the esteros.[77]

Unfortunately, despite plans to tap the drainage capacity of waterways, through the years, esteros were filled to give way to accommodate changes in the city's morphology. For example, Estero de Magdalena and Estero de San Lazaro, tributaries of Estero de la Reina, were once linked by a channel that eventually gave way to the extension of a street. The channel was "almost totally filled up except for a small shallow ditch that has replaced it."[78] Of course, "further encroachments on natural waterways, brought about by the construction of new structures and improvements, have made the problem more acute and its solution much more expensive."[79] As a result, the discharging capacity of Estero de la Reina—not to mention other minor esteros—had become insufficient.[80]

But at the same time, authorities behind the 1952 Drainage Plan were deaf to the situation of estero dwellers. The plan recommended the eviction of all estero squatters as "a mandatory step,"[81] without an accompanying framework to address or even to look into the housing inequalities that were at the root of this situation. In the 1952 master plan's list of things to do "For Immediate Action," topping the list is the eviction of estero dwellers: "Eviction of all estero squatters and strict enforcement of the laws prohibiting and penalizing the throwing of debris into estero channels and otherwise blocking the flow. In the Manila Watershed penalties for violation of this law should be made heavier to discourage continued violation."[82] Antonio A. Villanueva, chief flood control and drainage engineer, ranted:

> [In the] last forty years, more than 21 kilometers of estero channel have been filled up and appropriated by squatters with and without sanction by government authorities. Since no plan for an overall drainage system for the city had ever been prepared before, it may be said with impunity that the filling up of these esteros, where sanctioned by constituted authorities, was done arbitrarily. Furthermore, long sections of estero channels still existing, are now reduced by squatter encroachments to very narrow widths of two or three feet where they were formerly more than 20 feet wide. These estero channels are fast deteriorating and if not properly resuscitated and maintained will ultimately disappear with disastrous results.[83]

This discourse of evicting estero dwellers to obtain the "greater good" of making Manila flood-free was also seen in the mode of governance of other officials. Most notable among them was Manila Mayor Arsenio Lacson. Lacson boasted in a 3 January 1958 speech about his "gigantic project for the improvement of the Estero de Sampaloc as a main drainage facility."[84] However, the flipside of this was Lacson's ruthless eviction of household dwellers.[85] For Manila's local officials, slum dwellers were themselves a socioeconomic hazard "not only to morality but also to health and lives of the members of the community."[86] Their visibility and vulnerability made them the convenient scapegoat for the flood woes, with their propensity to throw garbage in the esteros often highlighted.[87]

Ripple Effects

These large-scale attempts at eviction, however, did not prevent the persistence and growth of estero communities in the 1950s and 1960s. Displaying resilience amid state-led evictions, low-income communities mushroomed along the banks of Estero Tripa de Gallina and Estero Concordia in Santa Ana district.[88] At the same time, middle-class housing in the form of multistory apartments rose along Estero de la Reina.[89] In Tondo, industries clustered along the banks of the esteros. These industries included, for instance, sawmills and lumber yards that lined the banks of Estero de la Reina and Estero de Vitas. Moreover, since the vicinity of Estero de Vitas had been pinpointed as an area for heavy industries in zoning plans, factories dominated its landscape.[90]

The rise of residential and industrial establishments in these areas led to unsurprising results. According to the foreign consultancy firm Black & Veatch: "Wastewater disposal is dependent on the water ways in the absence of an adequate sewerage system. Septic tank effluent and domestic wastewater empty into creeks which flow to the larger streams and rivers. This pollution has destroyed most of the animal life in the streams and the Pasig River."[91]

The resulting situation created by the abovementioned conditions led to the overall decay of the downtown districts that housed esteros. One important effect of this ecological decay was suburban flight: affluent households left downtown Manila for the growing suburban towns just outside it. Suburbanization defined the postwar geography of an emerging Metro Manila, but one that was different from the typical notion of suburbanization in Western society. To an extent, it produced a "crabgrass frontier" typical of US suburbia; towns just outside Manila became home to posh subdivisions for the well-to-do. Certainly, these residential developments were not affordable to the poor, who had to make do with makeshift dwellings.[92] But largely suburbanization meant the geographical expansion of blight into the erstwhile peripheries of Manila. One important aspect of this was the replication of estero communities in satellite towns such as Caloocan, where settlers had encroached unto fishponds, Pasay City, and Quezon City.[93]

Urbanization and suburbanization also meant the construction of more permanent structures along the waterways and high-density settlements, leading to further pollution and loss of drainage capacity in this growing metropolis. The encroachment of settlements in the forested areas also contributed to the problem, as seen in the "indiscriminate deforestation of the mountains of Marikina and Montalban above Quezon City in Rizal province,"[94] a problem already apparent in the previous decades. Equally important is the fact that suburbanization was a process that was facilitated not by transit but by automobiles. It meant more paved surfaces for thoroughfares and less drainage capacity. This combination reduced both the drainage capacity of waterways and the capability of urban land to absorb rainwater, ultimately leading to increased frequency and ferocity of floods in the metropolis from the 1950s to the 1970s. Navotas, Malabon, Caloocan, Pasay, San Juan, and Quezon City were not spared.[95]

By the 1960s, the estero ecology seen in Manila was already visible in the suburbs. Indeed, the most badly hit during floods were the informal settlers that had mushroomed along the waterways of the suburbs, as exemplified by the disastrous floods of 28 May 1960. Almost 200 casualties were recorded, with most of them coming from Manila and its neighboring towns of Quezon City, San Juan, Caloocan, Navotas, and Malabon, among others. Many of the victims lived in "Shanties occupied mostly by squatters in depressed areas traversed by the creeks and rivers."[96] As one reporter put it: "Nearly all the casualties were poor people who had to live as squatters on the banks of creeks and dry river beds because there was no other place for them to build their shanties."[97] Their bleak existence in the city was made even more apparent in these words:

> Meanwhile, many victims who had nowhere else to go began building temporary shelters and shanties in the same places on the banks of creeks and rivers from which their old homes were swept away by the flood.[98]

Conclusions

The Second World War turned Manila from the Venice of the East to the Warsaw of Asia in a span of just three years. However, lost in this narrative is the reality that many of Manila's physical faults were already apparent prior to the Japanese occupation. Focusing on the city's waterways in asserting this point, this chapter has shown that in Manila's rapid urban growth from the nineteenth century to the early twentieth century, esteros served multiple purposes for the capital and in the process endured physical degradation. Aside from acting as Manila's natural drainage system, these estuaries provided the city a network for the conveyance of passengers, produce, and even night soil and garbage.

Still, one cannot deny that the esteros' status as an environmental liability became far more pronounced after the war. The Japanese occupation dealt a shock to these waterways, as well as other nonhuman actors, similar to that felt by the city's human population. To a certain extent, the esteros and the households that relied on them for shelter suffered the same fate. Both were marginalized in a car-dependent, suburbanizing metropolis, based on US models of town planning. Whereas esteros were burdened by postwar demographic pressure, such pressure resulted from the accumulated actions of low-income households struggling against an income squeeze while fighting for their place in a metropolis, whose space is constrained by the dictates of geography and the state. The environmental and socioeconomic repercussions were serious and self-perpetuating. The uncontrolled urban growth contributed to further pollution and loss of drainage capacity in the built-up areas, factors that only served to worsen the situation of the esteros and the dwellers living near them. As a result, those with the political and economic capacity simply left the waterways to rot (as seen in the suburban flight, the eventual loss of the esteros' capacity as a transportation artery, and state neglect of the waterways) along with the families who had no other option but to cling to such marginal urban spaces. Floods, every now and then, make this sad reality the headline of the day, only to submerge it again and again in the depths of Metro Manila's collective unconscious.

In this light, there is a marked contrast between the aquatic shock that has left most of Manila's esteros biologically dead and the resilience demonstrated by Manila's social underside as they withstood not only the shock of war, but also the impact of natural, political, and economic blows. The war's aftermath forced the city's marginalized to make the most out of an erstwhile environmental asset that the war rendered obsolete. This distinction, however, should not obscure the reality that such resilience is but a survival mechanism for those drowning daily in near-death poverty.

Notes

1. This chapter is based on a research project funded by the Commission on Higher Education-Philippine Higher Education Research Network (CHED-PHERNET). I also want to thank Leo Angelo Nery for his assistance in this project.
2. Charles H. Storms, "The Esteros of Manila," *The Philippine Craftsman* 3, no. 2 (1914): 97–108.
3. Xavier Huetz de Lemps, "Waters in Nineteenth Century Manila," *Philippine Studies* 49, no. 4 (2001): 488–517, esp. p. 495.
4. William E. Reynolds and Evelyn J. Caballero, "The Changes through Time in Quiapo's Esteros," in Fernando Nakpil Zialcita, ed., *Quiapo: Heart of Manila* (Quezon City: Cultural Heritage Studies Program, Department of Sociology and Anthropology, Ateneo de Manila University and Metropolitan Museum of Manila, 2006), 70–95.
5. Beryl Hughes, "Manila's Gondolas and Gondoliers," *American Chamber of Commerce Journal* 6, no. 4 (1931): 7.
6. Daniel F. Doeppers, *Feeding Manila in Peace and War, 1850–1945* (Madison: University of Wisconsin Press, 2016), 254–256.
7. Huetz de Lemps, "Waters in Nineteenth Century Manila"; Reynolds and Caballero, "The Changes through Time," 76.
8. Rolando B. Mactal, *Kalusugang pampubliko sa kolonyal na Maynila (1898–1918): Heograpiya, medisina, kasaysayan* [Public Health in Colonial Manila (1898–1918): Geography, Medicine, History] (Quezon City: University of the Philippines Press, 2009), 156–157. The Philippine–American War began in 1899 and officially ended in 1902, although guerrilla warfare continued to disturb the environs of Manila until 1910. Although Manila did not directly experience war-related armed encounters, its residents felt the war's effects through the ecological disruptions caused by the conflict. For example, scholars have noted the role the war played in the outbreak of rinderpest and malaria, coeval catastrophes that negatively affected Manila's food supplies and public health situation, Doeppers, *Feeding Manila*, 233–234.
9. US Philippine Commission, *Annual Report of the Philippine Commission*, part 1 (Washington: Government Printing Office, 1901), 29, 52.

10. Municipal Board of Manila, *Annual Report of the Municipal Board of the City of Manila for the Fiscal Year 1909* (Manila: Bureau of Public Printing, 1910), 73.
11. J. R. McNeill, "Introduction: Environmental and Economic Management," in Alfred W. McCoy and Francisco A. Scarano, eds., *Colonial Crucible: Empire in the Making of the Modern American State* (Quezon City: Ateneo de Manila University Press, 2010), 475–478. Quote in p. 475.
12. Dean Worcester, *A History of Asiatic Cholera in the Philippine Islands* (Manila: Bureau of Printing, 1909), 83; I. V. Mallari, "Our Esteros," in Office of the Mayor [City of Manila], *City Gazette: Published Fortnightly by the Office of the Mayor of the City of Manila* 2, no. 5, 1 March (Manila: Office of the Mayor, 1943), 168.
13. *New York Times*, "Manila No Longer Venice of Orient," 5 July 1931: 38. Online, https://www.nytimes.com/1931/07/05/archives/manila-no-longer-venice-of-orient-twelve-canals-in-heart-of-city.html, accessed 26 Mar. 2018.
14. Mallari, "Our Esteros," 168.
15. Daniel Burnham and Peirce Anderson, "Report on Proposed Improvements at Manila," in US Philippine Commission, *Sixth Annual Report of the Philippine Commission, 1905*, part 1 (Washington, D.C.: Government Printing Office, 1906), 627–635.
16. Leonardo Q. Liongson, "The Esteros of Manila: Urban Drainage a Century Since," in *Pressures of Urbanization: Flood Control and Drainage in Metro Manila*, ed. Leonardo Q. Liongson, Guillermo Q. Tabios III, and Peter P. M. Castro (Quezon City: University of the Philippines, Center for Integrative and Development Studies, 2000), 1–16.
17. Reynolds and Caballero, "The Changes through Time," 79–85.
18. Mallari, "Our Esteros," 168. Cf. Felix Roxas, *The World of Felix Roxas: Anecdotes and Reminiscences of a Manila Newspaper Columnist, 1926–36*, trans. Angel Estrada and Vicente del Carmen (Manila: Filipiniana Book Guild, 1970), 260, 269; Alejo Aquino, "Esteros in Manila," in Office of the Mayor [City of Manila], *City Gazette: Published Fortnightly by the Office of the Mayor of the City of Manila* 2, no. 5, 1 March (Manila: Office of the Mayor, 1943): 165–167.
19. Manuel L. Quezon, *Messages of the President*, vol. 4, part 1 (Manila: Bureau of Printing, 1939), 270–271.
20. For the most authoritative secondary sources on the Philippines under Japan, see, for example, Teodoro A. Agoncillo, *The Fateful Years: Japan's Adventure in the Philippines, 1941–45*, 2nd ed., vols. 1 and 2 (Quezon City: University of the Philippines Press, 2001); A. V. H. Hartendorp, *The Japanese Occupation of the Philippines*, vols. 1 and 2 (Manila: Bookmark, 1967); Thelma Kintanar, Clemen C. Aquino, Patricia B. Arinto, and Ma.

Luisa T. Camagay, *Kuwentong Bayan: Noong Panahon ng Hapon (Everyday Life in a Time of War)* (Quezon City: University of the Philippines Press, 2006).
21. Doeppers, *Feeding Manila*, 307–330.
22. Hartendorp, *The Japanese Occupation*, vol. 1, 192.
23. Agoncillo, *The Fateful Years*, vol. 2, 512.
24. Joaquín L. García, *It Took Four Years for the Rising Sun to Set (1941–1945): Recollections of an Unforgettable Ordeal* (Manila: De La Salle University Press, 2001), 37.
25. Kintanar, et al., *Kuwentong Bayan*, 150–162.
26. Aquino, "Esteros in Manila," 166.
27. Ibid., 166.
28. Ibid., 167.
29. Mallari, "Our Esteros," 168.
30. Ibid., 168.
31. Bureau of Public Works (BPW), *Plan for the Drainage of Manila and Suburbs*, vol. I (Manila: Bureau of Public Works, 1952), 1.
32. García, *It Took Four Years*, 65; Gladys Savary, *Outside the Walls* (New York: Vantage, 1954), 106; Benito J. Legarda Jr., *Occupation: The Later Years* (Manila: De La Salle University Press, 2007), 44.
33. "Normal Water Supply for City Assured," *Tribune*, 18 Nov. 1943: 1; "Laurel Tours Area Affected by Flood," 19 Nov. 1943: 1, 6; Legarda, *Occupation*, 43.
34. Marcial P. Lichauco, *"Dear Mother Putnam": A Diary of the War in the Philippines* (Manila: n.p., 1949), 140.
35. Legarda, *Occupation*, 43.
36. A. V. H. Hartendorp, *The Santo Tomas Story* (New York: McGraw Hill, 1964), 189–191; A. V. H. Hartendorp, *The Japanese Occupation of the Philippines*, vol. 2 (Manila: Bookmark, 1967), 28–30.
37. Gerard Lico and Lorelei D. C. de Viana, *Regulating Colonial Spaces (1565–1944): A Collection of Laws, Decrees, Proclamations, Ordinances, Orders and Directives on Architecture and the Built Environment during the Colonial Eras in the Philippines* (Manila: National Commission for Culture and the Arts, 2017), 357–358.
38. BPW, *Plan for the Drainage of Manila*, vol. I, 1; Black & Veatch International Consulting Engineers, *Master Plan for a Sewerage System for the Manila Metropolitan Area: Final Report* (Manila: Black & Veatch, 1969), 10-8.
39. Alfonso J. Aluit, *By Sword and Fire: The Destruction of Manila in World War II, 3 February – 3 March 1945* (Makati City: Bookmark, 1994), 398–399.

40. Ibid.; Jose Ma. Bonifacio M. Escoda, *Warsaw of Asia: The Rape of Manila* (Quezon City: Giraffe, 2000).
41. Aluit, *By Sword and Fire*, 227, 249–50, 365, 399.
42. Escoda, *Warsaw of Asia*, 224–226.
43. "Can Manila Clean Up?," *Manila Times*, 27 Sept. 1947: 28–29.
44. Sergio Osmeña, *Ten Months of President Osmeña's Administration: A Review of Work Done under Unprecedented Difficulties* (n.p.: n.p., n.d.), 14.
45. Lyd Arguilla, "Pattern of a Modern City," *Manila Times*, 19 Sept. 1945.
46. Renato Arevalo, "Out of the Ruins," *Manila Times*, 10 June 1945: 6.
47. Jorge B. Vargas, City and National–Regional Planning in the Philippines: Its Nature, Problems, 7 Mar. 1953, p. 1 (Unpublished manuscript, Jorge B. Vargas Papers, SG: National Planning Commission, Folder 2, Jorge B. Vargas Museum and Filipiniana Research Center, Quezon City).
48. "Beautiful, Modern Manila to Rise from Ruins of Old, States Croft," *Manila Times*, 1 Oct. 1945: 1, 4.
49. Kenneth T. Jackson, *Crabgrass Frontier: The Suburbanization of the United States* (New York: Oxford University Press, 1985), 238–245.
50. Louis P. Croft, General Plan of Major Thoroughfares, Metropolitan Manila: Preliminary Report Prepared by the President's Office on City Planning, 13 June 1945, pp. 5–10 (Unpublished manuscript available at the American Historical Collection, Rizal Library, Ateneo de Manila University, Quezon City).
51. Arevalo, "Out of the Ruins," 6.
52. I. V. Mallari, "That Kind of a City," *Manila Times*, 18 Nov. 1945: 6.
53. Louis P. Croft, "Present Status of the City Planning Program," *American Chamber of Commerce Journal* 23, no. 8, (1947): 264–266, 291–294.
54. BPW, *Plan for the Drainage of Manila*, 52.
55. "Meralco Will Import More Busses for P.I.," 15 Sept. 1945: 4.
56. Osmeña, *Ten Months*, 14.
57. "City Conditions Greatly Improved, Survey Shows," *Manila Times*, 22 July 1945: 5; Bernardino Ronquillo, "Cost of Living in City Six Times More Than Pre-War, Survey Shows," *Manila Times*, 22 Dec. 1945: 1, 16.
58. Delfin Ferrer Gamboa, "I Live in a Barong-Barong," *Philippines Free Press*, 8 Mar. 1947: 22–23.
59. Charles Abrams and Otto Koenigsberger, A Housing Program for the Philippine Islands, 16 (Unpublished report prepared for the United Nations Technical Assistance Administration, 14 January 1959, available at the School of Urban and Regional Planning Library, University of the Philippines-Diliman, Quezon City).
60. Leon O. Ty, "Many of Our Countrymen Are Homeless," *Philippines Free Press*, 18 Sept. 1947: 4–5, 49.

61. G. Viola Fernando, House Bill no. 2798, Submitted during the Third Session of the Second Congress of the Republic of the Philippines, [1953?], 1 (Unpublished manuscript, Jorge B. Vargas Papers, SG: National Planning Commission, Folder 2, Jorge B. Vargas Museum and Filipiniana Resource Center, Quezon City).
62. Filemon V. Tutay, "Squatter Trouble," *Philippines Free Press*, 3 May 1952: 6–7, 27, 50. Quote in p. 27.
63. BPW, *Plan for the Drainage of Manila*, vol. I, 54.
64. "Can Manila Clean Up?," 28–29.
65. Tutay, "Squatter Trouble," 27.
66. C. V. Pedroche, "Slums Can Be Licked," *Sunday Times Magazine*, 27 Oct. 1946: 6–7.
67. Fernando, House Bill no. 2798, 1.
68. Leon O. Ty, "The Case of the Common *Tao*," *Philippines Free Press*, 29 May 1948: 2–3.
69. Ibid., 3.
70. BPW, *Plan for the Drainage of Manila and Suburbs*, vol. II (Manila: Bureau of Public Works, 1952), 275; Greg Bankoff, *Cultures of Disaster: Society and Natural Hazards in the Philippines* (London; New York: Routledge Curzon, 2003). It was also likely that the drive toward immediate reconstruction after the war played a role in the rampant deforestation and quarrying activities in Manila's immediate hinterland. However, further research is needed to verify and measure the extent of this causal link.
71. Doeppers, *Feeding Manila*, 175.
72. BPW, *Plan for the Drainage of Manila*, vol. II, 292.
73. Ibid., 269.
74. BPW, *Plan for the Drainage of Manila*, vols. I and II.
75. C. M. Hoskins, "Drainage Plan for the Greater Manila Area," *American Chamber of Commerce Journal* 28, no. 6 (1952): 219.
76. Marikina Project Coordinating Committee (MPCC), *Report on Marikina River Multi-Purpose Project: Proposed Development Plan and Evaluation* (Manila: National Economic Council, 1954).
77. A. A. Villanueva, "Memorandum of the Director of Public Works, 25 Apr. 1952," in BPW, *Plan for the Drainage of Manila and Suburbs*, vol. II, Appendix, 1–5 (Manila: Bureau of Public Works, 1952), 4.
78. BPW, *Plan for the Drainage of Manila*, vol. I, 52.
79. BPW, *Plan for the Drainage of Manila*, vol. II, 266.
80. BPW, *Plan for the Drainage of Manila*, vol. I, 54.
81. Villanueva, "Memorandum of the Director," 4.
82. Ibid., 3.
83. Ibid., 4.

84. City of Manila, Management and Planning Division, *Manila City Government* (Manila: Office of the Mayor, 1958), 5.
85. Tutay, "Squatter Trouble," 7.
86. City of Manila, *Manila City Government*, 18.
87. City of Manila, *City Government of Manila, 1952–55* (Manila: Office of the Mayor, [1955?]), 15.
88. Black & Veatch, *Master Plan*, 5-23.
89. Ibid., 5-20.
90. Ibid., 5-14–5-15.
91. Black & Veatch, *Master Plan*, 3-4.
92. Pedroche, "Slums Can Be Licked," 14.
93. Black & Veatch, *Master Plan*, 5-21, 5-25.
94. Filemon V. Tutay, "Worst Flood in Years," *Philippines Free Press*, 4 June 1960: 6–7, 42–43, 79. Quote in p. 6.
95. MPCC, *Report on Marikina River*, 22; Bankoff, *Cultures of Disaster*, 100–102.
96. Tutay, "Worst Flood in Years," 6.
97. Leon O. Ty, "'Why didn't I also die?'" *Philippines Free Press*, 4 June 1960: 4, 77. Quote in p. 77.
98. Tutay, "Worst Flood in Years," 79.

CHAPTER 13

Apocalyptic Urban Future

Atomic Cities and Cinema

Kimmo Ahonen, Simo Laakkonen, and William M. Tsutsui

INTRODUCTION

This is the D-Day. Total destruction by nuclear weapons, and from this hour forward the world as we know it no longer exists. And over all the lands and waters of the earth, hangs the atomic haze of death. Man has done his best to destroy himself. But there is a force more powerful than Man, and in His infinite wisdom He has spared a few.

With these words, the sonorous voice of the narrator begins Roger Corman's science fiction (sci-fi) movie *The Day the World Ended*, released in 1955. It depicted the world after the nuclear holocaust: the D-Day of Hiroshima was transposed into the D-Day of the entire planet, into global Armageddon. Corman's exploitation movie portrays how the last survivors

K. Ahonen (✉)
Tampere University, Tampere, Finland

S. Laakkonen
University of Turku, Turku, Finland

W. M. Tsutsui
Hendrix College, Conway, AR, USA

of the post-apocalyptic world struggle against the mutated monsters created by atomic radiation, the odorless, flavorless, and invisible force that can deform living cells. The film forced viewers, who lived in a world threatened with nuclear war, to imagine how humanity could survive in an environment ravaged by nuclear war.

This film was one among many outcomes of a single day, August 6, 1945, which changed human and environmental history forever. Altogether, the atomic bombing of Hiroshima and the attack on Nagasaki three days later caused the death of over 200,000 people. With the complete physical devastation wrought by the atom bomb, Hiroshima and Nagasaki became true shock cities. However, it took a long time before humankind came to understand what had actually happened in the blinding flashes of light that consumed those Japanese cities.

The unforgettable photograph of the huge mushroom cloud, taken by Enola Gay crew member George Caron, was one of the few images of Hiroshima that were shown in the immediate aftermath of the bombing. *Life* was the first American magazine to publish the picture of the towering cloud in its August 20, 1945 issue.[1] Such aerial images of Hiroshima and Nagasaki were very influential because they established the iconography of an atomic blast in the global consciousness. They became universal symbols of total warfare—observed from a safe distance, sublime, majestic, and haunting—but they did not reveal anything about the actual destruction that the atomic bombs caused to cities and their inhabitants. In addition, military officials tried to suppress the circulation of information about atomic radiation and its effects. With the human and environmental costs of nuclear weaponry obscured by the euphoria of victory and official censorship, major newspapers celebrated the atomic bombs as emblematic not just of the American triumph over Imperial Japan, but of America's postwar global ascendance.[2]

This "positive" public perception of nuclear arms changed drastically after the journalist John Hersey wrote a long article entitled "Hiroshima," which appeared in *The New Yorker* in fall 1946 and was published soon thereafter as a book. In his journalistic masterpiece, Hersey portrayed how six survivors experienced the atomic bombing of the city. He was the first who gave a voice to the victims, that is, the Japanese inhabitants of the devastated city. By reading Hersey's compelling work, the American public received a completely different view of the Hiroshima and Nagasaki bombings from what had previously been presented in newsreels and in the newspapers.[3] Indeed, photographs of the destructive effects of the bomb—particularly of the human victims—had not yet been shown in the media. The first photographs of Japanese killed and injured in Hiroshima and Nagasaki were not published until 1952 in *Life* magazine.[4]

Official censorship was even tighter for documentary footage than for photographs, as the American authorities apparently recognized the power of moving images of such vast urban destruction. Both US soldiers and Japanese journalists filmed the aftermath of the bombings of Hiroshima and Nagasaki but American civilian and military officials succeeded in suppressing virtually all of it for decades. However, when the censors closed the doors on such documentary records, they unintentionally opened a cultural Pandora's box: the absence of actual footage of the atomic bombings lifted the lid on apocalyptic imagination, releasing the creative energies and profound anxieties of writers, artists, and (perhaps above all) filmmakers around the world.[5]

A global boom produced hundreds of science fiction movies between 1948 and 1962, as postwar audiences seemed never to tire of sensational blockbusters featuring alien invasions, giant irradiated monsters, and terrifying nuclear mutants. These films, which poured from Hollywood backlots and Tokyo soundstages (but were also produced elsewhere in Asia and through Western Europe), were often quickly and inexpensively made, formulaic, and superficial in addressing complex contemporary issues.[6] But they were also extraordinarily revealing about their time and their audiences, addressing the contradictions of modern science, growing popular concerns about environmental degradation, and the unresolved traumas of the war just past, the dawning atomic age, and the superpower confrontations ahead. These science fiction movies, many of which are now regarded as classics of the genre, transformed the disturbing realities of Hiroshima and Nagasaki, so long hidden from the public, into chilling cinematic visions of urban devastation, disruption of the natural world, and post-apocalyptic survival.

This chapter explores the ways science fiction cinema made in the United States, as well as in Japan and around the world, dealt imaginatively with the threat of nuclear war, rising environmental concerns, and the destruction of cities after World War II.

SCIENCE FICTION, POPULAR CULTURE, AND ATOMIC WORLD WAR II

The aim of Nazi Germany, which had started the race to develop nuclear weapons, was to drop atomic bombs on the major cities of the United States. The Allies, however, succeeded in developing the bomb first and the new technology was used first by the United States not on Germany, which had already surrendered, but on Imperial Japan, which was still at war. Both Nazi Germany and the United States were looking beyond the current conflict to a possible future confrontation with the Soviet Union. As it turned out, however, the advantage achieved for the United States with the atomic bomb lasted for only a few years, since the Soviet Union responded to the challenge by exploding its own nuclear device in September 1949. After this, a psychological shift occurred in American domestic debates over the nuclear specter, in which the United States was changed from perpetrator into potential "victim," despite the fact that for many years it maintained an overwhelmingly superior nuclear arsenal.[7]

Only limited information on the international arms race seeped into the public domain, and the general public around the world long remained bewildered by the events in Japan in August 1945. The most difficult thing to comprehend was the destruction caused by those first atomic bombs: how was it possible that a single weapon could destroy an entire city? What was the impact of radiation on such a scale? What did a city struck by a nuclear weapon look like? When the destruction caused by the atomic blasts in Hiroshima and Nagasaki was discussed in the American press, in some cases a hypothetical comparison with New York was used: the aim was to understand what the consequences would be of dropping an equivalent bomb on a major US city.[8] To an American public that had not experienced widespread conventional bombing of cities (as the populations of Europe and Japan just had), comprehending the inconceivable scale of urban destruction from an atomic attack was an exercise in analogy and imagination.

In his 1946 essay "You and the Atom Bomb," George Orwell contemplated the effects of America's new superweapon on warfare and international politics. Orwell, who was among the first to embrace the name "Cold War" for the evolving postwar international order, noted that the idea of the atom bomb was not entirely a product of the recent war, because it had already been imagined decades earlier:

> For forty or fifty years past, Mr. H. G. Wells and others have been warning us that man is in danger of destroying himself with his own weapons, leaving the ants or some other gregarious species to take over. Anyone who has seen the ruined cities of Germany will find this notion at least thinkable.[9]

As Orwell argued, science fiction literature had addressed and even predicted the development of the atomic bomb and other weapons of mass destruction. In fact, Wells had presented the idea of the atom bomb even before World War I in his novel *The World Set Free* (1913–1914). For Wells, the power of atomic energy was a tool not for world peace, as was widely asserted during the Cold War, but for bringing about global destruction.[10]

Science fiction speculation took place at a time when the theoretical and technical knowledge necessary for developing the atom bomb was still in the future. But such fictional visions may also have shaped the creation of that fateful knowledge and the weapons that grew from it. Hungarian-born nuclear physicist Leo Szilard read Wells's book in 1932, and in the following year he conceived the idea of a neutron chain reaction.[11] During World War II, Szilard was one of the key figures in the Manhattan Project. At the same time, certain science fiction writers were speculating about future weapons of mass destruction. Imagination and reality combined dramatically in Cleve Cartmill's novella *Deadline*, which appeared in the pulp magazine *Astounding Science Fiction* in March 1944. The story reflected on whether the Allies would use an atomic bomb against the Axis powers. The publication was clearly read by people other than science fiction devotees, because the FBI soon investigated the editor of the magazine and Cartmill for endangering national security.[12] At a time of high stakes and high tensions, it was impossible to draw the line between science fiction and reality.

In the post-Hiroshima world, the belief in technological progress, which had held sway in science fiction since the beginning of the century, began to give way to darker dystopias. Philip Dick and Frederik Pohl, for example, described the near future of the United States in bleak terms in their science fiction novels and short stories of the 1950s. In the decade's science fiction cinema one can also find several thematic patterns which subvert the dominant triumphalist narrative of the time—that America was the richest and most powerful nation in the advancing Free World— with a more desolate vision.[13] First, alien encounters, which usually meant violent invasion, were one of the major topics in the science fiction films of the time. Another recurring theme was the attack of gigantic monsters and avenging, mutated creatures from the natural world, born from scientific experiments or atomic radiation. Third, it is possible to distinguish certain subgenres, such as post-apocalyptic films, which depicted the survival of a small group of people after the nuclear holocaust, including *Five* (1952) and *The Day the World Ended* (1955).[14] All of these three narrative patterns dealt with, to a greater or lesser degree, the question of the survival of the American way of life—and America's cities—after World War II and the beginning of the atomic age.

According to Winston Wheeler Dixon, who has studied images of the apocalypse in sci-fi cinema, visualizing the destruction of one's own life and civilization is not only entertaining for moviegoers but also oddly and unexpectedly comforting. Total devastation, Dixon notes, is a great social leveler: everybody is equal, regardless of their class in society or their ethnic background.[15] Joanna Bourke, who has studied the cultural history of fear, also stresses that cultural products offer individuals and society a channel for structuring their hopes and fears.[16] The images of aliens and monsters in science fiction films were connected to apocalyptic fantasies—both terrifying and liberating—which have long been part of the science fiction literary tradition of the West.[17] In his book about atomic bomb movies, Jerome Shapiro defines this tradition as the "apocalyptic imagination."[18] Hence, in these films, both contemporary and time-honored anxieties and fantasies met. According to Paul Boyer, who echoed the mainstream scholarly analysis of American popular culture of the immediate postwar era, American science fiction of the 1940s and 1950s can be read transparently as a contemporary commentary. Thus science fiction writing, sci-fi comic strips, and sci-fi movies dealt with the threats of nuclear war, totalitarianism, and invasion.[19]

THE WAR OF THE WORLDS AND THE ATOMIC APOCALYPSE

See New York disappear! See Seattle blasted! See San Francisco in flames! See paratroops take over the capital!

This is how the trailer for the science fiction movie *Invasion U.S.A.* (1952) revels in images of the destruction of major cities by atomic bombs.[20] American metropolises came under attack in many science fiction films of the 1950s, not however by Soviet troops as in *Invasion U.S.A.*, but more often by invading aliens, as for example, in *The War of the Worlds* (1953) or *Earth* vs *The Flying Saucers* (1956). However, in the fertile imaginations of filmmakers at the time, the agents of destruction could also be scientists whose experiments had gone badly wrong, nuclear war, monsters generated by radiation, or meteors colliding with Earth. Mainly science fiction movies offered audiences an imaginative portrayal of what would happen if the next world war were to start.

As characterizes the larger genre of science fiction, the relationship between sci-fi movies and science and technology was very ambivalent. At the beginning of *The Day the Earth Stood Still* (1951), an alien, Klaatu, lands in a spaceship in Washington, with the aim of warning humankind about the dangers of nuclear war. The use of nuclear weapons and nuclear power endangers the safety of "other planets" and is unacceptable to the advanced society from which Klaatu comes. As a character, Klaatu is a combination of Jesus Christ and an Old Testament prophet: he delivers an ultimatum, which forces the leaders of the major powers to make peace. In *The Day the Earth Stood Still*, aliens were portrayed as representatives of a higher civilization and the scientists sympathetic toward them are shown as heroes. The strife-torn nations of Earth are equipping themselves for a new world war, and it is only the intervention of extraterrestrials that brings hope for a better future.[21] *The Day the Earth Stood Still* was, however, an exception in its sanctimonious critique of the arms race and the xenophobia of postwar American society.

The majority of 1950s American science fiction films presented aliens as enemies attacking humanity, the nuclear family, and American values. At the same time, Hollywood's aliens diverted the public attention away from domestic conflicts toward a more immediate, identifiable other on the big screen. Mainstream invasion films also legitimized the Cold War arms race by demonstrating how scientists and the military defended the United States in a crisis. Not surprisingly, the United States Army actively strove to influence what kind of images Hollywood movies presented of soldiers and the armed forces. Its influence was partly indirect, since the studios needed military equipment when they made war movies. In any case, Hollywood and the Pentagon had a close relationship during World War II and at the start of the Cold War.[22] In a way, 1950s science fiction films depicted the rise of "the military-industrial complex," the sprawling agglomeration of America's armed forces and private businesses, about which President Dwight Eisenhower warned US citizens in his 1961 farewell address. In these films, however, it was the alliance of the military and science which saved the community—or, in many cases, betrayed it, having first created the threat to society through scientific experimentation gone wrong. The schizophrenic Dr. Jekyll and Mr. Hyde nature of the atomic age was apparent to everyone who went to a double feature or a drive-in during the 1950s.

In 1965, essayist Susan Sontag wrote an influential article entitled "The Imagination of Disaster" which dealt with the iconography and aesthetics of 1950s science fiction films. Sontag focused on the genre's pervasive imagery of physical destruction: science fiction films, she argued, were less about science than about the spectacle of destruction, which held a peculiar appeal and beauty. According to Sontag, the viewer "can participate in the fantasy of living through one's own death and more, the death of cities, the destruction of humanity itself."[23] Thus in many films the monuments of Washington, D.C., and the skyscrapers of New York came under attack as aliens and monsters laid waste to the cities and their most iconic structures. In the aftermath of Hiroshima and Nagasaki, and under the threat of Cold War nuclear holocaust, the modern vision of an urban environment as a technological utopia was literally demolished by science fiction films that reveled in the destruction of cities.[24]

The War of the Worlds, based on the 1898 novel by H. G. Wells, described a global invasion of aliens, climaxing with a fleet of spaceships destroying Los Angeles. At the beginning of the film version directed by Byron Haskin, the narrator connects the Martian attack to international politics and the development of weapons technology. Mention of the two world wars places the alien attack in the continuum of twentieth-century military history. During the narrator's prologue, which had not been part of Barry Lyndon's screenplay, images are shown of soldiers and rockets.[25] The newsreel-like narrative style brings a semi-documentary flavor to the film's beginning and links the alien threat to the fear of a third world war.

Mainstream science fiction movies, in which the armed forces and scientists fought together against invading aliens and monsters, were actually war films set in the near future. World War II remained a model for the mass invasion of aliens, and the war was re-staged with atomic weapons or with new military hardware. Both *The War of the Worlds* and *Earth* vs. *the Flying Saucers* included montage sequences, in which the global scale of the battle and the annihilation were depicted in a documentary style. Special effects created an illusion, but the documentary approach linked science fiction films to the reality of their time.[26] In particular, the use of stock footage shots of American postwar atomic bomb tests became an important genre-specific characteristic, connecting science fiction cinema to contemporary images of the Cold War threat.[27]

In a way, science fiction films dealt with the resilience of American society and its institutions in situations of crisis and warfare. Although science fiction films reveled in the destruction of human civilization, at the same time they offered a way out, enjoyment at the return—at least partially—

of the social order. Barry Keith Grant points out that, in these films, the restorers of social normalcy are scientists and soldiers, and that the critique of scientific-technological culture often turns into a paean of praise.[28] Grant's claim is mostly correct, but also too one-sided: very few films present science and technology *alone* as the solutions to humanity's existential crisis. For example, in *The War of the Worlds*, no weapons can stop the Martians; it is only the collective spiritual humbling of the nation that brings victory over the enemy. In the 1950s, the United States was still a very conservative and relatively rural society, and in science fiction movies, religion could be as potent a force as science.

Notably, even the darkest works of cinematic science fiction generally ended on crowd-pleasing notes of optimism: Hollywood's fictional apocalypses usually concluded with hope for a better tomorrow, promising redemption for humanity rather than nihilistic oblivion. In Roger Corman's dystopic *The Day the World Ended*, for instance, the leading man and woman find love in the end and the mutant produced by radiation dies when the rain falling from the skies "melts" it. The heroine of the movie states that "Man created it, God destroyed it." It appears that the rain is free of radiation and will wash away the remaining contamination, giving the surviving couple a chance to make a fresh start in the post-apocalyptic world. Instead of the customary "THE END," the film closes, almost cheerily, with the words "THE BEGINNING."

In Hollywood's poetics of destruction, the end of the world is actually also its beginning.

Science fiction movies played on the fears and insecurities of postwar American society, but to a surprising degree they also explored the utopian dreams that bloomed in the wake of World War II. First, they gave concrete form to the threat of a third world war by depicting the annihilation of cities and entire nations, and presenting imaginative visions of a post-apocalyptic world. Second, they affirmed that science and scientists had a decisive role in the new world order. Science was depicted as both a destructive and a redeeming force, and scientists were both heroes and potential villains. Third, they predicted a space race, where conquering extraterrestrial environments would become part of superpower rivalries and technological progress. Finding an alternative habitable planet in the place of an Earth rendered uninhabitable by radiation or destroyed by a third world war was established as one of humankind's great new missions of the future.

Nature Strikes Back: Radiation-Born Monsters

Hollywood's alien invasion movies addressed popular fears of the unknown and of hostile others in an era of rapid technological change, the proliferation of weapons of mass destruction, and tense superpower confrontations. Another imaginative response to the anxieties of the time was giant monster films, which flourished in the United States from the early 1950s into the 1960s and in Japan up to the 1970s, and which were also made in Britain, the Soviet Union, Denmark, South Korea, and China. Gigantic cinematic insects, spiders, and reptiles were almost invariably portrayed as the unintended consequences of scientific experimentation or innovation—pesticides and other toxic chemicals, nuclear weapons, new uses of radiation—that rendered fauna (and the occasional flora) huge and destructive. These creatures all seemed to seek revenge on humanity for disrupting the natural order and, like alien invaders, they focused their attacks on the greatest symbols of civilization, the world's cities. Visually arresting and entertaining, Cold War giant monster features enjoyed global popularity, playing on audiences' uneasiness about powerful but little-understood technological advances, a nagging sense of human overreach in scientific research, and a nascent environmental consciousness.[29]

The 1933 Hollywood classic *King Kong* was re-released internationally in 1953 and proved a box-office hit. Its success inspired filmmakers in Hollywood and in Japan to begin work on new "monster on the loose" pictures. The 1953 Warner Brothers production *The Beast from 20,000 Fathoms*, considered the first giant monster movie to feature radiation in the storyline, traced the rampage of a dinosaur through New York City after it was freed from hibernation in an Arctic glacier by nuclear weapons testing. The following year, the studio released an even bigger blockbuster, *Them!*, the cleverly conceived story of ants from the nuclear testing grounds of New Mexico, mutated to extraordinary size by exposure to residual radiation, that menace the desert Southwest before swarming into the storm sewers of Los Angeles and creating a giant nest. A television reporter breathlessly (and prematurely) declares that "Man as the dominant species of life on Earth will become extinct!" before soldiers with flamethrowers incinerate the ant colony, restoring order and reasserting the dominance of humankind over nature.[30]

Them! established basic patterns and themes that would become conventions in giant monster movies made in Hollywood and internationally in subsequent decades. First, the film's premise that radiation could render living creatures gigantic came to be accepted as plausible and even commonsensical by moviegoers, despite the lack of any scientific basis for such a belief. Second, *Them!* inaugurated an unlikely boom in "big bug" movies, with dozens of features involving gargantuan, destructive insects and arachnids produced over the subsequent decade. Third, it charted a simple but compelling narrative that would be followed, with slight variations, by most of the films in the genre: a monster appears or is created in a rural area; local authorities are unable to control the unprecedented threat; the creature then turns its sights on a major city, terrifying the populace and causing considerable damage; and the monster is finally vanquished by scientific ingenuity, brute military force, or an expedient plot twist, thus restoring peace and order to a traumatized but enduring humankind. Fourth, *Them!* (like Hollywood's alien invasion pictures) strongly affirmed the tradition of "secure horror," which allowed viewers to confront their real-world fears through cinematic sci-fi fantasies, but still leave movie theaters reassured that all would be right with society and the natural world.[31]

Hollywood's parade of giant, marauding arthropods that followed on the success of *Them!*—Warner Brothers' highest grossing film in 1954— was striking, creative, and frequently bizarre. In *Beginning of the End* (1957), locusts rendered huge by consuming experimental crops treated with radioactive fertilizer terrorize the Illinois countryside before descending on Chicago. Mysterious lab experiments produced the giant spider in *Tarantula!* (1955) and aliens seek to use huge, irradiated insects and reptiles to overcome humanity in *Killers from Space* (1954). In *The Deadly Mantis* (1957), a series of geological and climatic disasters frees a huge praying mantis from the Arctic and it attacks New York City. Mothra, a colorful irradiated moth with a 250-meter wingspan, appeared in a series of movies from Japan's Toho Studios starting in 1961 and laid waste to numerous cities, including Tokyo.[32] Eventually the cinematic "monsters on the loose" came to include a wide range of species and substances— from a bird in *The Giant Claw* (1957) to an octopus in *It Came from Beneath the Sea* (1954) to sinister blobs, slimes, and even giant stones— and were embraced by filmmakers globally. Dinosaurs and other gigantic lizards were perennial favorites and they terrorized London in *Gorgo* (1961) and *Behemoth, the Sea Monster* (1959), Copenhagen in *Reptilicus* (1961), Seoul in *Yongary, Monster from the Deep* (1967), and the solar system in the Soviet *Planeta Bur* (1962).

Giant monster movies reached levels of imaginative sophistication and cultural ubiquity in Japan even greater those in the United States in the 1950s. *Gojira* (1954) introduced the character later known internationally as Godzilla, a saurian survivor of the Jurassic period mutated into a gargantuan, vengeful monster by US hydrogen bomb testing in the South Pacific. The global appeal of *Gojira*, released in America in 1956 as the heavily edited *Godzilla King of the Monsters*, spawned a series of sequels (ultimately running to more than 30 films produced in Japan and Hollywood) and an entire cinematic ecosystem of other enormous monsters including a radioactive turtle (Gamera), a giant flying squirrel (Varan the Unbelievable), and all sorts of dinosaurs (from the pteranodon Rodan to the ankylosaurus Anguirus).[33]

Japan's rich and enduring genre of *kaijū eiga* (literally "strange beast movies") tapped into what the artist and critic Murakami Takashi has called "the twofold violence Japan experienced [in World War II] as both victimizer and victimized, as well as its fear of the Cold War."[34] *Gojira*, in particular, directly addressed Japan's atomic age traumas (Godzilla's furrowed skin resembled the keloid scars of the survivors of Hiroshima and Nagasaki[35]) and the health, environmental, and global security concerns created by Cold War nuclear testing. The repeated assaults on Tokyo by Godzilla and other creatures (most notably the somber and agonizing destruction of the city in *Gojira*) drew on audiences' still raw memories of wartime firebombing and the atomic attacks, and chillingly emphasized the fragility of urban civilization at a time of proliferating doomsday weaponry and in a nation caught between two hostile superpowers.[36] Moreover, *kaijū eiga* reflected growing popular concerns about the natural consequences of nuclear development and the environment degradation caused by accelerated industrialization in Japan and globally. In addition to exploring the radioactive contamination of Japan and the South Pacific through an imaginative lens, Japanese monster movies spoke to issues of air and water pollution in the early 1970s and, later, to the environmental and social costs of untrammeled economic growth.[37] Godzilla, above all, was presented as a cinematic conscience for humankind and a warning for a planet seemingly bent on its own destruction: as one character solemnly muses at the close of *Gojira*, "I can't believe that Godzilla was the only surviving member of its species. But if we keep on conducting nuclear tests, it's possible that another Godzilla might appear somewhere in the world again."[38]

Reality and fiction blended and merged in both Hollywood's big bug pictures and Japanese *kaijū eiga*, as filmmakers tapped the headlines for convenient plot devices and compelling storylines sure to engage (and perhaps alarm) audiences. In *The Deadly Mantis*, for instance, the monster was first detected by the DEW (distant early warning) Line, a network of US radar stations designed to give early warning of Soviet attacks across the North Pole. The radioactive fertilizer experiments responsible for creating the mammoth locusts in *Beginning of the End* actually were conducted, as American agricultural scientists apparently bought into Hollywood's fantasy that radiation could make living organisms larger.[39] The appearance of Godzilla in the series' first film drew directly on the Lucky Dragon incident, in which a Japanese tuna fishing boat was exposed to massive amounts of radiation in the Castle Bravo hydrogen bomb test on Bikini Atoll.[40] Moreover, as documented by William Tsutsui, Hollywood's giant insect movies addressed very real environmental concerns, "a widespread unease about insect infestation and humankind's ability to control it during the 1950s and 1960s."[41] Not only did films like *Them!* echo a media-fueled frenzy over invasive species like fire ants and Japanese beetles, but these seemingly unsophisticated special-effects features also channeled growing popular misgivings about the safety and effectiveness of chemical insecticides, particularly DDT.[42]

While postwar giant monster movies often seem humorous, even ridiculous, by twenty-first century standards, many resonated strongly with audiences at the time.[43] *The Saturday Review* praised *Them!* "for being as persuasively realistic a horror film as one can possibly imagine."[44] And even if some people were not terribly horrified by images of giant turtles or spiders rampaging through cities, many moviegoers were at least left uneasy by the disturbing "what if?" thoughts that these films evoked. A key message of "creature features" was that no one knew, with certainty, the outcomes—intentional and unintentional—of Western technoscientific progress. Questions and insecurities lingered and, after Hiroshima and Nagasaki, it was impossible to ignore catastrophic scenarios for humanity.[45] If governmental laboratories had been able to develop in secrecy bombs that could wipe out an entire city, then how sanguine could anyone be about the growth of the military-industrial complex or the fruits of scientific research that seemed even more outlandish than the fantasies on cinema screens? How might nature and natural forces be perverted, accidentally or intentionally, as threats to human beings or indeed to planet Earth itself? The science fiction movies of the 1950s treated both humankind and nature as potential victims in an unpredictable narrative of progress driven by international rivalries, the logic of capitalism, and an unchained atomic Prometheus.

Today we know that postwar science fiction filmmakers and their audiences may well have been justified in at least some of their darker suspicions. As Jacob Darwin Hamblin, Edmund Russell, and others have shown, the collaboration of the American military, scientists, and industry gradually extended the concept of total war into the natural world.[46] For example, during World War II the development of the A-bomb was paralleled by research on a cyborg-type B-bomb: the "bat bomb" was a cylinder filled with hibernating bats carrying individual incendiary devices, designed to be dropped by plane on an enemy city. In theory, the bats would wake up in the lower, warmer altitudes and fly from the cylinder to the attics and eaves of buildings to hide, where the timed bombs would explode. In 1943, the bomb-carrying bats successfully burnt a model of a Japanese town built in the Utah desert.[47] Weather was manipulated by both the Allies and the Axis powers.[48] During the Cold War, the US Navy funded, for example, an entomologist's study of army ants[49] and scientists continued to develop methods to utilize earthquakes, typhoons and other weather events, bacteria, and viruses for future warfare.[50] In retrospect, Hollywood studios were no match for the secret R & D laboratories of the new superpowers that World War II had created.[51] In the atomic age, truth was often stranger than fiction, fantasy was often intertwined with reality, and collective dreams and nightmares of an imagined future collided not just in science fiction films on drive-in screens, but in government laboratories and military bases, in the world's cities and through the natural world.

CONCLUSIONS

In our obsession with the antagonisms of the moment, we often forget how much unites all the members of humanity. Perhaps we need some outside, universal threat to make us recognize this common bond. I occasionally think how quickly our differences worldwide would vanish if we were facing an alien threat from outside this world. And yet, I ask you, is not an alien force already among us? What could be more alien to the universal aspirations of our peoples than war and the threat of war?

In 1987, President Ronald Reagan, speaking to the General Assembly of the United Nations, surprised his listeners by spontaneously reflecting on the consequences of an encounter with space aliens.[52] Reagan's speech about the threat of aliens and total war is eerily reminiscent of the science fiction movie *The Day the Earth Stood Still*. Reagan was a prototypical "Cold Warrior," who advocated a hardline foreign policy and stoked the nuclear arms race in his first presidential term. However, as a private person, he deeply feared nuclear conflict and its dire global implications.

This fear, which he shared with all humankind, was something new, born of war and scientific discovery in the mid-twentieth century. Belief in the end of the world had existed since the dawn of time but had originated in superstition, myths, religious faith, or prophecy.[53] After World War II, Armageddon was linked to manmade technology for the first time. With all its violence—from concentration camps, genocide, and the firebombing of cities to the horror of atomic warfare—that global cataclysm led to the birth of a new kind of fear, fear of a manmade end to the world, an anthropogenic apocalypse. Since World War II, and especially since two days in early August 1945, environmental catastrophism has been part of our environmental awareness.[54]

Science fiction movies were a meaningful part of the first wave of environmental catastrophism, which swept the world in the 1950s. This wave driven forward by atmospheric atomic bomb tests, increasingly critical scientists, growing political movements, and a culture saturated with alien invaders and irradiated monsters forced governments to act. The result of these global protests was the signing in 1963 of the Partial Test Ban Treaty, which prohibited nuclear weapons testing or any other nuclear explosions in the atmosphere, underwater, or in outer space. The goal was to protect both human lives and the environment from radiation exposure.[55] After that, average levels of radioactivity in the atmosphere began to decrease and the topic of radioactive fallout eventually disappeared from the mass media. The broad activism against nuclear testing paved the way, however, for subsequent environmental issues and movements, such as the rising concern about pesticide use that culminated in the publication of *Silent Spring*. The Partial Test Ban Treaty also diminished interest in movies about nuclear cataclysms, post-apocalyptic survival, and giant radioactive monsters. But it did not drive them to extinction.

In fact, during the following decades, science fiction films continued to thrive and over time the variety of monstrous creatures expanded. Celluloid biodiversity achieved new heights. Completely new species took to the screen in the final decades of the twentieth and early twenty-first centuries. Dinosaurs and arachnids continued to be favored by filmmakers, along with flies and bees, but mutated amphibians and cockroaches only made surprisingly belated debuts. Bacteria, viruses, mollusks, crustaceans, and rodents were also fresh to sci-fi fantasies, as were fish species like eels, piranhas, and above all the shoals of sharks that took on major roles over the following decades. Radioactive zombies re-emerged after the millennium to terrify audiences, but biology and gene technology gradually superseded radiation as Hollywood's preferred mutagen. Nevertheless, the "radioactive bogeyman" has enjoyed a long half-life in the global imagination.[56]

In brief, when nuclear weapons targeted cities, urban popular culture targeted nuclear weapons. The resulting "nuclear monster" movies of the 1950s started a completely new genre in the history of film, eco-horror. According to Andrew Newman, now that human society has changed everything, from the chemical composition of the atmosphere on down, we have everything to fear.[57] Consequently, science fiction movies played a significant role in postwar cultural resilience. The invention of imaginary futures, both the utopias and dystopias of human civilization, was a long-standing tradition of Western political thought and culture.[58] Envisioning the worst possible eco-scenarios provided structure and focus to society's problems, which in turn enabled the articulation of potential solutions. They portrayed both a vision of apocalypse as well as the possibility of salvation. Postwar nuclear cinema, with fantastical narratives and arresting images that foregrounded the dangers of contemporary military strategy and technology, opened the eyes of the general public to the possible futures of global civilization. And all of those futures appeared monstrous, not only for humanity's great cities, but for the world as a whole.

In the final analysis, one might well conclude that everything on Earth has reason to fear just one particularly aggressive species. As an actor as well as a Cold Warrior, Ronald Reagan realized that the real monster—"the alien inside," as he defined it—was the new human being that World War II had spawned and which had created the atomic future that we must live with today. Even more than half a century after Hiroshima, with hundreds of science fiction movies and celluloid monsters to shock us and compel us to action, the nuclear missiles are still there—targeted on towns and cities.

NOTES

1. Paul Boyer, *By the Bomb's Early Light: American Thought and Culture at the Dawn of the Atomic Age* (Chapel Hill, NC: University of North Carolina, 1994), 8. In the *Life* article, the Hiroshima and Nagasaki atomic bombs were depicted as an extension of the U.S. air raids on Japan.
2. Jiyoon Lee, "A Veiled Truth: The U.S. Censorship of the Atomic Bomb," *Duke East Asia Nexus* 3, no. 1 (2011); Janet Farrell Brodie, "Radiation Secrecy and Censorship after Hiroshima and Nagasaki," *Journal of Social History* 48, no. 4 (June 2015): 842–864.
3. John Hersey, *Hiroshima. A New Edition with a Final Chapter Written Forty Years After the Explosion* (New York: Vintage Books, 1989).
4. Dorothee Brantz, "Landscapes of Destruction: Creating Image through Photography," in Michael Geyer and Adam Tooze, eds., *The Cambridge History of the Second World War, Volume 3. Total War: Economy, Society and Culture* (Cambridge: Cambridge University Press, 2017), 725–748.
5. Jerome Shapiro, *Atomic Bomb Cinema: The Apocalyptic Imagination on Film* (New York: Routledge, 2002), 10–15, 192.
6. See, among the many sources on 1950s science fiction cinema, Joyce Evans, *Celluloid Mushroom Clouds: Hollywood and the Atomic Bomb* (Boulder, CO: Westview Press, 1998); Bill Warren, *Keep Watching the Skies! American Science Fiction Movies of the Fifties, Volume 1, 1950–1957* (Jefferson, NC: McFarland, 1982); and Shapiro, *Atomic Bomb Cinema*.
7. Mick Broderick and Robert Jacobs, "Nuke York, New York: Nuclear Holocaust in the American Imagination from Hiroshima to 9/11," *Asia-Pacific Journal* 11, No 6 (March 2012): 3.
8. Broderick and Jacobs, "Nuke York, New York," 4–5.
9. George Orwell, "You and the Atomic Bomb," first published in *Tribune* (London, October 19, 1945).
10. H. G. Wells, *The World Set Free* (1914), Project Gutenberg eBook 1059, release date, February 11, 2006. See also Spencer R. Weart, *The Rise of Nuclear Fear* (Cambridge: Harvard University Press, 2012), 12.
11. Richard Rhodes, *The Making of the Atomic Bomb* (New York: Simon & Schuster, 1986), 23–25.
12. See also John Clute, "Cartmill, Claude" in John Clute and Peter Nicholls, eds., *The Encyclopedia of Science Fiction* (New York: St. Martin Griffin, 1993), 202–203.
13. The narrative patterns of 1950s science fiction films were at least partly dictated by financial realities: successful films spawned imitators. See, for example, Victoria O'Donnell, "Science Fiction Films and Cold War Anxiety" in Peter Lev, ed., *The Fifties: Transforming the Screen 1950–1959, History of the American Cinema Vol. 7* (Berkeley: University of California Press, 2003), 169–195.

14. On post-apocalyptic films of the 1950s, see Kim Newman, *Apocalypse Movies: End of the World Cinema* (New York: St. Martin's Press, 2000), 99–106.
15. Winston Wheeler Dixon, *Visions of the Apocalypse: Spectacles of Destruction in American Cinema* (New York: Wallflower Press, 2003), 2–3.
16. Joanna Bourke, *Fear: A Cultural History* (London: Virago, 2005), 6–7, 74–75, 353–354.
17. The first movie where a dinosaur attacked a city was probably *The Lost World* (1925), which was based on Arthur Conan Doyle's novel (1912) of the same name. In this film a living Brontosaurus forcefully taken from the Amazon ravages through London.
18. Shapiro, *Atomic Bomb Cinema*, 10–15, 192.
19. Boyer, *By the Bomb's Early Light*, 258.
20. *Invasion USA, Theatrical Trailer* 1952. See also *Invasion U.S.A. Screenplay by Robert Smith*. From a Story by Robert Smith and Franz Spencer. UCLA Collection of Motion Picture Scripts, Box 521, University of California, Los Angeles.
21. On *The Day the Earth Stood Still*, see Mark Jancovich, *Rational Fears: American Horror in the 1950s* (Manchester: Manchester University Press, 1996), 141–146; Tony Shaw, *Hollywood's Cold War* (Edinburgh: Edinburgh University Press, 2007), 142–144.
22. Errol Vieth, *Screening Science: Contexts, Texts and Science in Fifties Science Fiction Film*. PhD, Griffith University, Faculty of Arts, School of Film, Media and Cultural Studies (1999), 75.
23. Susan Sontag, "The Imagination of Disaster," *Commentary* (Oct. 1965).
24. Vivian Sobchack, "Cities on the Edge of Time: The Urban Science Fiction Film" in Sean Redmond, ed., *Liquid Metal: The Science Fiction Film Reader* (New York: Wallflower Press, 2004), 80–81.
25. The prologue was not in Barré Lyndon's screenplay (updated in January 1952) at all. According to the screenplay, the film would have started with an image of the cover of Wells' book. Barré Lyndon, *The War of the Worlds. Revised Final Write*, January 11, 1952, 1–2, AMPAS Script Collection, Margaret Herrick Library. Academy of Motion Pictures Arts and Sciences, Los Angeles.
26. Vieth, *Screening Science*, 129.
27. Patrick Lucanio, *Them or Us: Archetypal Interpretations of Fifties Alien Invasion Films* (Bloomington: Indiana University Press 1987), 69–71.
28. Barry Keith Grant, "'Sensuous Elaboration': Reason and the Visible in Science Fiction Film" in Sean Redmond, ed., *Liquid Metal. The Science Fiction Film Reader* (New York: Wallflower Press, 2004), 21.
29. The scholarly literature on giant monster movies is substantial. For a source which focuses on one subgenre but which places postwar "creature features" in an environmental context, see William Tsutsui, "Looking Straight

at *Them!* Understanding the Big Bug Movies of the 1950s," *Environmental History* 12, no. 2 (April 2007): 237–253. See also Jancovich, *Rational Fears*, and Peter Biskind, *Seeing is Believing: How Hollywood Taught Us to Stop Worrying and Love the Fifties* (New York: Pantheon, 1983).
30. See also Cyndy Hendershot, "Darwin and the Atom: Evolution/Devolution Fantasies in 'The Beast from 20,000 Fathoms,' 'Them!,' and 'The Incredible Shrinking Man,'" *Science Fiction Studies* 25, No. 2 (July 1998): 319–335, and Tsutsui, "Looking Straight at *Them!*"
31. See Andrew Tudor, *Monsters and Mad Scientists: A Cultural History of the Horror Movie* (Oxford: Blackwell, 1989), 215.
32. Tsutsui, "Looking Straight at *Them!*"
33. Among the many works on Godzilla and Japanese monster movies, see William Tsutsui, *Godzilla on My Mind: Fifty Years of the King of Monsters* (New York: Palgrave Macmillan, 2004).
34. Murakami Takashi, *Little Boy: The Arts of Japan's Exploding Subculture* (New York: Japan Society, 2005), 205.
35. Tsutsui, *Godzilla on My Mind*, 33.
36. On Japan's postwar "imagination of disaster," particular as related to cities, see Susan Napier, "Panic Sites: The Japanese Imagination of Disaster from Godzilla to Akira," *Journal of Japanese Studies* 19, no. 2 (Summer 1993): 327–351, and William Tsutsui, "Oh No, There Goes Tokyo! Recreational Apocalypse and the City in Postwar Japanese Popular Culture" in Gyan Prakash, ed., *Noir Urbanisms: Dystopic Images of the Modern City* (Princeton: Princeton University Press, 2010), 104–126.
37. Tsutsui, *Godzilla on My Mind*, 61, 72–73.
38. Quoted in Tsutsui, *Godzilla on My Mind*.
39. Richard Leskosky, "Size Matters: Big Bugs on the Big Screen" in Eric Brown, ed., *Insect Poetics* (Minneapolis: University of Minnesota Press, 2006), 329; Tsutsui, "Looking Straight at *Them!*," 242.
40. Tsutsui, *Godzilla on My Mind*, Chapter 1.
41. Tsutsui, "Looking Straight at *Them!*," 246.
42. See, for instance, Edmund Russell, *War and Nature: Fighting Humans and Insects with Chemicals from World War I to Silent Spring* (Cambridge: Cambridge University Press, 2001).
43. See Weart, *The Rise of Nuclear Fear*.
44. Quoted in Adams, Evans, *Celluloid Mushroom Clouds*, 103.
45. Katy Waldman, "The Nuclear Monsters That Terrorized the 1950s," *Slate*, January 31, 2013, http://www.slate.com/articles/health_and_science/nuclear_power/2013/01/nuclear_monster_movies_sci_fi_films_in_the_1950s_were_terrifying_escapism.html

46. Jacob Darwin Hamblin, *Arming Mother Nature: The Birth of Catastrophic Environmentalism* (New York: Oxford University Press, 2013), 12, 21; Russell, *War and Nature*.
47. "Bat Bomb" project member Jack Couffer's memoirs are probably the most quoted publication on the subject. Jack Couffer, *Bat Bomb: World War II's Other Secret Weapon* (Austin: University of Texas Press, 1992).
48. See James Fleming, *Fixing the Sky: The Checkered History of Weather and Climate Control* (New York: Columbia University Press, 2010).
49. Derek Leebaert, *The Fifty-Year Wound: The True Price of America's Cold War Victory* (Boston: Little, Brown, 2002), 30–31.
50. See Hamblin, *Arming Mother Nature*.
51. See Hendershot, "Darwin and the Atom," 323.
52. Ronald Reagan, *Address to the 42nd Session of the United Nations General Assembly in New York*, September 21, 1987. According to presidential historian Lou Cannon, Reagan personally added the paragraph to his speech. Lou Cannon, *President Reagan: The Role of a Lifetime* (New York: Simon & Schuster, 1991), 63–64.
53. See, for example, David Adams Leeming, *Creation Myths of the World: An Encyclopedia* (Santa Barbara, California: ABC-CLIO, 2010).
54. Simo Laakkonen, Richard Tucker, and Timo Vuorisalo, "Hypotheses: World War II and Its Shadows," in Simo Laakkonen, Richard Tucker, and Timo Vuorisalo, eds., *The Long Shadows: A Global Environmental History of the Second World War* (Corvallis: Oregon State University Press, 2017), 324–325.
55. See http://www.state.gov/t/isn/4797.htm#treaty. Accessed February 5, 2018; John McCormick, *Reclaiming Paradise: The Global Environmental Movement* (Bloomington: Indiana University Press, 1989), 51–55.
56. Andrew Newman, "The Persistence of the Radioactive Bogeyman," *Bulletin of Atomic Scientists*, October 23, 2017, (no page numbers): https://thebulletin.org/bio/andrew-newman-0
57. David Ropeik, "Godzilla vs. Technology: How the Radioactive Monster Explains All Our Modern Environmental Debates," *Slate*, June 5, 2014, http://www.slate.com/articles/health_and_science/science/2014/06/godzilla_gojira_and_the_hydrogen_bomb_how_a_movie_monster_framed_the_environmental.html?via=gdpr-consent; Brian Merchant, "The Evolution of Eco-Horror, from Godzilla to Global Warming," *Storyboard*, November 14, 2014, https://motherboard.vice.com/en_us/article/xyy473/the-evolution-of-eco-horror-from-godzilla-to-global-warming
58. An important contribution to this critical tradition is Oswald Spengler's *Untergang des Abendlandes* which was published in two volumes between 1918 and 1923. It appeared in English as *The Decline of the West* in 1926.

SECTION V

Conclusions

CHAPTER 14

Epilogue: What Makes a City Resilient?

Simo Laakkonen, J. R. McNeill, Richard P. Tucker, and Timo Vuorisalo

INTRODUCTION

Some months after Stalin and Hitler had divided Europe between themselves in the so-called Molotov-Ribbentrop Pact on August 23, 1939, Soviet bombers appeared in the sky over Helsinki. But they also appeared above other Finnish towns and cities including Viipuri—the second biggest city in the country. Before the war, Viipuri had been an international, prosperous, and beautiful Finnish city. Unluckily, it was located near the Russian border, and consequently, Finnish and Soviet troops fought over it several times during World War II. More than two-thirds of Viipuri's buildings were ultimately damaged or destroyed by bombings, artillery

S. Laakkonen (✉)
University of Turku, Turku, Finland

J. R. McNeill
Georgetown University, Washington, DC, USA

R. P. Tucker
University of Michigan, Ann Arbor, MI, USA

T. Vuorisalo
Department of Biology, University of Turku, Turku, Finland

© The Author(s) 2019
S. Laakkonen et al. (eds.), *The Resilient City in World War II*, Palgrave Studies in World Environmental History, https://doi.org/10.1007/978-3-030-17439-2_14

barrages, or battles as well as the several ensuing fires. Only eight percent of its structures remained completely intact.[1] At the end of the war, Finland had to cede the capital of Karelia Province to the Soviet Union. Before the cession the entire Finnish population emigrated to other parts of Finland. The country miraculously survived World War II as an independent state and as a Nordic democracy.

Aside from stray dogs, Viipuri was a practically empty city when Soviet officers, soldiers, and settlers entered its streets at the end of the war and moved into the still habitable homes of the previous Finnish inhabitants. The occupied and ruined city was now renamed Vyborg (Выборг), and was left behind the Iron Curtain as a peripheral Soviet border town that remained closed to foreigners including its former inhabitants. Due to the destruction of war, the evacuation of the original population, the arrival of the Red Army, and the subsequent period of dereliction under Soviet power, the loss of Viipuri was full and final. Instead of facing a flourishing future as a free, democratic, and prosperous city, Viipuri became a marginalized outpost of the vast Soviet empire. If we adapt the words of urban historian Lewis Mumford,[2] Soviet Vyborg became for its original inhabitants a tyrannopolis as well as a nekropolis.

Meanwhile Helsinki, the capital of Finland, became known, along with other northern cities such as Stockholm, Copenhagen, or Oslo, as a Nordic model city with sophisticated municipal services including high-level public education, public transport, efficient environmental policy, and clean air, waters, and soil.[3] This tale of two cities clearly shows how in some cases World War II dictated and differentiated the fates of towns and cities. Viipuri may seem to have been an extreme case, but many towns and cities around the world shared its fate during and after World War II.[4] The consequent commemoration of lost homes and hometowns became above all in East Asia and Europe a long and painful process that still continues.

However, the socio-environmental stress brought about by war was there irrespective of whether towns or cities were on the winning or losing side of the war. While many towns and cities were ruined or damaged, for others the war years meant increasing industrial expansion and extremely rapid population growth due to immigration of new inhabitants and/or refugees. Such cities could be found, for instance, along the coastlines of North America, where, for instance, Seattle benefitted from its strategic location by the Pacific Ocean; in western Siberia beyond the Ural Mountains Sverdlovsk and other industrial cities grew when relocated factories intensified the exploitation of regional iron ore and coal reserves; and in the interior of China where Chongqing is an illustrative example. They also could be called shock cities due to the decisive impact that World War II had on their development. In contrast, there were also towns and cities where life continued in a relatively normal way despite the war. In brief, war had negative, positive, neutral, and controversial impacts on wartime towns and cities.

Urban Elements and Resilience

We have chosen resilience as a key concept in this collection of essays because resilience has been consistently linked with shocks. World War II was, if anything, a global shock. It was at the same time a demographic, economic, social, political, and cultural turning point. It affected directly or indirectly nearly every nation, city, and town in the world. Consequently, people all over the world still today regard World War II as by far the most important event in world history.[5]

The concept of resilience was put forward in natural sciences by the Canadian ecologist C. S. Holling, who defined it as a measure of the persistence of ecological systems and of their ability to absorb shocks and disturbance and still maintain the same relationships between state variables.[6] Holling's original biological definition made (and still makes) sense in the ecological context, but is for several reasons not directly applicable to social sciences or humanities. Natural scientists and even some urban ecologists have long neglected the fact that *Homo sapiens* is, almost by definition, the dominant species in towns and cities.[7] Consequently, modern urban communities cannot recover after shock to any previously undisturbed state; instead they change intentionally and constantly toward new states due to internal and external dynamics.[8]

Resilience has recently attracted lots of attention in the social sciences and humanities.[9] In both approaches and thereby also in historical studies, human agency is embraced as the most critical issue in resilience.[10] Therefore urban resilience refers here to the wartime capacity of towns and cities to function and maintain realistic living opportunities for human beings and other species, no matter what natural or anthropogenic problems they may encounter.[11] In a resilient city individuals and communities have the socio-environmental capacity to cope with wartime stress and shocks. Naturally no city can stand overwhelming hardships, and therefore the niche of resilience should be related to the scale of prevailing problems and availability of potential resources. Because cities are organized in ways that both produce and reflect underlying socio-economic disparities, some parts are much more resilient than others. Uneven or "partial" resilience, as Nicole Barnes defines it, threatens the ability of cities to function as a whole.[12] Although location, economy, culture, and history certainly matter, urban resilience is formulated and reformulated constantly by political choices. Essays in this volume also point out that power and related inequalities and interests are important approaches both in the development and in the evaluation of urban resilience. It is important to note that resilience is more a process than a feature of a given city.

Urban Infrastructure

Each town has a story to tell. The authors of the essays in this book endeavor to see how World War II affected urban inhabitants, infrastructure, and environment in different parts of the world. In the following pages we present the main themes of the book in a wider framework. We not only base our views on the experiences of the towns and cities that have been addressed in previous essays, but take into consideration experiences of other towns and cities as well. Some of the generalizations are of hypothetical nature, due to the lack of studies in this new subfield, that is, urban environmental history of wars in general and World War II in particular.

All industrial cities rely on critical infrastructures including water supply and sanitation, food production, transport and storage, energy production and distribution, electric power generation and transmission, telecommunications, and transportation. Extensive and functioning urban infrastructure is vital for urban resilience because it forms the pumping heart and veins of towns and cities by transforming and transporting people, food, materials and goods, water and wastewater, energy, and information. These metabolic analogies were earlier widely used in urban studies. E. W. Burgess already in 1925 suggested that the state of metabolism of a city might be measured by its "mobility," which he defined as a change of movement in response to a new stimulus or situation; this concept comes rather close to resilience.[13] In addition to internal infrastructure, all urban centers depend on the hinterland in transport of labor force and natural resources (Fig. 14.1).

Fig. 14.1 Weimer Pursell's poster (1943) for the Office of War Information shows how crisis awareness and public power stimulated innovative thinking even in American car culture. (Source: National Archives (NWDNS-188-PP-42), via Wikimedia Commons)

The case studies in this collection reflect two major water-related changes in urban resilience that took place in practically all industrialized countries by the early twentieth century. By this time most Western towns and cities had been able to construct large-scale water supply networks and water purification plants, which provided safe drinking and household water for the majority of their inhabitants. Another outcome of this reform and the structural modernization of industrial cities had led, in practice, to an end to urban fires afflicting more than one block of houses. A striking exception to these rules was World War II, when old urban nightmares concerning insufficient quantity and quality of drinking and household water reappeared in many formerly "water-secure" cities, and catastrophic urban fires returned to the modern and presumably "fire-secure" cities.

Many cities had great plans and initiatives in the interwar period to develop critical urban infrastructure. However, such initiatives in the fields of municipal waste management, sewerage and wastewater treatment and also air pollution control, as Peter Brimblecombe shows, suffered from deterioration owing to labor and materials shortages and wartime destruction of vehicles, networks, and plants.[14] Consequently, urban-industrial environmental protection suffered and revived only laboriously after the war. Urban environments were part of municipal infrastructures as well. The lack of maintenance during the war years also affected urban nature including soil, waterways, and air. The deterioration of *esteros* or wetlands in occupied Manila due to the lack of maintenance is an illustrative example of such large-scale socio-environmental changes.

URBAN AGRICULTURE

Food was often in short supply during the war, due to disruption of regional and international commerce or production, processing, storage, and delivery. The failure to provide food security was manifested in besieged Leningrad, in occupied Athens, and also in Hanoi where 1–2 million people perished in the famine of 1945.[15] The impetus for allotment gardening in European cities was primarily food shortages during wars. Even Sweden, which escaped both world wars, experienced a rapid increase in community gardening, reflected by the rise from 30,000 gardens prior to the World War II to 130,000 during its peak, and providing perhaps ten percent of all vegetables consumed in Sweden. In Britain the number of allotment or victory gardens, as they were called, rose from 800,000 to 1,400,000 during the World War II peak, providing British citizens with 1,300,000 tons of food. In Germany, the number of allotments rose from 450,000 during the economic crises of the early 1930s to 800,000 at the war's end. In most countries, the area of allotment gardens decreased only slowly after the war.[16]

Rauno Lahtinen addresses the importance of urban agriculture and animal husbandry for food security during periods of military crises. But provision of urban food security was not a simple thing. It required at least three critical factors; first, availability of potentially arable public lands— private lands were seldom used extensively; second, additional watering of generally dry urban soils; and third, practical knowledge of subsistence farming.[17] Fortunately, a big proportion of the wartime urban population consisted of ex-farmers, or people who had grown up on farms. An innovative cultural heritage was a crucial ingredient of urban agriculture and of urban resilience.

Another important but so far unexamined factor for successful urban agriculture during the exceptionally cold war years is the urban heat island, a term that refers to the fact that towns and cities tend to have, due to human influence, higher mean temperatures than the surrounding rural areas.[18] This phenomenon was already described in the early nineteenth century for London and the nearby countryside.

Urban Fauna and Flora

Considering the importance of birds and plants for the urban way of life, and the high number of bird-watchers and amateur field naturalists who must have personally experienced the war in their home towns or cities, surprisingly little is known about the effects of warfare on urban fauna and flora.[19] We are not aware of any systematic wartime or historical study on the topic. This is not a surprise, considering the fact that until the 1970s practically all ecological studies concentrated on more or less pristine natural habitats, and cities were considered unworthy of study as human-dominated ecological anomalies. However, the origins of urban ecology as a discipline are deeply rooted in World War II. Vuorisalo and Kozlov point out that many pioneer works on botany and zoology were performed in bombed cities and related wastelands, which today would be called brownfields. In isolated West Berlin, the lack of accessible countryside forced botanists and zoologists to continue investigating urban habitats and develop urban ecology. Since then, urban ecology studies have gradually gained practical importance in urban green planning in cities all over the world.[20]

Urban wildlife showed considerable resilience in all examined cities. Early successional plant species rapidly colonized bombed sites, and along with vegetation, rather diverse animal life thrived among the ruins. The black redstart (*Phoenicurus ochruros*), a bird species, became a symbol of war-damaged urban habitats in many European cities during and after World War II. It was, however, a flower, the oleander that became the most impressive biological symbol for urban natural resilience. It was widely believed that nothing would grow in the atom bomb-scorched soil of Hiroshima for the next 75 years. But as spring 1946 arrived, the surviving inhabitants were amazed to see the devastated terrain dotted with red oleander flowers. The oleander (*Nerium oleander*) and six surviving gingko trees (*Ginkgo biloba*) dispelled the concerns that the destroyed city had lost all of its fertility, and revived the hope of recovery for the population of Hiroshima. Oleander was chosen the official flower of the city.[21] The first sign of urban resilience after the first atom bombing on Earth was thus urban nature. While poppies became symbols of the destruction of countryside during World War I, oleander symbolized urban killing fields of World War II.

Urban biodiversity covers not only wild animals and plants but also pets, domestic animals, and zoo animals. Cities and towns constituted, paraphrasing the title of George Orwell's famous novel, a strange "urban animal farm" during World War II. Because of the expansion of urban agriculture, the number of production animals increased considerably in towns and cities. While urban agriculture took place mostly in public lands, chickens, rabbits, and pigs were reared in private backyards and sometimes even indoors to guarantee provision of protein and extra incomes. Due to the lack of animal power, fodder, and medical care, the surviving urban horses, mules, and cows were overworked and undernourished. Yet public and private pets faced perhaps the worst fate during the war. Altogether probably millions of pets including birds, cats, and dogs were culled because their owners did not have enough food to share, or they had to go off to war, or their homes were bombed. In starving cities people even had to eat their pets—often after swapping them in order to avoid eating their own cats or dogs.[22] Pets were also released, and at the end of the war the high number of stray cats and dogs constituted a particular urban and rural problem that was addressed by shooting and poisoning them.

War, bombing, and battles killed not only innocent people in towns and cities but also the even more innocent animals. In addition to private pets, animals in public zoos were adversely affected. Only a minority of zoo animals were evacuated from cities. Due to the lack of staff, rationing of food and fuel, and fear for air raids potentially dangerous animals such as elephants, large predators, venomous snakes, and constrictors and arachnids were either euthanized or simply destroyed in hundreds in zoos all over the world.[23] Warfare destroyed or badly damaged famous zoos of, for example, Warsaw, Hamburg, and Vienna. However, the remaining "public" zoo animals substituted the lost "private" animals and became important symbols for survival for the inhabitants in London, Berlin, Tokyo, and Leningrad as well. Members of the staff of Leningrad Zoo miraculously managed to keep some animals, including Belle the hippo, alive in the starving city and the zoo open during the siege.[24] Mieke Roscher and Anna-Katharina Wöbse indicate that zoo animals, some of which were kept alive with great sacrifices, contributed significantly to the populations' morale and endurance. Consequently despite all killing and culling, a particular form of urban resilience developed in wartime cities, that is, interspecies resilience.

Urban Society and Environmental Health

The social structures of Eurasian cities, already unstable in the depression years, were fractured by the war, becoming more turbulent and improvisational than ever. Many cities had to absorb massive arrivals of rural people torn from the land by military movements. Public administration faced a Herculean task. It could be severely disrupted, as social services in peacetime were diverted to the military and its priorities. Where sanitation systems were disrupted, disease threatened. Where regional transport systems were badly damaged or diverted to military priorities, as in Munich, the movement of food supplies from farm to city was curtailed. This left the urban poor, whether local or immigrant, dependent on unreliable relief operations or scavenging to feed themselves. Malnutrition, even famine, intensified the likelihood of epidemic disease. Malaria was a constant danger in southern Europe and tropical Asia, in Hanoi, Manila, Rangoon, and elsewhere. Typhus was a common threat also in European cities, from Naples to Munich to Warsaw.

In the three war-torn cities of Asia discussed in this book, the influx of refugees from conflict zones caused crises in environmental health that bore down disproportionately on the poor and dislocated populations. Breakdowns of water and power supplies and medical facilities resulted in severe public health crises, especially for the newly arrived. Sudden, radical population changes in these cities revealed the wartime regimes' social biases, which protected the wealthy and powerful and largely marginalized or punished the poor.[25]

In the Nationalist Chinese headquarters of Chongqing, the wartime population nearly tripled, producing massive turmoil, confusion, and public health breakdowns. Refugees from Japanese-controlled regions improvised bamboo shelters in squatter colonies along river banks. But this was risky; the flood of spring 1938 left 30,000 people homeless. From then onward, they suffered poor sanitation, seasonal epidemics, polluted air in the mountain valley setting, and even lung disease in the weapons factory. Some medical care was provided in various hospitals and clinics, including three mission hospitals. But as the war continued, those facilities ran low on staff and equipment. More seriously, the military administration consistently favored the wealthy and even criminalized the poor and vagrant.

Hanoi struggled through the war under an administration inherited from colonial French times, supervised by the Japanese military. Under the hierarchy of social privilege, landlords, Chinese rice merchants, and French bureaucrats were relatively secure; vast numbers of poor were marginalized. Aerial bombardment of transport networks, especially railways, disrupted the movement of supplies into Hanoi, as in many European and East Asian cities. Troubles in the crowded city intensified following the American bombing that began in 1943, cutting railways from rice-surplus areas far to the south. Then nature, the typhoon and flood of early 1945, drowned rice lands in the Red River lowlands; weather was a natural force, but social and environmental dislocations intensified the impact. Medical treatment collapsed almost entirely. Famine struck Hanoi that March, when the Japanese took direct control of the city. More than a million Vietnamese, many already weakened by malnutrition and disease, died by August when the war ended. Ho Chi Minh's administration that took over after that faced an enormous challenge to rebuild tolerable living conditions in the capital city.

In Manila, dismal health conditions confronted the poor majority, who lived primarily in coastal wetlands and meandering waterways that were increasingly polluted. Wartime neglect meant that public services in those areas deteriorated. In November 1943, a disastrous typhoon and subsequent flooding compounded the environmental stress of the lowlands; many people fled to rural areas. The Filipino administration under Japanese military overlords favored the well connected, protecting them from flood, pollution, and shortages. In all Asian examined cities, urban resilience was socially a partial process.

The main target of industrial warfare was civilian populations. The dropping of atom bombs on Hiroshima and Nagasaki was only one stage in the short history of terror bombings of towns and cities from the air that had started during the Great War, expanded in the Spanish Civil War, and then escalated during World War II.[26] Terror bombing culminated in atomic warfare, which characterized the urban nature of World War II.

The 1945 atom bombs turned a new and more sinister page in the history of humankind. It took a long cultural process to think of the unthinkable and understand the understandable, not to speak of what could happen next in targeted towns and cities. The existing and potential dangers of atomic weapons, and the new techno-scientific urban-industrial culture as a whole, were increasingly addressed in society and popular culture. Manipulated and mutilated nature emerged all over industrialized nations in movies, television, books, magazines, newspapers, and comics.[27] The new technology developed during World War II[28] created a new world order of eco-horror, which from the 1950s onward compelled industrial nations to seek salvation in international environmental policy, cooperation, and agreements.

Resilience Lessons from Wartime Cities: Creative Urban Devolution

Inhabitants of a town or a city experiencing a severe military crisis ultimately have three options: to die, survive at home, or migrate elsewhere. Survival or migration, in the military context, both mean some degree of resilience or adaptation to adverse conditions. As our essays show, all these options were realized full-scale in World War II urban areas. The literature indicates three mechanisms for urban resilience: persistence, transition, and transformation. *Persistence* simply refers to the ability of technical systems, organizations, and individuals to resist disturbance. While retaining basic urban functions (urban metabolism) is an important part of most definitions of resilience, many definitions of urban resilience refer to *transition*, that is, the ability to adapt incrementally, or more radically even to *transform*, where different structural or administrative parts or elements of the city assume entirely new functions.[29] A transition may be a short-term and local reaction to some acute crisis, while a transformation may require more centralized decision-making and reorganization of urban infrastructure and practices.

One key concept to understand wartime urban transitions and above all transformations is *evolutionary* resilience[30] or rather *devolutionary* resilience. The basic stages of human history that resulted in contemporary post-industrial towns and cities consist of the modification of natural sites by hunter-gatherers, agricultural communities, manufacturing sites, and industrial centers. The key drivers of this change toward the present industrialized world were expanding commerce, constant flow of innovations, increasing division of labor, higher productivity, and ever-heavier reliance on external natural resources. These broad categories of socio-cultural development represent the diverse ways by which societies may adapt to a variety of environments based on the available natural resources, technologies, and knowledge.

In some countries, above all in the United States, World War II signified accelerated urban-industrial evolution. At the end of the war, its "arsenal of democracy" produced about half of global industrial production. Its dramatic wartime economic development helped the country to move beyond the Great Depression, but it also greatly changed towns and cities. Expanding arms and other industries made millions of people move to towns and cities, which accelerated suburban sprawl by creating green neighborhoods for affluent white middle-class families. The new suburbs in turn created new and strong social bases for environmentalism for postwar North America.[31] But on a structural level, war accelerated the development of consumer culture, car-dependent cities, interstate highways (begun in the mid-1950s but inspired by wartime concerns), and metropolitan regions that required massive amounts of fossil energy and other natural resources and produced massive amounts of pollution. Peter Brimblecombe's essay on air pollution shows well the consequences of such development in different locations. Thereby World War II became one of the factors creating perhaps the most environmentally harmful model of town planning and urbanization that world had yet seen and spreading this model to other regions and continents as well (Fig. 14.2).

On the other hand, across Europe and Asia, in cities that suffered most directly from warfare, urban-industrial development was halted or even reversed, in which case devolutionary resilience was achieved by a return to some earlier phase of socio-cultural development. In such cities, the volume of industrial production decreased and focused on a few strategic sectors. The first step toward urban devolution was to shift emphasis from

Fig. 14.2 The power of persuasion. During world wars the best artists were engaged in national campaigns to reduce consumption, spare natural resources, and reuse and recycle for the public good. After the war, national medias were harnessed by private interests to increase consumption and use of natural resources. (Source: Poster designed by Vanderlaan. National Archives (NWDNS-79-WP-103), via Wikimedia Commons)

industrial toward manufacturing and domestic production by means of using more simple tools and processes, saving energy whenever possible and reusing or recycling all possible raw materials and tools.

The second widespread transformation toward urban devolution was resorting to agriculture in towns and cities. As mentioned earlier, this required providing access to additional farmland by putting public parks, private lands, bombed lots, and urban wastelands under cultivation. Access to water was essential for urban agriculture; in some places urban inhabitants dug new wells or put old channels, pipes, wells, or natural watercourses to new uses. Municipal and domestic production of vegetables, root vegetables, meat and skins became a significant mental and material factor in wartime urban resilience. In brief, urban inhabitants shifted from being essentially only food consumers to local food producers, and occasional urban horticulture and gardening developed into serious and continuous urban farming.[32]

The third transformation took place in more desperate situations where inhabitants of towns and cities had to resort even to the original strategy of human existence, that is, the hunter-forager economy. Municipalities often relaxed rules to allow more extensive and intensive use of urban wildlife including hunting, fishing, and gathering. Possibilities for hunting of large mammals were naturally limited within and near cities; therefore hunting focused mostly on available small-sized mammals like hares, rabbits, squirrels, stray dogs or cats, and birds including ducks, domestic pigeons, crows, jackdaws, and in extreme cases even small passerine birds. Such solutions to protein scarcity were inevitably transient, as in most wartorn cities edible mammals and birds vanished rapidly. In towns and cities with abundant watercourses, fishing provided somewhat better possibilities to secure additional food. Also gathering became important. It included picking of available natural berries, fruits, and/or mushrooms depending on their availability in urban areas and on the local traditional diet. Wartime urban gathering was extended to collection of firewood not only from urban and nearby forests but also from parks and scavenging of combustible materials such as planks, furniture, or books among ruins of destroyed or deserted buildings. People and on the frontline soldiers also scavenged dumps or bombed neighborhoods and industrial sites in order to find something edible or otherwise useful, or valuable materials. Organized private or municipal gathering became important in some specific cases.[33]

In order to utilize previous stages of urban development, wartime cities and their populations had to be creative and look for and develop retrovations.[34] Modern adaptions of old innovations and knowledge were required, in order to use all potentially available urban natural resources including changing seasons, day and night, air and wind, rain, surface and ground water, properties of soil, arable land, biomass, fish and game. Using modern terminology, socio-ecological transformations brought about in wartime towns and cities created reversed modernization, which implied a significantly reduced ecological footprint or rather wartime "bootprint" of these urban areas.[35] Heating and the average temperature of premises were reduced. Daylight saving that was first adopted during World War I was also reintroduced during World War II to save energy. The more natural diurnal cycle in blacked-out cities was probably good for both urban inhabitants and nature. Saving, reuse, and recycling of various raw materials became widespread in households, enterprises, and public institutions. Decreased use of cars and increased use of collective transport, combined with walking and cycling made many cities more active, healthy, clean, and quiet.[36] Increased use of local natural resources including certain natural cycles, urban soil, green space, game and fish, and rain, ground and surface waters made people more aware of urban natural resources and deepened their relation with urban nature. Tom Arnold reveals how *Müncheners* adapted by turning back the clock to pre-industrial stages of development, showing that years of city life had not weakened their relationship to nature in any manner. In wartime towns and cities, the number of members in nature protection movements also increased consistently, which indicates the importance of nature for urban inhabitants during wartime.[37] Naturally, not all people supported these activities. The imagined resilience of American wartime collection campaigns, described by Sarah Frohardt-Lane, enlisted civilians on the home front to help in war efforts, while industries avoided rationing. Nevertheless, from today's point of view these developments provide grounds to call many wartime towns and cities eco-cities.

There was of course a dark side to these developments—or several dark sides (Fig. 14.3). Reversed modernization in shock cities resulted in increased self-reliance, decreasing division of labor, lower productivity and consumption. Such a rapid large-scale reverse of urban development was, as Annika Björklund has explained, an anomaly and it came with a price.[38] The well-known by-products of reversed urban development included poverty, malnutrition, heightened rates of infectious diseases, worsening hygiene,

14 EPILOGUE: WHAT MAKES A CITY RESILIENT? 295

Fig. 14.3 The long shadows of war. A World War II veteran and his son looking at the Golden Gate Bridge in spring 1945. (Source: San Francisco History Center, San Francisco Public Library)

emergence of pests, abuse of animals and non-human life in general, and overexploitation of local natural resources. In the long run, resorting to the previous stages of socio-economic development was not enough to support large urban human populations. In desperate situations, inhabitants had no other alternative but to interact first with the countryside to gain fuel and food, and then to move there to be able to grow food.[39] In practice, wartime urbanization of the countryside and ruralization of cities enabled nature to take over certain parts of cities, in practice first the most badly devastated and emptied areas, and then gradually others. In principle, urban devolution signified returning to the origins of urbanization—to nature.

Future Uses for Wartime Practices?

World War II made it clear that public power, national governments, states, and municipalities were able to achieve, if necessary, fundamental changes in urban order and practices in a notably short time period. This observation has prompted several environmental scholars and activists to ask what kind of implications this wartime experience may have for the contemporary era characterized by accelerating climate change.[40] If democracy is too slow and inefficient system to react in crisis, could centralized government be summoned for "climate mobilization," to gather and compel citizens once more to change their production and consumption habits and bring about a wholesale industrial retooling, this time against environmental risks? Such drastic measures may be needed in the future, but they are not yet a viable option. Democratic governments temporarily acquired exceptional powers only because such violent ideologies as Communism, Fascism, and Nazism threatened world peace and democratic nations. Despite warning signs, climate crisis has not yet acquired such an aura of menace that would make people concede to democratic governments exceptional power. Today sustainable urban futures require more carefully constructed climate-based mitigation and adaptation agendas than centralized decision-making systems.[41] However, mitigation and adaptation to climate change in urban areas might obviously benefit from wartime experiences in resilience-promoting urban practices.

What, then, were the main components of urban resilience during World War II? We maintain that the key to the utilization of potential natural resources in extreme wartime conditions was the capacity for reversed modernization, which in practice meant a return to survival strategies typical for manufacturing, agricultural, or even hunter-gatherer societies. Nevertheless, previous studies and essays in this book show that wartime urban societies included various macro- and micro-level components of resilience that have been addressed above. But what were the common basic prerequisites that supported these sub-properties and wartime urban resilience as a whole?

We return to the conceptual starting point of this book, to the three basic elements of urban communities—nature, technology, and culture. The more nature, and the more varied nature, there was left in the city, the more resilient a wartime city was in principle able to be. Similarly, the more varied historical layers a city had, the more possibilities it had to develop different strategies to utilize existing urban natures. However, the human agency was the critical factor for wartime urban resilience as well. In practice, the efficient use of potential urban natural resources and critical urban infrastructures required rich cultural heritage, that is, creative utilization of necessary knowledge, skills, and tools originating from industrial, manufacturing, agricultural, and hunter-gatherer economies. Hence, in addition to urban nature and infrastructure, urban resilience was based on the memories, values, interests, and capabilities of urban inhabitants and their institutions. In brief, urban resilience required natural, urban, and cultural diversity.

Nevertheless, it is good to bear in mind that towns and cities that did not participate in the war were probably the most resilient ones. Therefore active peace-building by means of consolidating local, national, and international democracy, socio-economic justice, and functioning public institutions is a crucial part of progressive urban culture that may make the world a better place in the future as well.

Notes

1. Tapio Hämynen and Yuri Shikalov, *Viipurin kadotetut vuodet 1940–1990* [The Lost Years of Viipuri, 1940–1990] (Helsinki: Kustannusosakeyhtiö Tammi, 2013), 77, 86, 105, 110–116.
2. Lewis Mumford, *The City in History: Its Origins, Its Transformations, and Its Prospects* (New York: Harcourt, Brace & World, 1961), 234.
3. The criteria and ratings change somewhat annually, but in the twenty-first century the top ten green cities in the world have been constantly located in the Nordic countries, Central Europe (e.g. Amsterdam, Berlin, Vienna, Zurich), or Canada (Vancouver). It is not a coincidence that these cities have also had the most equal socio-economic structures. See, for example, Global Green Economy Index, Global Sherpa, and European Urban Ecosystem Survey. These rankings are not comparable because they address different criteria, although not always explicitly.
4. See, for example, Steven Brakman, Harry Garretsen and Marc Schramm, "The Strategic Bombing of German Cities during World War II and its Impact on City Growth," *CESifo Working Paper*, no. 808 (2002), 1–38. http://hdl.handle.net/10419/76256

5. James H. Liu et al., "Social Representations of Events and People in World History Across 12 Cultures," *Journal of Cross-Cultural Psychology* 36, no. 2 (March 2005): 185. For reasons of importance of wars in collective memory, see Magdalena Bobowik et al., "Beliefs about history, the meaning of historical events and culture of war," *Revista de Psicología* 28, no. 1 (2010): 115–116.
6. C. S. Holling, "Resilience and Stability of Ecological Systems," *Annual Review of Ecology and Systematics*, Vol. 4 (1973): 1–23; Thomas Elmqvist, "*Urban resilience thinking*," *Solutions* 5, no. 5 (September 2014): 26–30; Akira S. Mori, "Resilience in the Studies of Biodiversity-Ecosystem Functioning," *Trends in Ecology & Evolution* 31, no. 2 (February 2016): 87–89.
7. Indeed, many ecological studies conducted in urban settings have simply considered humans as agents of disturbance. See Nancy E. McIntyre, K. Knowles-Yánez and D. Hope, "Urban ecology as an interdisciplinary field: differences in the use of "urban" between the social and natural sciences," *Urban Ecosystems* 4, no. 1 (January 2008): 11, 17.
8. David E. Beel et al., "Cultural resilience: The production of rural community heritage, digital archives and the role of volunteers," *Journal of Rural Studies* 54 (2017): 459–461. See Kristen Magis, "Community Resilience: An Indicator of Social Sustainability," *Society & Natural Resources* 23, no. 5 (2010): 402.
9. See, for example, Jeremy Walker and Melinda Cooper, "Genealogies of Resilience: From Systems Ecology to the Political Economy of Crisis Adaptation," *Security Dialogue* 42, no. 2 (2011): 143–60; Muriel Cote and Andrea J. Nightingale, "Resilience thinking meets social theory. Situating social change in socio-ecological systems (SES) research," *Progress in Human Geography* 36, no. 4 (2011): 475–489. See also Tobias Plieninger and Claudia Bieling, eds., *Resilience and the Cultural Landscape: Understanding and Managing Change in Human-Shaped Environments* (Cambridge: Cambridge University Press, 2012).
10. Marjolein Spaans and Bas Waterhout, "Building up resilience in cities worldwide – Rotterdam as participant in the 100 Resilient Cities Programme," *Cities* 61 (January 2017): 111; Beel et al., "Cultural resilience," 460.
11. Our concept that focuses on wartime cities is near definitions presented in Ian P. McCarthy, Mark Collard, and Michael Johnson, "Adaptive Organizational Resilience: An Evolutionary Perspective," *Current Opinion in Environmental Sustainability* 28 (October 2017): 1.
12. Lawrence J. Vale, "The politics of resilient cities: whose resilience and whose city?" *Building Research & Information* 42, no. 2 (2013): 191–201.

13. Ernest W. Burgess, "The Growth of the City: An Introduction to a Research Project," originally published 1925, reprinted in John M. Marzluff, Eric Shulenberger, Wilfried Endlicher et al., eds., *Urban Ecology. An International Perspective on the Interaction Between Humans and Nature* (New York: Springer, 2008), 71.
14. Raymond G. Stokes, Roman Köster, and Stephen C. Sambrook, *The Business of Waste: Great Britain and Germany, 1945 to the Present* (Cambridge: Cambridge University Press, 2013); Simo Laakkonen, Sari Laurila, eds., *Harmaat aallot. Ympäristönsuojelun tulo Suomeen* (Helsinki: SHS, 1999).
15. Polymeris Voglis, "Surviving hunger: life in the cities and the countryside during the occupation," in Robert Gildea, Olivier Wieviorka and Anette Warring, eds., *Surviving Hitler* and *Mussolini: daily life* in *occupied Europe* (Oxford and New York: Berg, 2006), 16–41.
16. Modified on base of Stephan Barthel, John Parker and Henrik Ernstson, "Food and Green Space in Cities: A Resilience Lens on Gardens and Urban Environmental Movements," *Urban Studies* 1, no. 18 (2013): 4–5.
17. Ibid., 2; Stephan Barthel, Sverkel Sörlin, and John Ljungqvist, "Innovative memory and resilient cities: echoes from ancient Constantinople," in Paul J. J. Sinclair, Gullög Nordquist, Frands Herschend and Christian Isendahleds, eds., *The Urban Mind: Cultural and Environmental Dynamics* (Uppsala: Uppsala University Press, 2010), 391–406.
18. Bernhard Klausnitzer, *Verstädterung von Tieren* (Wittenberg Lutherstadt: Die Neue Brehm-Bücherei, 1988), 12–13.
19. Impacts of World War II on wildlife were, however, studied in some less inhabited areas of the global war theater. On Midway Atoll, the so-called moaning birds (e.g. wedge-tailed shearwaters) that had a depressing voice were disliked and persecuted by soldiers, while in another Pacific Ocean island military construction benefited local birds; the Papua New Guinea swiftlets started to use abandoned Japanese tunnels as their nesting sites. See Harvey I. Fisher and Paul H. Baldwin, "War and the birds of Midway Atoll," *The Condor* 48, no. 1 (January–February 1946): 3–15, and Harry L. Bell, "Occupation of urban habitats by birds in Papua New Guinea," *Proceedings of the Western Foundation of Vertebrate Zoology* 3, no. 1 (1986): 17.
20. For origins of urban ecology, see Herbert Sukopp, "Stadtforschung und Stadtökologie in Vergangenheit und Gegenwart," *Geobotanische Kolloquien* 11 (1994): 3–16; Jens Lachmund, "Exploring the city of rubble: botanical fieldwork in bombed cities in Germany after World War II," in Sven Dierig, Jens Lachmund, and Andrew Mendelsohn, eds., Science and the City, a special issue, *Osiris* 18 (2003): 234–54. For a wider picture, see Jens Lachmund, *Greening Berlin: The Co-Production of Science, Politics and Urban Nature* (Cambridge, MA: MIT Press, 2013).

21. Dan Listwa, "Hiroshima and Nagasaki: The Long Term Health Effects," University of Columbia, Center for Nuclear Studies, August 9, 2012, online article, https://k1project.columbia.edu/news/hiroshima-and-nagasaki, accessed June 15, 2018. See also William M. Tsutsui, "Landscapes in the Dark Valley: Toward an Environmental History of Wartime Japan," in Richard P. Tucker and Edmund Russell, eds., *Natural Enemy, Natural Ally: Toward an Environmental History of War* (Corvallis: Oregon State University Press, 2004), 195–216.
22. Claire Campbell, *Bonzo's War: Animals Under Fire 1939–1945* (Glasgow, Scotland: Little, Brown Book Group, 2013); Hilda Kean, *The Great Cat and Dog Massacre: The Real Story of World War Two's Unknown Tragedy* (Chicago: University of Chicago Press, 2017); Anna Reid, *Leningrad: Tragedy of a City Under Siege, 1941–44* (London: Bloomsbury, 2011).
23. For Japan, see Mayumi Itoh, *Japanese Wartime Zoo Policy: The Silent Victims of World War II* (New York: Palgrave Macmillan, 2010); for Munich Zoo, see Michael Kamp and Helmut Zedelmaier, eds., *Nilpferde an der Isar. Eine Geschichte des Tierparks Hellabrunn in München* (München: Buchendorfer, 2000); for Leningrad, see E. E. Denisenko, *Ot zverincev k zooparku. Istorija Leningradskogo zooparka* (Sankt Peterburg: Iskusstvo-SPb, 2003).
24. Frederick S. Litten, "Starving the Elephants: The Slaughter of Animals in Wartime Tokyo's Ueno Zoo," *The Asia-Pacific Journal* 38, no. 3 (2009): 1–17.
25. For further background on refugees in Asian war zones, see Micah Muscolino, "Conceptualizing Wartime Flood and Famine in China," and Richard P. Tucker, "Environmental Scars in Northeastern India and Burma," in *The Long Shadows*, 101–9, 120–22. For similarly massive refugee movements in wartime Europe, see, for example, Michael R. Marrus, *The Unwanted: European Refugees in the Twentieth Century* (New York and Oxford: Oxford University Press, 1985); Malcolm J. Proudfoot, *European Refugees 1939–52* (Evanston: Northwestern, 1956); Andrew Paul Janco, *Soviet Displaced Persons in Europe, 1941–1951*, PhD dissertation, University of Chicago, 2012.
26. See Tom Arnold's chapter in this volume. For related moral questions, see Charles S. Maier, "Targeting the city: Debates and silences about the aerial bombing of World War II," *International Review of the Red Cross* 87, no. 859 (September 2005): 429–44, and Conrad C. Crane, *American Airpower Strategy in World War II: Bombs, Cities, Civilians, and Oil* (Kansas City: University Press of Kansas, 2016).
27. See, for example, Ferenc M. Szasz and Issei Takechi, "Atomic Heroes and Atomic Monsters: American and Japanese. Cartoonists Confront the Onset of the Nuclear Age, 1945–80," *The Historian* 69, no. 4 (Winter 2007): 728–752.

28. William E. Burrows, *This New Ocean: The Story of the First Space Age* (New York: Random House, 1998), 99–100.
29. Sara Meerow, Joshua P. Newell, and Melissa Stults, "Defining urban resilience: A review," *Landscape and Urban Planning* 147 (March 2016): 44; Patrick Martin-Breen and J. Marty Anderies, "Resilience: A Literature Review," in *Bellagio Initiative* (Brighton: IDS, 2011), 8. See also Michael Burayidi, ed., *City Resilience*, 4 vols. (London: Routledge, 2015).
30. See Simin Davoudi, "Resilience: A Bridging Concept or a Dead End?" *Planning Theory & Practice* 13, no. 2 (June 2012): 302.
31. Adam Rome, *The Bulldozer in the Countryside: Suburban Sprawl and the Rise of American Environmentalism* (New York: Cambridge University Press, 2001); Christopher Sellers, *Crabgrass Crucible: Suburban Nature and the Rise of Environmentalism in Twentieth-Century America* (Chapel Hill: University of North Carolina Press, 2012).
32. Mia Kunnaskari, "Kaupunkiviljely joustavuuden koekenttänä," *Terra* 128, no. 4 (2016): 221; Simo Laakkonen, "Polemosphere. The War, Society, and the Environment," in Laakkonen, Tucker, Vuorisalo, *The Long Shadows*, 22.
33. For example, when there were not enough chicken eggs available for a children's hospital in Helsinki, they were substituted in the summertime by collection of gulls' eggs from the islets outside the city. Eero Haapanen et al., *Lukuja luodoilta. Helsingin saaristolinnut nyt ja ennen* (Helsinki: Helsingin kaupungin ympäristökeskus, 2017), 210–215.
34. Jaakko Suominen and Anna Sivula, "Retrovation – the Concept of a Historical Innovation," *WiderScreen* 3–4 (2016): online (http://widerscreen.fi/numerot/2016-3-4/retrovation-the-concept-of-a-historical-innovation/).
35. Simo Laakkonen, "War – An Ecological Alternative to Peace," in *Natural Enemy, Natural Ally*, 190. Generally, the ecological footprint refers to the biologically productive area needed to provide for all resources needed by people in a particular area, and for processing their wastes. See Mathis Wackernagel and William E. Rees, *Our Ecological Footprint: Reducing Human Impact on the Earth* (Gabriola Island, BC: New Society Publishers, 1996). For an innovative long-term assessment, see Helmut Haberl, Karl-Heinz Erb and Fridolin Krausmann, "How to calculate and interpret ecological footprints for long periods of time: the case of Austria 1926–1995," *Ecological Economics* 38, no. 1 (July 2001): 35.
36. Naturally, life in Leningrad or Warsaw did not become healthy. However, in Leningrad, people paid attention to the exceptionally clean air and silence in the besieged city. Lidia Ginzburg, *Leningradin piirityksen päiväkirja* (A diary of the blockade of Leningrad, translated from Russian by Kirsti Era) (Helsinki: Into Kustannus oy, 2011), 27.

37. John Sheail, "War and the Development of Nature Conservation in Britain," *Journal of Environmental Management* 44, no. 3 (June 1995): 267–83; Raymond H. Dominick III, *The Environmental Movement in Germany: Prophets and Pioneers, 1871–1971* (Bloomington: Indiana University Press, 1992), 101, 114.
38. Annika Björklund, *Historical Urban Agriculture. Food Production and Access to Land in Swedish Towns before 1900* (Stockholm: Stockholm University, 2010), 214–215.
39. A good example of this phenomenon is Japanese cities. The population of Tokyo decreased from 6.8 million in 1940 to 2.8 million in 1945 while the populations of Osaka, Nagoya, and Yokohama were halved in the same period. The share of urban population decreased in wartime Japan as a whole from 37.7 percent down to 22.8 percent. Source: Demography of Imperial Japan, From Wikipedia, the free encyclopedia, accessed 15 June 2018, https://en.wikipedia.org/wiki/Demography_of_Imperial_Japan
40. See, for example, Laurence L. Delina and Mark Diesendorf, "Is wartime mobilisation a suitable policy model for rapid national climate mitigation?" *Energy Policy* 58 (2013): 371–380; Hugh Rockoff, "The U.S. Economy in WWII as a Model for Coping with Climate Change," *NBER Working Paper*, no. 22590, The National Bureau of Economic Research, September 2016; Bill McKibben, "We're under attack from climate change—and our only hope is to mobilize like we did in WWII," *The New Republic*, August 15, 2016.
41. Meghan Doherty, Kelly Klima and Jessica J. Hellmann, "Climate change in the urban environment: Advancing, measuring and achieving resiliency," *Environmental Science & Policy* 66 (2016): 310–313.

Index[1]

NUMBERS AND SYMBOLS
1952 Drainage Plan, 248

A
Agricultural Collective of Barcelona, 36
Agricultural involution, 221
Agriculture, 9, 29, 48, 105, 106, 108, 111–113, 115, 118, 120, 122, 125n39, 219, 233, 286–288, 293
Agro-ecological crisis, 221
Aigües de Barcelona Empresa Col lectivitzada, 35
Airborne particles, 78
Aircraft industry, 56
Airfields, 139
Air pollution, 69–79, 79n6, 292
Air Pollution Control Association, 75
Air raid, 3, 5–6, 13, 32, 36–38, 71, 134, 137–138, 143, 145, 152, 155, 157–160, 162–163, 169, 204, 206–213, 215n12, 216n20, 217n29, 223–224, 227, 288
Air Raid Precautions Department, 71
Aleppo, viii, xn5, xn6, 41n4, 175n100
Alkali Inspectorate, 71
Allotment gardens, 109, 121, 286
All-out Campaign Against Famine, 230
Alps, 127, 136, 145
Aluminum, 54–56
American Civil War, 10
American Fat Salvage Committee (AFSC), 86, 89–92, 94, 95, 97, 101n41
Anarcho-syndicalists, 30
Andreas-Friedrich, Ruth, 138
Anh, Nguyen The, 231, 234n2, 236n36
Animal abuse, 116
Animal husbandry, 107, 111, 113–115, 117, 120, 286
Animal protection, 4
Animal protection movement, 115

---

[1] Note: Page numbers followed by 'n' refer to notes.

© The Author(s) 2019
S. Laakkonen et al. (eds.), *The Resilient City in World War II*, Palgrave Studies in World Environmental History, https://doi.org/10.1007/978-3-030-17439-2

INDEX

Annam, 222
Antarctica, 12
Anthropomorphism, 155
Apocalyptic imagination, 261, 264
Apocalyptic urban future, 259–274
Aqueduct, 24
Aquifer, 24
Aquino, Alejo, 242, 253n18
Aragón, 30
Arkansas, 56
Army Air Forces, 54, 57, 65n17, 66n33
Arnold, Thurman, 91, 101n44
Arsenal of democracy, 47, 291
Arsenic, 60
Arxiu General d'Aigües de Barcelona (AGAB), 25
Athens, ix
Atlantic Ocean, 51
Atomic cities, 259
Atomic radiation, 260, 263
Augsburg, 136, 144
August Revolution, 230, 233
Aurajoki River, 117
Aurochs, 162, 164, 168

B

B-17 "Flying Fortress," 47
B-29 "Superfortress," 47
Bao Dai, 222
Barcelona, 23–25, 29–30, 34–40, 40n1, 41n6, 42n7, 42n12, 43n18, 43n21, 43n23, 43n24, 45n30–34, 46n35
Bataan Peninsula, 241
Battersea, 71, 79n5
Battle of Britain, 23
Battle of Manila, 243
Battle of the Ebro, 37
Bauxite ores, 56
Bavaria, 127, 132, 140, 143
Beijing, 205, 216n19
Belgium, 18n33, 71
Bengal famine, 220, 233
Berlin, 16n15, 16n17, 19n40, 105, 119, 122n2, 135, 145n3, 148n25, 149n52, 151–153, 156, 161–169, 173n50, 173n54, 173n55, 174n58–65, 174n71–74, 175n94, 175n101, 177, 187, 193, 195n2, 199n89, 287, 288, 297n3, 299n20
Beveridge Report, 7
Bikini Atoll, 271
Bilston, 75
Bings, 77, 79n4
Biodiversity, 273, 288
Bird-watching, 192
Black & Veatch, 249, 254n38, 257n88, 257n91, 257n93
Black market, 114, 224, 227
Black redstart, 192–194, 287
Black Tuesday, 72
Blaisdell, Thomas, 91
Blitz, 7, 145n3, 152, 171n11, 189, 190, 192
Boeing, 47, 51, 53, 56, 57, 59, 60, 63, 63n1, 64n4, 64n10, 65n17–19, 65n28, 66n29, 66n30, 66n32–34, 67n50
Bolshevism, 134
Bombed sites survey, 188, 190
Bonneville Dam, 52
Botanists, 177, 180, 191, 287
Bourke, Joanna, 264
Bowles, Chester, 92
Boyer, Paul, 264, 275n1
Breitbart, Myrna B., 29
Brownfields, 287
Building stock, 111
Building waste, 111
Bureau of Public Health, 207, 210, 214
Bureau of Smoke Control, 78

INDEX   305

Burgess, E. W., 284
Burnham, Daniel, 240, 245, 253n15
Burt, Jonathan, 159

C
Caloocan, 250
Cambodia, 225
Canal de la Reina, 245
Canales del Lozoya, 24–27, 32–34, 36–38, 42n8, 42n9, 43n16, 43n17, 44n26–28
Canal No. 5, 242
Canton, 220
Carossa, Hans, 136
Carrot, 109
Cartmill, Cleve, 263
Cascade Mountains, 48, 49
Catalan government, 30, 34, 35, 37, 45n34
Catalonia, 29, 34, 37, 45n31
Catastrophism, 273
Cats, 106
Cattle, 106, 114, 161, 168
Censorship, 108
Charcoal, 142
Chauvet, Paul, 224
Chemicals, 60, 268
Chen, King C., 230
Chennault, Claire, 226
Chetverikov, Sergey, 185
Chicago, 10, 15n10, 17n27, 48, 63n3, 73, 75, 82–86, 89, 90, 96, 98, 98n5, 99n8, 99n9, 99n11, 99n14, 99n17, 100n20, 100n22, 100n30, 101n37, 101n38, 101n39, 102n58, 102n64, 102n65–67, 146n7, 171n18, 214n1, 269, 300n22, 300n25
Chicago Commission of Civilian Defense, 84
Chicago Daily Tribune, 96, 98n5

Chickens, 90, 106, 141, 161, 288
Chimpanzees, 153, 157–159
China, 5, 13, 203–214, 214n1, 215n4, 215n6, 215n8, 215n9, 215n11, 215n12, 216n22, 216n23, 217n31, 218n42, 218n44, 222, 225, 236n32, 268, 283, 300n25
Chinese Medicine Hospital, 211, 218n36
Chlorination, 27
Cholera, 221
Chongqing, 8, 203–213, 214n1, 215n3, 215n6, 215n12, 216n17, 216n18, 216n20, 217n26, 217n29, 217n30, 217n32, 218n34, 218n39, 218n41–43, 289
Chongqing Municipal Hospital, 207
Churchill, Winston, 12, 19n38, 23, 40, 154, 155
City of Tomorrow, 7, 15n12
City Planning Office, 244
Civil engineers, 209
Civilian, viii, 5, 6, 24, 26, 40, 63, 82, 84, 86, 92, 96, 105, 112, 130, 205, 208, 220, 224, 243, 244, 261, 290
Civil servants, 206, 209
Civil war, 25, 204
Classical smog, 70
Clauss-Ehlers, C. S., 128
Clean Air Act of 1956, 78
Climate change, 13, 81–82, 97, 296
Coal, 37, 72, 73, 75–78, 132–134, 137, 140, 142–145, 217n31, 222, 283
Coal mines, 77
Coal thievery, 133
Coffee, 112, 227
Cold War, 4, 13, 63n1, 63n2, 122n3, 177, 262, 263, 265, 266, 268, 270, 272, 275n13, 276n21, 278n49

Colgate-Palmolive-Peet, 86, 91
Collectivizations' Decree, 35
Columbia River, 52, 55
Columbus, 76
Commonwealth government, 240, 241
Communist, 43n16, 204, 205, 214
Concentration camps, 148n30, 273
Condor Legion, 32, 34
Conservation, 4, 7, 10, 15n13, 75–76, 81, 82, 84, 87, 93–95, 97, 98, 106, 132, 133, 137, 189
Conservation campaigns, 81
Consumers, 78, 82, 92, 94, 96, 97, 209, 293
Continuation War, 111, 112, 115, 121, 123n13, 179
Cooperatives, 209
Co-Prosperity Sphere, 222
Corman, Roger, 259, 267
Corpses, 156, 167, 188, 189, 221, 227, 229
Countryside, 9–11, 112, 127, 141, 142, 144, 151, 177, 220, 223, 224, 231, 232, 234n2, 241, 269, 287, 295
Crew, Joseph C., 226
Cripplegate, 188, 190, 191, 193, 199n71, 199n72, 199n74, 199n75
Croft, Louis, 244, 245, 255n48, 255n50, 255n53
Cultural diversity, 297
Culturally-focused resilient adaptation, 128
Cultural resilience, 128
Currie, P. W. E., 192, 199n67

D

Dachau, 134, 148n30
Daily Mail, 126n61, 154, 171n8
The Daily Telegraph, 159

Damon, W. A., 71
D-Day, 259
Decoux, Jean, 226
Defense industries, 56
Deforestation, 247, 250, 256n70
Democratic Republic of Vietnam (DRV), 230
Denmark, 112, 268
Department of Scientific and Industrial Research, 70, 73, 75
Detroit, 4, 76
Deutch, I. A., 75
Devolutionary resilience, 291
DEW (distant early warning) Line, 271
Dick, Philip, 263
Diesel, viii, 78
Dig for Victory, 105
Dikes, 220–222, 225, 226, 228, 233
Dixon, Winston Wheeler, 264, 276n15
Dogs, 106
Domestic animals, 107, 108, 114, 117–122, 125n36, 288
Dotterweich, Helmut, 136
Douglas fir, 51
Douhet, Giulio, 129, 143, 147n12
Dresden, viii, 5, 8, 15n9, 167, 183
Dumps, 111
Dunkirk, 23
Dust bowl, 9
Dutch Guiana (Suriname), 56
Duwamish Valley, 49
Duwamish waterway, 49
Dystopias, 263

E

Eco-cities, 294
Eco-horror, 274
Ecological degradation, 239
Ecological footprint, 294
Economic autarky, 222
Eco-scenarios, 274

INDEX   307

Eggs, 141, 301n33
Eisenhower, Dwight, 265
Electricity, 19n34, 37, 52, 53, 55, 62, 71, 130, 133, 134, 136–139, 141, 144, 147n22, 209, 243
Elephant, 154, 157, 161, 164–167, 288
Elephant hawk-moth, 190
El Villar, 26, 27, 38
Ely, Sumner B., 76
Energy saving, 137
England, 51, 105, 115, 171n20
English Channel, 168
Enola Gay, 260
Environmental health, 289
Environmentalism, 292
Environmental management, 239
Environmental protection, 4, 10–11, 60, 286
Environmental shock, 233
Epidemic disease, 289
Epidemics, 210, 221, 239, 289
Epstein, Israel, 208
Erkamo, Viljo, 178–183, 195, 196n7, 197n25
Essen, 4
Esteros, 237–252, 252n2, 252n4, 253n12, 253n14, 253n16, 253n18, 254n26, 254n29
Estuarial creeks, 237, 238
Europe, viii
Eutrophication, 178
Evening Standard, 151, 159, 170n1, 172n29
Eviction, 246, 248, 249
Extinction, 162, 187, 194, 273

F

Fabritius, Leo J., 114
Factories, 5, 6, 30, 43n23, 47, 54, 55, 73, 129–132, 139, 189, 249, 282–283

Famine, 195, 220, 222, 227–230, 232, 233, 233n1, 234n2, 235n7, 235n14, 235n26, 286, 289, 290, 300n25
Fat salvage, 81–86, 89–96, 98n3–5, 99n6, 99n9, 99n16, 100n19, 100n24, 100n26–29, 100n31, 101n35, 101n36, 101n40, 101n41, 101n42, 101n45–47, 101n48, 102n50, 102n51, 102n52, 102n56, 102n57, 102n61
Feeding of birds, 194
Feral pigeon, 119, 182, 187, 192, 194–195, 293
Fertilizer, 107, 109, 122, 269, 271
Fiehler, Karl, 132
Fifteenth Air Force, 138
Finland, vii, ix, 105–126, 178–183, 194–195
Fire, 15n8, 15n9, 34–37, 42n7, 46n34, 134, 138–140, 146n9, 254n39, 255n41, 300n22
Fire-bombing, 32
Fishing, 48, 146n5, 271, 293
Fitter, Richard, 178, 188–190, 192, 194
Flood control, 233, 248
Flood Control Board, 243
Floods, 206, 213, 219, 222, 228, 230, 238, 239, 242, 243, 247, 250
Flora of bombed areas, 189
Fodder, 180, 288
Food Administration, 109
Food packers, 59
Food Rationing Division, 90, 98n3, 99n5, 99n6, 99n9, 100n24, 100n29, 101n31, 101n32, 101n41
Food security, 223, 230, 233, 286
Food shortages, 36
Fortress city, 47–63

14th Air Force, 226
Franco, Francisco, 13, 25, 30–34, 36–37, 39
Franco-Prussian War (1870–1871), 13
Fuel supplies, 127
Fundación Pablo Iglesias, 25, 42n8, 42n11, 42n14, 43n16, 44n28

G
Garbage collection, 241
Gas, 132–134, 137, 139, 141, 144, 147n20, 148n34
Gasoline, 83, 87, 242
Gdansk, vii, 134
Geertz, Clifford, 221, 235n6
Genetics, 185
Genocide, 273
Georgia, 56, 75
Gia Lam, 226
Gibraltar strait, 31
Giessler, Paul, 131
Gingko, 287
Giraffe, 160, 165
Glycerine, 82, 87, 91, 92, 95, 97
Goat, 163
Godzilla, 270, 271, 277n33, 277n35–38, 277n40, 278n57
Goebbels, Joseph, 133
Gorilla, 168
Göring, Hermann, 133, 162–164, 168
Gourou, Pierre, 220
Grand Coulee Dam, 52
Grant, Barry Keith, 267
Great Depression, 4, 60, 291
Great Famine, 226, 234n1, 234n2, 236n36
Great Northern Railway, 48
Great Smog, 78
Great tit, 187
Green spaces, 127, 134
Gregg, Victor, 5

Grzimek, Bernhard, 165
Guadarrama, 26
Guangxi, 231
Guernica, 40
Guerrilla forces, 204
Guerrilla gardening, 122
Guerrillas, 122, 204, 213, 241, 246, 252n8
Gulf Wars, 10
Gunbelt, 62

H
Ha Tinh, 221
Hamblin, Jacob Darwin, 122n3, 272, 278n46
Hamburg, 122n2, 135, 138, 141, 149n43, 167, 183, 288
Hanoi, 219–233, 233n1, 234n2, 235n9, 235n19, 235n20, 236n26, 236n31, 286, 289
Harbor Island, 51, 54, 57, 60, 67n47
Harbors, 225
Harlem River, 73
Hart, Basil Liddell, 129
Hartendorp, A. V. H., 243, 253n20
Hausenstein, Wilhelm, 136, 139, 141, 148n39
Haussmann Plan, 209
Haussner, Alfred, 141
Health officials, 71, 210, 212, 213
Heck, Lutz, 162, 164
Heinroth, Katharina, 165, 174n73, 174n82
Helsinki, vii, 3, 10, 14n1, 17n26, 109, 113, 114, 117–119, 123n13, 124n15, 125n36, 125n50, 126n54, 126n57, 179, 196n15, 281, 282, 297n1, 299n14, 301n33, 301n36
Hersey, John, 260, 275n3
Hidráulica de Santillana, 27, 37

Hiroshima, viii, 8, 183, 259–263, 266, 270, 271, 274, 275n1, 275n2, 275n3, 275n7, 287, 290, 300n21
Hitler, Adolf, 162, 281
Ho Chi Minh, 229, 230, 290
Holling, C. S., 178, 283
Hollywood, 154, 261, 265, 267–272, 274, 275n6, 276n21, 277n29
Holocaust, 231
Home Office, 71
Hong Kong, 206, 216n21, 235n4
Hongji Hospital, 210
Horses, 106–108, 113–117, 119, 160, 163–164, 180, 192, 288
Household waste, 109, 122
House sparrow, 187, 192
Hull, 7, 15n12
Hull Regional Survey, 7, 15n12
Human-animal relations, 152
Hup Thien graveyard, 229
Huxley, Julian, 153–155
Hydraulic pact, 221
Hydropower, 54–56
Hydropower dams, 39
Hygiene, 30, 107, 115, 117, 119, 120, 122, 294

I
Ickes, Harold, 51, 64n13
Illinois Central, 73
Indochina, 18n30, 220, 222, 225, 226, 230, 235n3, 235n5, 235n17, 235n19
Industrial cities, 282, 284, 285
Infectious Disease Hospital, 210–212, 218n38
Inflation, 206, 212, 213, 227, 233
Informal communities, 246
Infrastructure, viii, 5, 7, 10, 13, 32, 34, 35, 37, 38, 40, 41n3–5, 52, 54, 59, 61, 62, 127, 129, 130, 140, 210, 220, 222, 225, 232, 247, 284–286, 291, 297
Inner city, 139, 142, 223
Innsbruck, 140
International Brigades, 32, 34
Interspecies resilience, 159
Invasive plants, 180
Iron ore, 283
Isar River, 127, 134–135, 145

J
Japan, viii
Japanese Imperial Army, 203, 241

K
Kai-shek, Chiang, 203–206, 212, 214
Karelia, 180, 181, 282
Kean, Hilda, 157, 171n18, 300n22
Kelly, Ethel, 84–86, 98
Kenkichi, Yoshizawa, 224
Kim, Tran Trong, 222
King County Airport, 54, 56, 63
Kollaa battlefield, 180
Koltsov, Nikolay, 185
Koltushi, 185, 188
Komarov's Botanical Institute, 184
Konetzky, Margarete, 130, 131
Kunming, 226, 230, 231
Kupittaa field, 109
Kuwaiti oil fires, 69

L
Labbé, Danielle, 223, 233n1
Lacson, Arsenio, 249
Lagerström, Herman, 116
Landfill, 181
Laos, 225
Lapland War (1944–1945), 111
Laurel, José P., 241

310 INDEX

Lead, 29, 60, 73, 139, 148n28
Leningrad, vii, viii, 16n17, 40, 41n3, 178, 183–188, 195, 196n7, 197n31, 197n35, 197n36, 197n37, 197n38, 198n43, 198n44, 198n45, 198n47, 198n53, 286, 288, 289, 300n22, 300n23, 301n36
Leningrad Blockade, 183, 188
Lever Brothers, 86, 91, 101n42, 101n47
L'Herpinière, Michel, 227
Liége, 71
Limestone mountains, 204
Liquid fuels, 70
Llobregat River, 24, 45n31
Logging, 48, 49, 207
London, vi–vii, 5, 15n12, 15n14, 16n15, 16n17, 16n20, 17n23, 17n25, 17n27, 19n38, 41n3, 41n5, 42n9, 44n25, 63n2, 69, 71, 78, 79n1, 79n11, 80n15, 80n25, 105, 145n3, 147n14, 148n36, 151–163, 167, 170n4, 171n5, 171n11, 171n15, 171n21, 172n24, 172n27, 173n47, 173n57, 175n101, 178, 188–195, 198n57, 198n58, 198n60–65, 199n67, 199n69, 199n71, 199n72, 199n76, 199n77, 199n78–80, 199n78–80, 199n81–83, 199n81–83, 199n85–87, 200n90, 200n91–94, 216n20, 234n2, 235n3, 256n70, 269, 275n9, 276n16, 276n17, 287–288, 300n22, 301n29
London Natural History Society (LNHS), 190, 195, 199n78, 199n82
London Zoological Society, 158, 167
London's Natural History, 188

Los Angeles, 9, 17n24, 62, 63n3, 66n29, 67n51, 69, 72, 75–77, 79n2, 79n10, 80n22, 80n23, 266, 268, 276n20, 276n25
Lousley, J. E., 189
Luftwaffe, 152, 189
Lung disease, 210, 289
Luzon, 237, 241, 243, 246

M

Macau, 220, 235n4
MacLean, Ken, 224
Macpherson, Arthur Holte, 191
Madrid, 23–38, 40, 40n1, 41n6, 42n7–9, 42n10, 42n13, 42n15, 43n16, 44n25, 44n26, 44n27–29, 45n31, 45n32, 46n35
Magnesium bombs, 135
Malamud, Randy, 159
Malaria, 252n8, 289
Mallari, I. V., 242, 253n12, 253n14, 253n18, 254n29, 255n52
Mallorca, 36
Malnutrition, 290, 294
Mangroves, 247
Manila, 237–252, 252n2–8, 253n10, 253n11–13, 253n15, 253n16, 253n18, 253n19, 253n20, 254n21, 254n24, 254n26, 254n31, 254n32, 254n34, 254n36–39, 255n40, 255n43, 255n45, 255n46, 255n48, 255n50, 255n52, 255n54, 255n57, 256n63, 256n64, 256n70–72, 256n74–80, 257n84–87, 286, 289, 290
Manila Bay, 237, 238, 241
Manila Electric Railroad and Light Company, 245
Manure, 107
Manzanares River, 32

INDEX 311

Marikina, 248, 250, 256n76, 257n95
Marr, David, 231, 234n2
Marsh, Arnold, 75, 76, 78
Marx, Karl, 8
Mascots, 161, 163
Mass bombings, 4
Mass production, 4
Maternal and Children's Health clinics, 210
Maternity Hospital, 211, 212
Meat, 51, 89, 91, 92, 114, 115, 122, 141, 160, 167, 293
Medical care, 210, 213, 288, 289
Medicines, 82, 87, 92, 96
Mediterranean, 36
Meetham, A. R., 78
Megaherbivores, 162
Meiling, Song, 204, 206
Mekong delta, 222
Mercury, 60
Metropolitan Manila Planning Association, 244
Meuse Valley, 71, 79n6
Mice, 119
Mickelsen, Martin, 226
Middle East, viii, 6
Midway Island, 76
Migration, 62, 146n5, 246, 291
Militarized landscapes, 48, 63
Military-industrial complex, 48, 56, 62, 265, 271
Military Police, 209
Mines, 132
Mines Department, 71
Ming (Panda), 151, 154, 155, 161, 168, 171n11, 172n30, 173n48, 234n2
Mining, 24, 36, 48, 49
Minister of Health, 71
Ministry of Supply, 114
Ministry of Town and Country Planning, 189

Minsk, 163
Missionaries, 212
Missionary hospital, 212
Mitchell, William "Billy," 129
Model city, 6–8, 62, 282
Model City, 6
Molero, Federico, 34, 41n1
Molotov-Ribbentrop Pact, 281
Monster movies, 270
Montalban, 250
Mordant, Eugene, 226
Morocco, 25
Mumford, Lewis, 282, 297n2
Munich, 127–145, 145n1, 146n10, 147n23, 148n26, 148n27, 148n35, 149n52, 149n58, 164, 289, 300n23
Munich City Council, 132
Munich Railroad District, 140
Municipal Hospital, 211
Munitions, 87, 96, 106, 132
Munn, Orson, 95
Mushrooms, 141, 293

N
Nagasaki, 183, 260–262, 266, 270, 271, 275n1, 275n2, 290, 300n21
Nanjing, 205, 209, 212
Nationalist Party, 207, 213, 214
Nationalist State (1927–1949), 214
National parks, 9
National Resources Planning Board, 52, 64n5, 64n11, 64n12, 65n16, 65n24, 65n26, 66n39
National Revolutionary Army (NRA), 203
National Smoke Abatement Society, 73, 74, 76
National Urban Planning Commission, 244
Native Americans, 49

Natural environment, 145, 238
Natural gas, 78
Natural History Society, 190
Natural resources, 48, 51, 60,
    128–130, 132, 140, 145, 146n7,
    241, 284, 291, 292, 294–297
Nature conservation, 4, 7, 10, 15n13,
    302n37
New Deal, 52, 56, 62, 64n14
New Mexico, 268
New York, xn1, xn2, xn4, 15n8,
    15n14, 15n15, 16n19, 17n27,
    18n30, 18n32, 41n5, 48, 64n4,
    64n6, 64n10, 66n29, 67n51,
    72–73, 80n17, 86, 98n1, 99n10,
    99n12, 100n21, 102n58, 146n4,
    146n7, 146n9, 146n11, 148n30,
    171n5, 171n6, 172n25, 173n51,
    174n66, 197n34, 197n39,
    199n70, 215n7, 215n8, 215n11,
    215n12, 236n29, 239, 253n13,
    254n32, 254n36, 255n49,
    256n70, 262, 264, 266,
    268–269, 275n3, 275n5, 275n7,
    275n8, 275n11, 275n12,
    276n14, 276n15, 276n24,
    276n28, 277n29, 277n33,
    277n34, 278n46, 278n48,
    278n52, 297n2, 299n13,
    300n23, 301n28, 301n31
Nghe An, 221
Nitrogen oxides, 70
Nonrenewable resources, 9
Nordic cities, 106
North Africa, 31, 47, 87
Northern Pacific Railway, 48
Northern wheatear, 187
Northwest Power Pool, 55
Nuclear weapons, 259, 261, 265, 268,
    273, 274
Nuremberg, 144, 149n58
Nutrition, 195, 227

O
Oesch, Karl Lennart, 180, 196n14
Office of Civilian Defense, 84
Office of Price Administration, 87,
    98n3, 98n4, 99n5, 99n6,
    100n29, 101n32
Oleander, 287
Operation Barbarossa, 75
Opium Treatment Hospital, 210
Orangutans, 157
Orbeli, Leon, 185
Oregon, vii, 14n7, 56, 65n25, 66n38,
    100n25, 278n54, 300n21
Ornithological Laboratory, 185
Orwell, George, 262, 275n9, 288
Osaka, 4, 302n39
Osmeña, Sergio, 244, 255n44
OSS, 229, 231, 236n33

P
Pacific, 47, 48, 52, 54–57, 60, 63,
    63n1, 64n5, 64n6, 64n8, 64n11,
    64n12, 64n14, 65n16, 65n20,
    65n23, 65n24, 65n26, 66n37,
    66n39, 66n40, 171n5, 206, 243,
    270, 275n7, 282, 299n19, 300n24
Pacific Theater, 47
Parks, 30, 112, 120–122, 136, 139,
    142, 183, 194, 242, 293
Partial Test Ban Treaty, 273
Pasay City, 250
Pasig River, 237, 238, 244, 247, 249
Pasteur Institute, 227
Patti, Archimedes, 229
Pavlov, Ivan, 185
Pearl Harbor/Pearl Harbour,
    75, 206, 241
Pentagon, 265
People's Liberation Army, 214
Persistence, 291
Pesticide, 273

INDEX 313

Philadelphia, 73
Philippine–American War, 239, 252n8
Philippines, 225, 237–257, 252n8, 253n16, 253n20, 254n34, 254n36, 254n37, 255n47, 255n58, 255n59, 255n60, 256n61, 256n62, 256n68, 256n70, 257n94, 257n97
Photochemical smog, 9, 70, 76, 79
Physical environment, 13
Pigs, 106, 107, 113, 114, 118, 161, 188, 288
Pioneer plant, 189
Pittsburgh, 9, 16n17, 17n24, 48, 76–78, 79n2, 80n20, 80n24, 80n26, 122n1
Pneumothorax, 165
Pohl, Frederik, 263
Poland, 23, 118, 156, 164
Pollution, 7, 10, 12, 27, 60, 62, 69–74, 76–78, 79n4, 79n6, 79n7, 80n16, 80n20, 80n27, 117, 238, 239, 244, 249–251, 270, 285, 290, 292
Pondicherry, 223
Population boom, 59
Population growth, 282
Portland, 62, 65n16, 65n23, 65n28, 66n38
Potable water, 239, 243
Potato, 109, 111, 112, 119, 121
Power lines, 130
"Precision" raids, 140
Preis, Kurt, 138
Price of water, 37
Prinzregentenstrasse, 131
Prinzregentplatz, 131
Procter & Gamble, 86, 91
Promptov, Alexandr, 178, 185, 187–188, 195, 197n41
Propaganda, 6, 87, 92, 96, 131, 134, 147n23, 152, 153, 155, 162, 221

Protectorate Moroccan soldiers (*Regulares*), 31
Public health, 11, 70, 71, 105, 120, 213, 221, 239, 252n8, 289
Puentes Viejas, 26, 27, 38
Puget Sound, 47–49, 63, 67n49
Purification, 36
Pyrenees, 37, 39, 45n34

Q
Qing (1644–1911), 214
Quang Ngai, 221
Quang Tri, 225
Quezon, Manuel, 240, 241, 244, 253n19
Quezon City, 250

R
Rabbits, 106, 113, 114, 118, 288, 293
RAF, 130, 131, 133–135, 138, 139, 141, 148n37
Railroads, 64n6, 72, 76, 127, 129, 130, 139–141, 144
Rangoon, 289
Rationing, 89, 91, 92, 96, 97, 111, 112, 160, 194, 220, 288, 294
Rats, 111, 118–119, 188
Raw materials, 51, 54, 59, 62, 293, 294
Reagan, Ronald, 273, 274, 278n52
Recycling, 75, 293
Red Army, 142, 167, 282
Red Cross, 144, 207, 210, 300n26
Red River, 219
Red River delta, 219–221, 223, 226, 232, 233
Refugees, 32, 35, 37, 46n35, 128, 166, 203, 206, 209, 212, 213, 246, 282, 289, 300n25
Regensburg, 140, 149n58

Regent's Park, 153, 157, 161
Regional Railroad Authority, 140
Renton, 57, 61, 63, 66n33, 67n50
Reservoirs, 24–26, 32, 34, 38
Resilience, viii, ix, 3–14, 16n15, 16n16, 19n41, 25, 38, 40, 81–98, 127–130, 145, 146n4, 146n5, 151–170, 178, 194, 196n4, 196n5, 203–214, 215n7, 221, 246, 249, 251, 266, 274, 283–284, 287, 288, 291–297, 298n6, 298n8, 298n9, 298n10–12, 299n16, 301n29, 301n30
Resilient city, 8, 15n15, 283
Retrovations, 294
Réunion, 223
Reversed modernization, 294
Rice-based alcohol fuel, 227
Roads, 52, 130, 190, 208, 245
Rook, 187
Roosevelt, Eleanor, 105
Roosevelt, Franklin D., 52, 75
Rosebay willowherb, 181
Rosendorfer, Theo, 134
Rosenheim, 140
Roth, Eugen, 135
Roxas, Manuel, 244
Royal Air Force Coastal Command, 189
Royal Botanic Gardens at Kew, 189
Rubber, 77, 81, 222
Rubble, 131, 134, 138, 139, 141, 144, 145, 165, 167, 190, 191, 199n66, 244, 299n20
Ruhr, 17n25, 132, 145, 149n52
Ruralization, 295
Rutabaga, 109

S
Saigon, 224, 225, 231
Salinization, 24
Salisbury, E. J., 189

Sampaloc district, 243
Sanatorium, 212
San Francisco, 62, 67n51, 264
Sanitation, 26, 115, 117, 210, 245, 284, 289
Santa Ana district, 249
Satellite towns, 250
Sawmills, 49
Schmeider, Maria, 144
School gardens, 111
Schumpeter, Joseph, 8
Scully, Cornelius D., 77
Seattle, 17n24, 47–49, 51–54, 56, 57, 59, 60, 62, 63n1, 63n3, 64n4, 64n7, 64n8, 64n9, 64n11, 65n17, 65n18, 66n29, 66n32, 66n35–37, 66n42–45, 67n47, 67n49, 67n50, 122n1, 264, 282
Seattle Housing Authority, 59, 67n44
Seattle-Tacoma Shipbuilding Corporation, 54, 57
Secondary pollution, 70
Sen, Amartya, 220
Seville, 26, 31
Sewage, 60, 117, 139, 239, 246
Sewerage system, 239, 242, 249
Shantytowns, 206
Shapiro, Jerome, 264
Sheep, 106, 113, 114, 118, 161
Shelters, 36, 38, 115, 118, 134, 143, 157, 160, 165, 204, 206, 207, 209, 213, 215n12, 216n20, 250, 289
Shock city, 3–6, 8, 62
Shortages, 148n34, 206, 224–225, 233
Sichuan, 154, 203, 210, 214n2, 215n3
Siltation, 239
Sindicato de Obreros de Aguas de Barcelona, 29
Sitka spruce, 51
Smoke abatement, 70, 72, 73, 75, 76, 80n21

INDEX 315

Smoke Prevention Association, 72
Smugglers, 231
Smyntyna, Olena, 128, 146n5
Snakes, 156, 165, 288
Soap, 91
Social inequality, 206
Social service centers, 209
Sociedad General de Aguas de
    Barcelona, 24, 29, 37
Solar power, ix
Solidary organizations, 223
Sonck, Carl Eric, 182
Sontag, Susan, 265, 276n23
South Korea, 268
Soviet Union, viii
Söyrinki, Niilo, 180, 196n11
Spanish Civil War, 5, 13, 23, 25,
    34–38, 40, 42n7, 42n9, 43n18,
    44n25, 44n26, 44n27, 290
Spanish Foreign Legion (*Legionarios*), 31
Spanish Republic, 25, 31, 34
Speer, Albert, 137
SS, 162, 164, 168
Stalin, Josif, 16n21, 197n39, 281
Stalingrad, viii, 8, 40, 133
Starvation, 160, 161, 183, 184, 188,
    194, 224, 227, 230, 231
Steel, 59, 77, 145n1, 155
St. James's Park, 192
St. Louis, 17n24, 72, 78, 79n9
Stockholm, 16n21, 18n30, 118,
    123n4, 282, 302n38
Stockpiles, 220, 224, 232
Strategic Bombing Survey, 140,
    149n53, 149n58, 150n74
Streetcars, 83, 107, 141, 144
Subsistence farming, 286
Subterranean warfare, 34
Suburbanization, 249, 255n49
Suburbs, 291
Succession, 191
Sugar, 81, 109, 112

Sulfur dioxide, 71
Sustainable cities, 12
Sweden, 112, 115, 117, 118, 286
Syrian Civil War, viii
Szilard, Leo, 263

T
Terror bombing, 5
Textiles, 82
Thái Bình, 228
Thirty Years' War, 12
Throwaway society, 106
Thrush nightingale, 183
The Times, 161, 171n11, 172n30,
    173n48
Todd Shipyards Corporation, 53
Toilets, 246
Tokyo, ix, 16n15, 40, 118, 171n5,
    183, 261, 269, 270, 277n36,
    288, 300n24, 302n39
Tomb of Fascism, 34
Tonkin, 221, 226, 234n2
Total war, 133
Town planning, 10, 251, 292
Trans-Indochinese Railroad, 220
Transition, 77, 94, 128, 291
Transport system, 130, 142, 145
Trauma Hospital, 210, 217n29,
    218n37
Trench warfare, 129
Trenchard, Hugh "Boom," 129
Truman, Harry S., 229
Tuberculosis Sanatorium, 210
Tucker, Raymond, 72
Tucker, Richard P., 12
Turku, vii, ix, 106–109, 111–120,
    123n8, 123n9, 123n10, 124n14,
    124n17, 124n28, 125n35,
    125n36, 125n47, 125n49, 126n63
Turku Animal Protection Association,
    115, 125n49

Turku Friends of Animals, 115
Tutzing, 136, 139
Typhoid fever, 24, 37, 39, 46n35
Typhoon, 243, 290
Typhoon Gertrude, 247
Typhus, 289

U
Ude, Karl, 143
United Nations, viii, xn2, xn3, 255n59, 273, 278n52
United States, vii
University City, 32, 34, 44n27
Urban agriculture, 106, 126n67
Urban ecology, 177, 178, 195n1, 196n3, 287, 299n20
Urban gardening, 36
Urban heat island, 286
Urban History Association, 9
Urbanization, viii
Urban metabolism, 38
Urban nature, 6, 7, 10, 13, 177–200, 286–287, 294, 297
Urban resilience, viii, 13, 14, 24, 39, 122n2, 152, 283–288, 290, 291, 293, 296, 297, 301n29
Urban society, 7, 13
Urban stress, 24
Urban wastelands, 293
USAAF, 130, 133, 138–142
USAFFE, 241, 243
US Navy, 51, 272

V
Vaccination campaigns, 212, 221
València, 32
Vancouver, Washington, 55
Vavilov's All-Union Institute of Plant Industry, 184
V-E day, 90

Vegetable gardens, 36
Vichy, 219, 220, 222, 225, 226, 235n17
Victory garden, 105
Vienna, 16n17, 164, 288, 297n3
Viet Minh, 221, 223, 228, 230–233, 234n2
Vietnam War, 10
Viipuri, 178–183, 187, 194, 195, 196n6, 197n24, 197n42, 281, 282, 297n1
Villanueva, Antonio A., 248
V-J day, 90
Vo Anh Ninh, 228–230
Volatile organic compounds, 70, 77

W
Wall Street, 51
War Department, 53, 235n8, 235n10, 235n15, 235n16
War Emergency Plan, 157
War Food Administration, 83
War industry, 52
War of Resistance, 204, 206, 207, 212, 214n1, 215n2, 215n3
War Production Board, 55, 65n27, 83, 86, 91, 92, 101n44, 102n52
Warner Brothers, 268, 269
Warsaw, 16n15, 41n3, 164, 193, 199n89, 244–249, 251, 255n40, 255n42, 288, 301n36
Waste coal, 71
Waste management, 117, 285
Wastewater treatment, 249, 285
Water-borne diseases, 238
Water consumption, 35
Water management, 24
Water pipes, 130, 139, 142
Water purification plants, 285
Water supply, 24

Watford, 189
Weber, Franz, 141
Wehrmacht, 138, 143, 163, 173n56
Welfare state, 7
Wells, H. G., 262, 266, 275n10
Wet-field rice, 220
Whipsnade, 155, 157, 161, 171n11, 172n34, 173n43, 173n45, 175n89
White, Mary Webster, 98
Wichita, 57
Wilderness, 9, 10, 152
Wildlife, 177–195, 195n2, 228, 287–288, 293, 299n19
Winter War (1939–1940), 111
Wisent, 162, 164, 168
Wolves, 160, 162, 163
Wood, 17n28, 66n39, 143
Wood as a fuel, 142
Woodpigeon, 194
Work camps, 224
Wrighton, F. E., 190–191

X
Xiangxi, Kong, 206

Y
Yersin hospital, 226
Yilin, Mei, 207, 211
Yokohama, 302n39
Yorkshire, 7
Yugoslav Wars (1991–2001), 13
Yunnan, 215n8, 225, 231

Z
Zebras, 160, 166
Zoo, 151–169, 170n4, 171n5, 171n7, 172n24, 172n27, 172n31, 174n72, 175n85, 175n86, 175n100, 175n101, 185, 186, 195, 197n37, 198n43, 288, 300n23, 300n24
Zoophilia, 164